**Tutorials, Schools, and Workshops
in the Mathematical Sciences**

This series will serve as a resource for the publication of results and developments presented at summer or winter schools, workshops, tutorials, and seminars. Written in an informal and accessible style, they present important and emerging topics in scientific research for PhD students and researchers. Filling a gap between traditional lecture notes, proceedings, and standard textbooks, the titles included in TSWMS present material from the forefront of research.

More information about this series at http://www.springer.com/series/15641

Radosław A. Kycia · Maria Ułan ·
Eivind Schneider
Editors

Nonlinear PDEs, Their Geometry, and Applications

Proceedings of the Wisła 18 Summer School

 Birkhäuser

Editors
Radosław A. Kycia
The Faculty of Science
Masaryk University
Brno, Czech Republic

Faculty of Physics Mathematics and
Computer Science
Cracow University of Technology
Kraków, Poland

Maria Ułan
Baltic Institute of Mathematics
Warszawa, Poland

Eivind Schneider
Department of Mathematics and Statistics
UiT The Arctic University of Norway
Tromsø, Norway

ISSN 2522-0969   ISSN 2522-0977   (electronic)
Tutorials, Schools, and Workshops in the Mathematical Sciences
ISBN 978-3-030-17030-1   ISBN 978-3-030-17031-8   (eBook)
https://doi.org/10.1007/978-3-030-17031-8

Library of Congress Control Number: 2019935998

Mathematics Subject Classification (2010): 35Q30, 45Gxx, 35Q79, 35J96, 35K96

This book is published under the imprint Birkhäuser, www.birkhauser-science.com by the registered
company Springer Nature Switzerland AG
The registered company address is: Gewerbestrasse 11, 6330 Cham, Switzerland

We would like to dedicate this book to the speakers and scientific committee: A. Kushner, V.V. Lychagin, I. Roulstone, V. Rubtsov, J. Slovák, S.N. Tychkov, M. Wolf, and organizing committee R.A. Kycia, J. Szmit, M. Ułan. Extra acknowledgements to O.A. Laudal, U. Persson.

Foreword

The Summer School Wisla 18: Nonlinear PDEs, Their Geometry, and Applications was organized by the Baltic Institute of Mathematics and took place in a beautiful mountain region of Wisła city in the south of Poland between 20 and 30 of August 2018. The city is located in the Beskid Mountains and is surrounded by mountains that are covered by wild forest. The name of the city comes from the biggest river in Poland which has its source near the town. In Poland, it is a well-known good place for hiking, biking, and skiing and even ski-jumping in winter. Wisła is also a spa. The city is located near the border with the Czech Republic and a few kilometers from the border with Slovakia, with the unique triple point where the borders of Poland, The Czech Republic, and Slovakia meet.

The school was devoted to geometric theory of differential equations and applications to physics. A special place is occupied by the theory of the Monge–Ampère equations and their applications in meteorology. There was also a session for learning how to use the Computer Algebra System Maple® in practical computations. The second part of the school was devoted to participant's contributions. This splitting into lectures and contributions of participants is reflected in this book.

Fig. 1 Sunset over the mountains surrounding Wisła (photography by Eivind Schneider)

Brno, Kraków, Oslo, Tromsø, Warszawa   Maria Ułan
2019   Eivind Schneider
                                                         Radosław A. Kycia

Preface

This book is a summary of *The Summer School Wisla 18: Nonlinear PDEs, Their Geometry, and Applications* that took place on 20–30 August 2018 in Wisła, Poland. It is divided into two complementary parts: The first part the book presents geometric methods in nonlinear differential equations and their application to physics (including thermodynamics and meteorology) is presented. They are sorted in the ascending order from introductory to more advanced. It is required that the reader has basic knowledge of differential geometry (at the level of standard university course of Global Analysis and/or Differential Geometry) and some knowledge in physics for lectures touching upon this subject. The second part of the book contains participant's contributions. They are original research articles that show how to apply the methods present in the first part. Some of the papers are complementary to the lectures and present different viewpoints on the same subject. This part is more advanced and concise, however a motivated reader should not have big problems with following them after reading the lecture notes from the first part of the book.

The first part of the book contains lectures given at the School and focuses on various subjects from the geometric theory of differential equations.

The first lecture set in the book by Valentin V. Lychagin explains interesting connections between the theory of measurements, contact geometry, and thermodynamics. Its novel approach should be interesting to anyone who wants to investigate the intriguing connection between probability and differential geometry.

The second set of lectures by Alexei Kushner, Valentin V. Lychagin, and Jan Slovák contains an introduction to geometric theory of differential equations, reformulation of the Monge–Ampère equations in this language, and examples of use of Maple CAS to computations. It should be interesting for anyone new to the subject.

The following lectures by Volodya Rubtsov take the reader deeper into the Monge–Ampère equations and related structures including complex, Kähler and hyperkähler. The last chapter in this series can be treated as an introduction to the equations used in meteorology.

The final set of lectures by Sergey N. Tychkov introduces, in the form of a hands-on approach, symbolic computations in differential geometry using Maple®.

The second part of the book contains research papers by participants of the school. They are more concise and the reader can treat them as an exercise in understanding the material from the first part. They describe various subjects and therefore they give reader a glimpse of current research in these areas.

We hope that the mix of lectures and original research articles will give you, dear reader, a good starting point in your journey to the world of geometric theory of differential equations and their applications, and that they will help you improve your skills in this field.

Brno, Kraków, Oslo, Tromsø, Warszawa   Maria Ulan
January 2019   Eivind Schneider
                                                        Radosław A. Kycia

Acknowledgements

The organizers of the school would like to thank all participants of the school for a great, friendly atmosphere and strong motivation to learn.

The editors would like to thank the Birkhäuser Mathematics, Springer Nature crew, especially Chris Tominich, and Samuel DiBella for assistance.

Contents

Contributors

Bertrand Banos Faculté des sciences et sciences de l'ingénieur, Université de Bretagne Sud, Lorient, France

Anton A. Gorinov Institute of Control Sciences of Russian Academy of Sciences, Moscow, Russia

Stanislav Hronek Faculty of Natural Sciences, Masaryk University, Brno, Czech Republic

Alexei Kushner Lomonosov Moscow State University, Moscow, Russia; Moscow Pedagogical State University, Moscow, Russia; V. A. Trapeznikov Institute of Control Sciences of Russian Academy of Sciences, Moscow, Russia

Radosław A. Kycia The Faculty of Science, Masaryk University, Brno, Czech Republic; Faculty of Physics Mathematics and Computer Science, Cracow University of Technology, Kraków, Poland

Valentin V. Lychagin Department of Mathematics, UiT Norges Arktiske Universitet, Tromsø, Norway; V. A. Trapeznikov Institute of Control Sciences of Russian Academy of Sciences, Moscow, Russia

Mikhail D. Roop Institute of Control Sciences of Russian Academy of Sciences, Moscow, Russia; Lomonosov Moscow State University, Moscow, Russia

Ian Roulstone Department of Mathematics, University of Surrey, Guildford, Surrey, United Kingdom

Volodya Rubtsov Maths Department, University of Angers, Angers, France; Theory Division, Mathematical Physics Laboratory, ITEP, Moscow, Russia

Eivind Schneider UiT the Arctic University of Norway, Tromsø, Norway

Jan Slovák Department of Mathematics and Statistics, Masaryk University, Brno, Czech Republic

Sergey N. Tychkov V. A. Trapeznikov Institute of Control Sciences of Russian Academy of Sciences, Moscow, Russia;
Institute of Control Sciences of Russian Academy of Sciences, Moscow, Russia

Maria Ulan Baltic Institute of Mathematics, Warszawa, Poland

Acronyms

CAS	Computer Algebra System
MA	Monge-Ampère equations
MDI	The Principle of Minimum Discrimination Information
SMAE	Symplectic Monge-Ampère equations

Part I
Lectures

This part presents the lectures on geometric theory of differential equations and its applications.

Chapter 1
Contact Geometry, Measurement, and Thermodynamics

Valentin V. Lychagin

1.1 Preface

This paper has a long story and goes back to the middle of 80s but its recent version is based on the series of lectures I gave during the Summer school Wisla 18.

It is difficult now to say what kind of students and their backgrounds I had in mind. Possibly, and it reflects my own background, I had in mind students working in differential geometry, differential equations, and mathematical physics.

My friend, when I asked his opinion about thermodynamics, just said to me: consider physics as a big family with brothers, sisters, cousins, etc. Then thermodynamics shall be witch among them.

This is provocative and possibly too strong image, but something like this I feel every time when I read books on thermodynamics, although nobody will deny that this is the more important part of physics.

Thus, the idea underlying these lectures was to show, first of all, that some kind of thermodynamics is presented in practically all acts of measurement and second, that it is strongly related to splendid mathematics, especially to contact and symplectic geometry and to the theory of singularities.

The paper is organized in the following way. In the first part, we give a short exposition of topics from probability theory that will be important for us. All of this more or less standard with, the exception of discussion on coherence conditions for dependency, which is heavily dependent on the Mac Lane coherence condition ([18]).

V. V. Lychagin (✉)
Department of Mathematics, UiT Norges Arktiske Universitet, Postboks 6050,
Langnes 9037, Tromsø, Norway
e-mail: valentin.lychagin@uit.no; lychagin@math.uit.no

V. A. Trapeznikov Institute of Control Sciences of Russian Academy
of Sciences, 65 Profsoyuznaya street, Moscow 117997, Russia

The second part is central to the paper. Here we discuss the measurement procedure for random vectors based on the principle of minimal information gain. We use this principle instead of principle of maximum entropy because the notion of entropy is not appropriate and not well defined in this situation.

The main result of this chapter and main geometrical image that stands behind any measurement procedure might be formulated as a Legendrian or Lagrangian manifold in contact or symplectic space equipped with an additional Riemannian structure.

At the end of this chapter, we apply these results to measurement of velocities in classical mechanics and more generally to measurement of physical fields. We show that variational methods in mechanics and classical field theory are just a continuation of the principle of minimal information gain and the functions of information gain are just Lagrangians in the corresponding variational problems.

In the final chapter, we apply the previous results to classical thermodynamics and illustrate them for the case of ideal and real gases in order to show how to get the corresponding Lagrangian and Legendrian manifolds, and then describe phase transitions and thermodynamical processes in the corresponding gases.

1.2 A Crash Course in Probability Theory

1.2.1 Measure Spaces and Measurable Maps

We begin with a motivation to consider σ-algebras, given by Giuseppe Vitali.

Let's consider subsets $X \subset \mathbb{R}$ and let's try to measure their "size" $\mu(X)$, where μ is a positive function, $\mu(X) \in [0, \infty]$.

Assume that we, for intervals, have the standard measure $\mu([a, b]) = b - a$, $\mu(\varnothing) = 0$, and let us also require that $\mu(X) \leq \mu(Y)$, if $X \subset Y$.

Moreover, assume that $\mu(\cup_i X_i) = \sum_i \mu(X_i)$, for disjoint countable unions: $X_i \cap X_j = \varnothing$, for $i \neq j$, and $\mu(X + t) = \mu(X)$, i.e., "size" does not change under translations of subsets, $t \in \mathbb{R}$. These are more or less natural expectations for the "size-function".

We will now show that such function does not exist. To this end, consider the factor group $\mathbb{R}/\mathbb{Q}$ consisting of classes $x \bmod \mathbb{Q}$ of real numbers $x \in \mathbb{R}$ module rationals.

Due to the axiom of choice, it is possible to choose representative $\bar{x} \in \mathbb{R}$ for any class $x \bmod \mathbb{Q}$, and therefore, one gets a map $\phi: \mathbb{R}/\mathbb{Q} \to \mathbb{R}$, such that $\pi \circ \phi = \mathrm{id}$, for the natural projection $\pi: \mathbb{R} \to \mathbb{R}/\mathbb{Q}$.

Taking fractional part of ϕ, we'll assume, that $Z = \mathrm{Im}\,\phi \subset [0, 1]$ and therefore, $c = \mu(Z) \leq 1$.

On the other hand, $\mathbb{R}$ is a disjoint union $\cup_{t \in \mathbb{Q}}(Z + t)$, and therefore, $c > 0$.

Consider now the following disjoint union: $Z' = \cup_{t \in \mathbb{Q} \cap [0,1]}(Z + t) \subset [0, 2]$. Then, $\mu(Z') \leq 2$ and $\mu(Z') = \sum_{t \in \mathbb{Q} \cap [0,1]} c$.

This contradiction shows that subset Z is not *measurable*. However, we still have enough *measurable* subsets which we can construct from intervals by taking their unions, intersections, and complements. This argument leads us to the following notion.

A *measurable space is a pair* $(\Omega, \mathcal{A})$, where Ω is a set and $\mathcal{A}$ is a σ-algebra, i.e., a subalgebra of the power algebra $\mathcal{P}(\Omega)$ $(= 2^\Omega)$ of all subsets of the set Ω closed with respect to *countable* unions and intersections as well as taking complements, and which satisfies $\varnothing, \Omega \in \mathcal{A}$.

In probability theory, elements of $\mathcal{A}$ are usually called *events* and while points $\omega \in \Omega$ are called *experiments*. Then unions $\bigcup X_i$, where $X_i \in \mathcal{A}$, correspond to the statement that at least one event X_i holds, and intersections $\bigcap X_i$ correspond to the statement that all events X_i hold.

Compare the notion of measurable space with the notion of *topological* spaces, where one defines a topology $\tau \subset \mathcal{P}(\Omega)$ as a collection of τ of open sets which is closed with respect all possible unions and finite intersections, and which satisfies $\varnothing, \Omega \in \tau$.

Thus, for a given topological space (Ω, τ) we'll consider the measurable space $(\Omega, \mathcal{B})$, where $\mathcal{B} = \mathcal{B}(\tau)$ is a σ-algebra generated by countable unions, intersections, and complements of open sets in τ. This algebra is called a *Borel* σ-algebra.

Any mapping $\phi \colon \Omega_1 \to \Omega_2$ of sets induces a mapping of power algebras

$$\phi^{-1} \colon \mathcal{P}(\Omega_2) \to \mathcal{P}(\Omega_1),$$

by taking preimages, and also

$$\phi^{-1}(X \cup Y) = \phi^{-1}(X) \cup \phi^{-1}(Y),$$
$$\phi^{-1}(X \cap Y) = \phi^{-1}(X) \cap \phi^{-1}(Y),$$
$$\phi^{-1}(X^c) = \left(\phi^{-1}(X)\right)^c,$$

for all subsets X, Y of set Ω_2, i.e., ϕ^{-1} is a morphism of the algebras.

For the mapping ϕ of given measurable spaces $(\Omega_1, \mathcal{A}_1)$ and $(\Omega_2, \mathcal{A}_2)$, we say that ϕ is *measurable (or random point)* if

$$\phi^{-1}(\mathcal{A}_2) \subset \mathcal{A}_1,$$

i.e., if ϕ^{-1} maps elements of σ-algebra $\mathcal{A}_2$ to $\mathcal{A}_1$ and preserves intersections, unions, and complements.

Once more, compare and recall that a mapping ϕ of topological spaces (Ω_1, τ_1) and (Ω_2, τ_2) is *continuous* if $\phi^{-1}(\tau_2) \subset \tau_1$, i.e., if ϕ^{-1} maps τ_2-open sets to τ_1-open sets. Hence, if we consider topological spaces as measurable with respect to the corresponding Borel σ-algebras, then continuous mappings are also Borel measurable.

Let now $(\Omega, \mathcal{A})$ be a measurable space. Then a *probability measure* p on $(\Omega, \mathcal{A})$ is a map $p \colon \mathcal{A} \to [0, 1]$ such that

- $p(\varnothing) = 0, \; p(\Omega) = 1$, and

- $p\left(\cup_i X_i\right) = \sum_i p\left(X_i\right)$, for all disjoint countable unions $X_i \cap X_j = \varnothing$, for $i \neq j$, and measurable sets $X_i \in \mathcal{A}$.

The triple $(\Omega, \mathcal{A}, p)$ is called a *probability space* and the value $p\left(X\right)$, where $X \in \mathcal{A}$, is called the *probability of the event X*.

Remark 1.1 (*Important!*) In these lectures, we consider only probability measures, but, for the sake of simplicity, we call them simply measures.

1.2.2 Operations Over Measures, Measure Spaces, and Measurable Maps

1. **Restriction and induction**.
   Let $(\Omega, \mathcal{A})$ be a measurable space and let $\Omega' \subset \Omega$ a be a subset. Then the σ-algebra $\mathcal{A}'$ generated by intersections $X \cap \Omega'$, for all $X \in \mathcal{A}$, defines a structure of measurable space on Ω'.
   The following is more general construction. Let $\phi \colon \Omega' \to \Omega$ be a map and $\phi^{-1}\left(\mathcal{A}\right)$ be the σ-algebra generated by all preimages of measurable subsets $X \subset \Omega$, $X \in \mathcal{A}$. Then $\left(\Omega', \phi^{-1}(\mathcal{A})\right)$ is a measurable space.
2. **Tensor product**.
   Let $(\Omega_1, \mathcal{A}_1)$ and $(\Omega_2, \mathcal{A}_2)$ be measurable spaces. Denote by $\mathcal{A}_1 \otimes \mathcal{A}_2$ the σ-algebra of subsets of the product $\Omega_1 \times \Omega_2$ generated by cylinders $X_1 \times X_2$, where $X_1 \in \mathcal{A}_1$ and $X_2 \in \mathcal{A}_2$. This algebra is called *tensor product* of σ-algebras $\mathcal{A}_1$ and $\mathcal{A}_2$.
3. **Measurable maps**.
   Let $\phi_1 \colon (\Omega_1, \mathcal{A}_1) \to (\Omega_2, \mathcal{A}_2)$ and $\phi_2 \colon (\Omega_2, \mathcal{A}_2) \to (\Omega_3, \mathcal{A}_3)$ be measurable maps. Then their composition $\phi_2 \circ \phi_1 \colon (\Omega_1, \mathcal{A}_1) \to (\Omega_3, \mathcal{A}_3)$ is measurable too. The constant maps as well as identity maps are measurable.
4. **Image of measure**.
   Let $\phi \colon (\Omega_1, \mathcal{A}_1) \to (\Omega_2, \mathcal{A}_2)$ be a measurable map and let p be a measure on $(\Omega_1, \mathcal{A}_1)$, then the formula

$$\phi_*\left(p\right)\left(X\right) \stackrel{\text{def}}{=} p\left(\phi^{-1}\left(X\right)\right)$$

defines the measure on $(\Omega_2, \mathcal{A}_2)$. It is easy to check that

$$(\phi_2 \circ \phi_1)_*\left(p\right) = \phi_{2*}(\phi_{1*}\left(p\right)).$$

5. **Tensor product of measures**.
   Let p_i, $i = 1, 2$, be measures on measurable spaces $(\Omega_i, \mathcal{A}_i)$, then the formula

$$(p_1 \otimes p_2)\left(X_1 \times X_2\right) \stackrel{\text{def}}{=} p_1\left(X_1\right) \cdot p_2\left(X_2\right),$$

where $X_1 \in \mathcal{A}_1$ and $X_2 \in \mathcal{A}_2$, defines the measure on $(\Omega_1 \times \Omega_2, \mathcal{A}_1 \otimes \mathcal{A}_2)$. This measure is called *(tensor) product of the measures* p_1 and p_2.

6. **Random vectors and random variables**.
   Measurable maps $\phi(\Omega, \mathcal{A}, p) \to \mathbb{R}$ or more generally $\psi: (\Omega, \mathcal{A}, p) \to E$, where E is a finite-dimensional vector space (over $\mathbb{R}$), and both spaces $E, \mathbb{R}$ are considered as measurable with respect to the Borel σ-algebras, and are called *random variable* and *random vector*, respectively.
   From the practical point of view, it means that preimages $\phi^{-1}(U)$ and $\psi^{-1}(V)$ are $\mathcal{A}$-measurable for all open sets $U \subset \mathbb{R}$ and $V \subset E$.

Remark 1.2 Almost all constructions that we will use in these lectures could be also applied in the case when E is an infinite-dimensional Banach or Hilbert space.

The set of all random vectors is closed with respect to addition and multiplication by numbers and therefore forms a vector space itself (often infinite-dimensional). The set of random variables is also closed with respect to product and therefore forms an algebra over $\mathbb{R}$.

7. The images $\phi_*(p)$ or $\psi_*(p)$ are probability measures on $\mathbb{R}$ and E, respectively. They are called *probability laws* for random variable ϕ or random vector ψ. We consider the values $\phi_*(p)(U)$ or $\psi_*(p)(V)$ as probabilities of finding a random variable ϕ or random vector ψ in the open set $U \subset \mathbb{R}$ or $V \subset E$.

1.2.3 The Lebesgue Integral

By *step functions* on a measurable space $(\Omega, \mathcal{A})$, we mean measurable functions that take only a finite number of values.

It is easy to see that any such function s could be presented in the form

$$s = \sum_{i=1}^{n} c_i \cdot \chi_{A_i},$$

where $A_i \in \mathcal{A}$ are measurable subsets, and χ_{A_i} are their indicator functions, i.e., $\chi_{A_i}(\omega) = 1$, when $\omega \in A_i$, and $\chi_{A_i}(\omega) = 0$ in the opposite case, and c_i are the values of the function $s : s|_{A_i} = c_i$ on A_i.

For a step function s, the definition of the (Lebesgue) integral with respect to the measure p is very natural:

$$\int_{\Omega} s \, dp \stackrel{\text{def}}{=} \sum_{i=1}^{n} c_i \, p(A_i).$$

For a nonnegative measurable function f this integral is defined as a limit of approximations of f by nonnegative step functions, or more precisely as

$$\int_{\Omega} f \, dp \overset{\text{def}}{=} \sup_{s} \left(\int_{\Omega} s \, dp, \text{ for all nonnegative step functions } s \text{ such that } s \le f \right).$$

(1.1)

It is easy to see that any measurable function f could be presented in the form: $f = f_+ - f_-$ for some measurable and nonnegative functions f_+, f_- and we'll define

$$\int_{\Omega} f \, dp \overset{\text{def}}{=} \int_{\Omega} f_+ \, dp - \int_{\Omega} f_- \, dp.$$

One can easily check (or read in textbooks) that the Lebesgue integral satisfies the most of familiar properties of integrals. Among them

1. The function $f \longmapsto \int_{\Omega} f \, dp$ is $\mathbb{R}$-linear in f.

2. For any nonnegative measurable function λ, such that $\int_{\Omega} \lambda \, dp = 1$, the formula

$$p_\lambda (A) = \int_{\Omega} \chi_A \lambda \, dp,$$

for $A \in \mathcal{A}$, defines a new probability measure on $(\Omega, \mathcal{A})$, and

$$\int_{\Omega} f \, dp_\lambda = \int_{\Omega} f \lambda \, dp$$

for any measurable function f.

3. If $\phi \colon (\Omega, \mathcal{A}) \to (\Omega', \mathcal{A}')$ is a measurable map and $p' = \phi_* (p)$ is the image of the measure p on Ω. Then

$$\int_{\Omega'} f' \, dp' = \int_{\Omega} \phi^* (f') \, dp,$$

for any measurable function f' on Ω'.

Here, $\phi^* (f') = f' \circ \phi$.

Remark 1.3 1. The difference between Riemann and Lebesgue integrals one can see and feel for the case when $\Omega = [0, 1]$ and $f = \chi_A$ is the indicator function of the set $A = [0, 1] \cap \mathbb{Q}$. Then the Riemann integral of f obviously does not exist but the Lebesgue integral equals zero.

2. Let's consider a measurable function $f \colon (\Omega, \mathcal{A}, p) \to \mathbb{R}$ as a random variable. Then the mean value (or average) $\langle f \rangle_p$ of f is the integral

$$\langle f \rangle_p \overset{\text{def}}{=} \int_{\Omega} f \, dp.$$

Let $p' = f_* (p)$ be the probability law for f. Then, due to the above property,

$$\langle f \rangle_p = \int_\Omega f \ dp = \int_\mathbb{R} x \ dp',$$

where x is the standard coordinate on $\mathbb{R}$.

1.2.4 The Radon–Nikodym Theorem

Let p and q be two measures on a measurable space $(\Omega, \mathcal{A})$. We'll write $q << p$ and say that measure q is *absolutely continuous with respect to measure* p, or *dominated* by p, if $q (A) = 0$ whenever $p (A) = 0$, for $A \in \mathcal{A}$.

If $q << p$ and $q << p$ we say that the measures p and q are *equivalent*.

The Radon–Nikodym theorem (see, for example, [25]) states that if $q << p$ then there is a nonnegative measurable function ρ such that

$$q (A) = \int_\Omega \rho \chi_A dp, \tag{1.2}$$

for any set $A \in \mathcal{A}$.

This function is called density q with respect to p, or *Radon–Nikodym derivative* $\frac{dq}{dp}$.

In this notation, formula (1.2) takes the more transparent form:

$$q (A) = \int_\Omega \chi_A \frac{dq}{dp} dp.$$

1.2.5 The Fubini Theorem

Let $(\Omega_1, \mathcal{A}_1, p_1)$ and $(\Omega_2, \mathcal{A}_2, p_2)$ be probability spaces. We define their product as the triple $(\Omega_1 \times \Omega_2, \mathcal{A}_1 \otimes \mathcal{A}_2, p_1 \otimes p_2)$, where $\mathcal{A}_1 \otimes \mathcal{A}_2$ is the σ-algebra generated by the products $A_1 \times A_2 \subset \Omega_1 \times \Omega_2$ of measurable sets $A_1 \in \mathcal{A}_1$ and $A_2 \in \mathcal{A}_2$ and the product of the measures $p_1 \otimes p_2$ is a measure on space $\Omega_1 \times \Omega_2$, such that

$$(p_1 \otimes p_2) (A_1 \times A_2) = p_1 (A_1) \cdot p_2 (A_2).$$

It is easy to see that the product $p = p_1 \otimes p_2$ is also probability measure: $(p_1 \otimes p_2) (\Omega_1 \times \Omega_2) = 1$.

We say that a measurable function f on $\Omega_1 \times \Omega_2$ is *integrable* if

$$\int_{\Omega_1 \times \Omega_2} |f| \, dp < \infty.$$

For integrable functions the Fubini theorem states that

$$\int_{\Omega_1 \times \Omega_2} f \, dp = \int_{\Omega_1} \left(\int_{\Omega_2} f \, dp_2 \right) dp_1 = \int_{\Omega_2} \left(\int_{\Omega_1} f \, dp_1 \right) dp_2. \tag{1.3}$$

Remark that functions given by the integrals $\int_{\Omega_1} f \, dp_1$ and $\int_{\Omega_2} f \, dp_2$ involved in the iterated integration is not necessarily defined everywhere, but the sets on which they are not defined are sets of measure zero.

1.2.6 Random Vectors

Let $X : (\Omega, \mathcal{A}, p) \to V$ be a random vector, $\dim_{\mathbb{R}} V = n < \infty$.
Define a *mean or average or expectation of* X as a vector $E(X) \in V$ such that

$$\alpha(E(X)) = \int_{\Omega} (\alpha \circ X) \, dp, \tag{1.4}$$

for any covector $\alpha \in V^*$.
Two remarks on this definition. First of all, if $e_1, ..., e_n$ is a basis in V and

$$X(\omega) = X_1(\omega) e_1 + \cdots + X_n(\omega) e_n$$

is a decomposition of X in this basis for some measurable functions $X_1, ..., X_n$, then (1.4) means that

$$E(X) = \langle X_1 \rangle_p e_1 + \cdots + \langle X_n \rangle_p e_n,$$

or

$$E(X) = \int_{\Omega} X \, dp,$$

if in the last formulae we mean coordinate wise integration.
Second, formula (1.4) allows us to define means of random vectors for infinite-dimensional Banach vector spaces also if we use the Bochner integral (see, for example, [4]) instead of the Lebesgue integral.

1.2.7 Conditional Expectation

Let $\phi: (\Omega_1, \mathcal{A}_1) \to (\Omega_2, \mathcal{A}_2)$ be a random point on Ω_2 and let p be a probability measure on $(\Omega_1, \mathcal{A}_1)$. Then the push-forward measure $p_\phi = \phi_*(p)$ on $(\Omega_2, \mathcal{A}_2)$ is the probability law, in the sense that $\phi_*(p)(B) = p(\phi^{-1}(B))$ is equal to the probability of finding the random point in set $B \in \mathcal{A}_2$.

Let now $X: (\Omega_1, \mathcal{A}_1) \to V$ be random vector. Then there is a random vector $E(X|\phi): (\Omega_2, \mathcal{A}_2) \to V$ (defined in the complement of a zero measure set) called *conditional expectation*, such that

$$\int_{\Omega_1} \phi^*(f) X dp = \int_{\Omega_2} f E(X|\phi) dp_\phi. \tag{1.5}$$

Taking indicator functions χ_B, for sets $B \in \mathcal{A}_2$, we can rewrite the above formula in equivalent form

$$\int_{\phi^{-1}(B)} X dp = \int_{B} E(X|\phi) dp_\phi. \tag{1.6}$$

The last formula allows us to consider conditional expectation as a "density" of local averages $\int_{\phi^{-1}(B)} X dp$ along preimages $\phi^{-1}(B)$.

To outline the construction of conditional probability we remark that due to definition (1.4) it is enough to consider only nonnegative random variables $X: (\Omega_1, \mathcal{A}_1) \to \mathbb{R}$.

For them, the left-hand side of (1.6) defines a measure on $(\Omega_2, \mathcal{A}_2)$. This measure is bounded and absolutely continuous with respect to probability law p_ϕ if $X \in L_1(\Omega_1, p)$. Therefore, at least in this case, conditional expectation $E(X|\phi)$ exists. For more details, see [10, 11].

Example

Let $\phi: \Omega_1 \times \Omega_2 \to \Omega_2$ be the projection $\phi(\omega_1, \omega_2) = \omega_2$ and let $p = \rho p_1 \otimes p_2$, where p_i are measures on Ω_i, respectively, and $\rho(\omega_1, \omega_2)$ is the density. Then

$$dp_\phi = \left(\int_{\Omega_1} \rho(\omega_1, \omega_2) dp_1 \right) dp_2,$$

and formula (1.5) gives us

$$E(X|\phi)(\omega_2) = \frac{1}{\int_{\Omega_1} \rho(\omega_1, \omega_2) dp_1} \int_{\Omega_1} X(\omega_1, \omega_2) dp_1.$$

1.2.8 Dependency, Coherence Conditions, and Tensor Product of Random Vectors

Let $X_1 \colon (\Omega_1, \mathcal{A}_1, p_1) \to V_1$ and $X_2 \colon (\Omega_2, \mathcal{A}_2, p_2) \to V_2$ be random vectors, $\dim_{\mathbb{R}} V_i = n_i < \infty$, $i = 1, 2$.

Define their tensor product as a random vector

$$X_1 \otimes X_2 \colon (\Omega_1 \times \Omega_2, \mathcal{A}_1 \otimes \mathcal{A}_2, p_{12}) \to V_1 \otimes V_2,$$

where

$$(X_1 \otimes X_2)(\omega_1, \omega_2) = X_1(\omega_1) \otimes X_2(\omega_2) \in V_1 \otimes V_2,$$

and p_{12} is a measure absolutely continuous with respect to product $p_1 \otimes p_2$.

Random vectors X_1 and X_2 are said to be *independent* if $p_{12} = p_1 \otimes p_2$.

Denote by $E(X_1)$ and $E(X_2)$ averages with respect to measures p_1 and p_2, respectively, and let $E(X_1 \otimes X_2)$ be the average with respect to measure p_{12}.

Then, due to the Fubini theorem,

$$E(X_1 \otimes X_2) = E(X_1) \otimes E(X_2),$$

if X_1 and X_2 are independent.

In general, the tensor

$$\mathrm{cov}\,(X_1, X_2) = E(X_1 \otimes X_2) - E(X_1) \otimes E(X_2) \in V_1 \otimes V_2.$$

is called the *cross-covariance* between random vectors X_1 and X_2.

In the special case when $X = X_1 = X_2$, $V = V_1 = V_2$, and $p_{12} = p \otimes p$ the tensor $\mathrm{var}(X) = \mathrm{cov}\,(X, X) \in V \otimes V$ is a symmetric tensor and is called the *variance* of the random vector X.

In order to measure various tensor products, we fix a probability space $(\Omega, \mathcal{A}, p)$ and call "dependency" a sequence of probability measures $p^{(k)}$ on products $(\Omega^k, \mathcal{A}^k)$, where $\Omega^k = \Omega \times \cdots \times \Omega$, $\mathcal{A}^k = \mathcal{A} \otimes \cdots \otimes \mathcal{A}$ are kth Descartes and tensor degrees, which are absolutely continuous with respect to the tensor product measure $p^k = p \otimes \cdots \otimes p$ and satisfy

$$\pi_{k,i}\left(p^{(k)}\right) = p^{(k-1)}. \tag{1.7}$$

Here $\pi_{k,i} \colon \Omega^k \to \Omega^{k-1}$ are the natural projections,

$$\pi_{k,i}(\omega_1, \ldots, \omega_k) = (\omega_1, \ldots \omega_{i-1}, \omega_{i+1}, \ldots, \omega_k)$$

for $\omega_i \in \Omega$.

Independence in these terms means that $p^{(k)} = p^k$ for all $k \geq 2$.

Conditions (1.7), as we'll see later on, uniquely define measures $p^{(k)}$, for $k \geq 3$, for given measures $p^{(1)} = p$ and $p^{(2)}$.

It is very similar to the Mac Lane coherence conditions for associativity of a tensor product (see [18]), where associativity of a product of three elements implies associativity of products of any number of elements.

For these reasons, we also call (1.7) *coherence conditions for probability measures*.

Let ρ_k be densities of measures $p^{(k)}$ with respect to measures p^k, then (1.7) means that

$$\int_{\Omega_i} \rho_k \, dp = \rho_{k-1}, \qquad (1.8)$$

where Ω_i are fibers of the projections $\pi_{k,i}$.

In order to show that such dependencies exist, we'll consider, at first, the case of finite-dimensional dependencies.

Namely, we'll assume that there is a finite-dimensional vector space W in the space of $\mathcal{A}$—measurable functions on Ω, $W \subset \mathcal{F}(\Omega)$, such that $1 \in W$ and all densities ρ_k, $k \geq 2$, belong to the kth tensor products $W^{\otimes k} = W \otimes \cdots \otimes W \subset \mathcal{F}(\Omega^k)$.

Here, we've identified the product of functions $f(\omega_1) \, g(\omega_2) \in \mathcal{F}(\Omega^2)$, where $f, g \in W$, with the tensor product $f \otimes g \in W \otimes W$.

Denote by $I: W \to \mathbb{R}$ the linear functional given by integral, $f \in W \longmapsto \int_{\Omega} f \, dp$, and let $e_0 = 1, e_1, \ldots, e_n$ be a basis in W, where $I(e_i) = 0$ if $i \geq 1$.

Assuming that $\rho_2 \in W \otimes W$ we represent it in the form

$$\rho_2 = \sum_{i,j \geq 0} c_{ij} e_{ij},$$

where $e_{ij} = e_i \otimes e_j$ and $c_{ij} \in \mathbb{R}$.

Denote by $I_1 = I \otimes \mathrm{id}$ and $I_2 = \mathrm{id} \otimes I$, we rewrite condition (1.8) in the form:

$$I_1(\rho_2) = 1 \in W \quad \text{and} \quad I_2(\rho_2) = 1 \in W,$$

or

$$\sum_{j \geq 0} c_{0j} e_j = 1 \in W, \qquad \sum_{i \geq 0} c_{i0} e_i = 1 \in W.$$

Therefore,

$$c_{00} = 1, c_{0j} = 0, c_{i0} = 0, \text{ if } i, j \geq 1,$$

and thus we get that

$$\rho_2 = 1 \otimes 1 + \sum_{i,j \geq 1} c_{ij} e_{ij}.$$

The similar calculations show that for the given ρ_2 of the above form, we get

$$\rho_3 = 1 \otimes 1 \otimes 1 + \sum_{i,j \geq 1} c_{ij} e_{ij}^{(3)}, \tag{1.9}$$

where

$$e_{ij}^{(3)} = 1 \otimes e_i \otimes e_j + e_i \otimes 1 \otimes e_j + e_i \otimes e_j \otimes 1,$$

and for general value of k we get

$$\rho_k = 1 \otimes \cdots \otimes 1 + \sum_{i,j \geq 1} c_{ij} e_{ij}^{(k)}, \tag{1.10}$$

where

$$e_{ij}^{(k)} = 1 \otimes \cdots \otimes 1 \otimes e_i \otimes e_j + \cdots + e_i \otimes e_j \otimes 1 \otimes \cdots \otimes 1.$$

Going back to the standard functional notations we'll rewrite these formulae in the following way.

Theorem 1.1 *For a given density function $\rho_2(\omega_1, \omega_2)$, the formula*

$$\rho_k(\omega_1, \ldots, \omega_k) = \sum_{i<j} \rho_2(\omega_i, \omega_j) - \frac{(k+1)(k-2)}{2} \tag{1.11}$$

gives the solution of coherence equation (1.8).

Remark 1.4 We saw the uniqueness of solution for densities generated by the finite-dimensional vector spaces W. It follows from the Stone–Weierstrass theorem that the uniqueness is also held in the case when Ω is a compact subspace of $\mathbb{R}^n$, $\mathcal{A}$ is the Borel algebra and we consider only continuous solutions of (1.8).

1.3 Measurement of Random Vectors

The problem of measurement of vectors, namely, the measurement of velocity vectors for gas particles, was initially investigated by Maxwell (1860) and later by Boltzmann (1872). They showed that the density of probability to get velocity vector $v \in T$ in a tangent space T with respect to the Lebesgue measure satisfies the normal law

$$\rho = \left(\frac{m}{2\pi k_B T} \right)^{3/2} \exp \left(-\frac{mv^2}{2k_B T} \right),$$

where k_B is the Boltzmann constant.

This distribution describes behavior only for ideal noble gases, for real ones the picture is much more interesting and much more complicated.

1.3.1 Entropy and the Shannon Formula

The notion of entropy as well as the name Entropie appears for the first time, due to Rudolf Clausius ([3]), in 1865.

In modern terms, it claims that the differential 1-form δQ which measure the change of heat is proportional to the differential of some function S, called *entropy*, i.e., $\delta Q = T dS$. The multiplier T is called temperature. Remark that in this setting the entropy function S is not uniquely defined. The absolute entropy function was defined later by using statistical approach and the third law of thermodynamics, or Nernst's postulate.

In statistical mechanics Boltzmann defined entropy as related to the number W of microscopic configurations of the system:

$$S = k_B \ln W. \tag{1.12}$$

Entropy in this picture is considered as a measure of uncertainty of our system after observation of the main macroscopic quantities of the system.

It was noted by Max Plank ([22]) that "The logarithmic connection between entropy and probability was first stated by L. Boltzmann in his kinetic theory of gases."

From a probabilistic point of view the Boltzmann formula could be written in the form $S = -k_B \ln p$, where p is a probability of taking a microscopic configuration, and in this form, it has very transparent interpretation.

Let (Ω, p) be a probability space, where $p(A)$ is the probability of an event $A \subset \Omega$.

We say that a function $s \colon A \mapsto s(A) \in \mathbb{R}$ is a *surprise function*, if it satisfies the following properties:

- $s(A) \geq 0$, $s(\Omega) = 0$, $s(\varnothing) = \infty$.
- $s(AB) = s(A) + s(B)$, if events A and B are independent.
- $s(A) = f(p(A))$, for a continuous function f.

Then, if we assume that the function $p \mapsto s(p)$ is continuous, we get that

$$s(A) = -\log_a p(A),$$

for some constant $a > 1$.

The base number a is used to give name for units of the surprise measurement: *nats, for $a = e$*; *bits*, for $a = 2$; and *hartleys*, for $a = 10$.

In what follows, we'll use the unit of nats.

Assume now that we have a finite probability space $\Omega = \{\omega_1, \ldots, \omega_n\}$ and let $p = \{p_1, \ldots, p_n\}$, where $p_i = p(\omega_i)$, is a probability measure. Then, due to Shannon, the average, or expected value, of the surprise function:

$$S(p) = -\sum_i p_i \ln p_i, \tag{1.13}$$

is called the *entropy of probability measure p*.

Now let's try to transfer this notion to the case of infinite probability spaces. Take, for example, $\Omega = \mathbb{R}$ and $p = f(x)dx$, where $f(x)$ is a probability density.

Then taking discretization $\mathbb{R} = \cup_i [x_i, x_{i+1}]$ and $\widetilde{p}_i = f(\xi_i)$, for some intermediate points $\xi_i \in [x_i, x_{i+1}]$, we expect that possible value of $S(p)$ is a limit of integral sums $S(\widetilde{p}) = -\sum_i \widetilde{p}_i \ln \widetilde{p}_i$.

We have

$$S(\widetilde{p}) = -\sum_i f(\xi_i) \ln f(\xi_i)\Delta_i - \sum_i f(\xi_i)\Delta_i \ln \Delta_i,$$

where $\Delta_i = x_{i+1} - x_i$.

The first term in this expression tends to $\int_{-\infty}^{\infty} f(x) \ln f(x)dx$, but the second one has no limit.

To resolve this problem, let's reconsider definition (1.13) by taking two probability distributions on Ω; say $p = \{p_1, \ldots, p_n\}$ and $q = \{q_1, \ldots, q_n\}$.

Assume that q is our initial distribution, then differences $(-\ln q_i) - (-\ln p_i)$ tell us about "the gain of information", and the average

$$S(p, q) = \sum_i p_i \ln \left(\frac{p_i}{q_i} \right)$$

is called the *gain of information* or the *Kullback–Leibler divergence* between probability distributions p and q.

Let's compare this gain of information with the previous notion of entropy. If we assume that $q_h = \{\frac{1}{n}, \ldots, \frac{1}{n}\}$ is the homogeneous distribution, "noise", then

$$S(p, q_h) = ln(n) - S(p).$$

In other words, the gain $S(p, q_h)$ is a difference between entropy of "noise", which is equal to $\ln(n) = \max_p S(p)$, and entropy $S(p)$ of the given distribution p.

Going now back to the case, when $\Omega = \mathbb{R}$, and assuming that $p = f(x)dx$ and $q = g(x)dx$, we get that

$$S(\widetilde{p}, \widetilde{q}) = -\sum_i f(\xi_i) \ln \frac{f(\xi_i)}{g(\xi_i)}\Delta_i \longrightarrow \int_{-\infty}^{\infty} f(x) \ln \frac{f(x)}{g(x)}dx \tag{1.14}$$

which now has sense.

1.3.2 Gain of Information

Let us now consider a general probability space $(\Omega, \mathcal{A}, p)$ with probability measure p and let q be another probability measure, which is equivalent to measure p, $p \sim q$, i.e., these measures have the same set of negligible (or measure zero) sets.

Then formula (1.14) leads us to the following notion of *information gain* or *Kullback–Leibler divergence* between probability distributions in the general case:

$$I(p, q) = \int_{\Omega} \ln\left(\frac{dp}{dq}\right) dp. \tag{1.15}$$

The main property of this function is positivity (*Gibbs inequality*) :

$$I(p, q) \geq 0, \text{ and } I(p, q) = 0 \text{ if and only if } p = q \text{ almost everywhere.}$$

It follows directly from Jensen's inequality:

$$\int_{\Omega} \ln\left(\frac{dp}{dq}\right) dp = -\int_{\Omega} \ln\left(\frac{dq}{dp}\right) dp \geq -\ln\left(\int_{\Omega} \frac{dq}{dp} dp\right) = -\ln\left(\int_{\Omega} dq\right) = 0.$$

Moreover, $I(p, q) = 0$ if $\frac{dq}{dp} = 1$, or $p = q$, almost everywhere.

Therefore, this function shows how to measure p diverges from measure q. Remark that this function is not a distance between two measures because it is not symmetric: $I(p, q) \neq I(q, p)$: Moreover, it does not satisfy the triangle inequality.

Formula (1.15) in terms of density ρ, $dp = \rho dq$, could be rewritten in the form

$$I(p, q) = \int_{\Omega} \rho \ln \rho \, dq,$$

which is similar (up to sign) to the entropy expression.

Applying, for example, this function to random points we would now be able to compare two random points $\phi_i : (\Omega, \mathcal{A}, p) \to (\Omega', \mathcal{A}')$, $i = 1, 2$, by taking $I(\phi_{2*}(p), \phi_{1*}(p))$, as well as estimate the dependency p_{12} between two measures p_1, p_2 on probability spaces $(\Omega_i, \mathcal{A}_i, p_i)$ by taking their *mutual information*: $I(p_{12}, p_1 \otimes p_2)$.

1.3.3 Principle of Minimal Information Gain

The *principle of maximum entropy* was formulated by Edwin Thompson Jaynes in 1957 ([6]). In these papers, he gave a very transparent relation between statistical mechanics and information theory.

Later Kullback ([14]) proposed the principle of minimum discrimination information (MDI) which we used here under the name *principle of minimal information gain*.

Consider a random vector $X\colon (\Omega, \mathcal{A}, q) \to V$ on a finite-dimensional vector space V. Then we consider the average

$$E(X) = \int_\Omega X \, dq$$

as a result of the measurement.

Indeed, assume that g is a metric on V and the result of measurement is a vector $c \in V$, such that expectation of length of vectors $X - c$, i.e., $E(g(X - c, X - c))$, is minimal.

Then, due to Jensen's inequality, we get that

$$E(g(X - c, X - c)) \geq g(E(X - c), E(X - c)) = g(E(X) - c, E(X) - c) \geq 0,$$

and it takes zero value if and only if $c = E(X)$.

Changing random vector X for $X - E(X)$, we'll assume that X is a *centered random vector*, i.e., $E(X) = 0 \in V$.

Let's consider the map $X\colon (\Omega, \mathcal{A}, q) \to V$ as some device that produces vectors of vector space V and uses probability measures $p \sim q$ on Ω as control parameters.

Namely, we assume that the result of the vector measurement for the given value of p is

$$E(X, p) = \int_\Omega X \, dp.$$

In order to get the vector $x \in V$ as a result of our measurement, we should adjust our device state $(\Omega, \mathcal{A}, q)$ to another $(\Omega, \mathcal{A}, p)$, where $p \sim q$, in such a way that

$$E(X, p) = x.$$

In terms of density functions, $dp = \rho dq$, this means that we are looking for non-negative and $\mathcal{A}$-measurable functions ρ on space Ω such that

$$\int_\Omega \rho \, dq = 1, \quad \int_\Omega \rho X \, dq = x. \tag{1.16}$$

Conditions (1.16) are too weak to define function ρ completely. So, and this is *the principle of minimal gain information*, we'll require in addition that the measure p is the closest one to the basic measure q in the sense that it minimizes the information gain:

$$I(p, q) = \int_\Omega \rho \ln \rho \, dq \Longrightarrow \min,$$

under conditions (1.16).

To this end, we consider the functional

$$\mathcal{L}(\rho) = \int_\Omega \rho \ln \rho \, dq - \lambda_0 \left(\int_\Omega \rho \, dq - 1 \right) - \left\langle \lambda, \int_\Omega \rho X \, dq - x \right\rangle,$$

where $\lambda_0 \in \mathbb{R}$ and covector $\lambda \in V^*$ are the Lagrange multipliers.

Then, the condition

$$\frac{\delta \mathcal{L}}{\delta \rho} = \ln \rho + 1 - \lambda_0 - \langle \lambda, X \rangle = 0,$$

gives

$$\rho = \exp(\lambda_0 - 1 + \langle \lambda, X \rangle),$$

and the first condition of (1.16) gives us

$$\rho = \frac{1}{Z(\lambda)} e^{\langle \lambda, X \rangle}, \tag{1.17}$$

where

$$Z(\lambda) = \int_\Omega e^{\langle \lambda, X \rangle} \, dq, \tag{1.18}$$

is the so-called *partition function*.

Remark that this integral may not exist for all values of λ.

Let's fix a connected and simply connected domain $D \subset V^*$ where the function $Z(\lambda)$ is defined and smooth, and assume that $0 \in D$.

Then the differential of this function at a point $\lambda \in D$ is a vector of the cotangent vector space $T^*(D)$ which we, using the affine structure on V^*, shall identify with $(V^*)^* = V$. In other words, we'll assume that $d_\lambda Z \in V$.

Remark 1.5 The hidden probability space Ω could be eliminated from the formula for the partition function and the last formula could be written in terms of vector space V only:

$$Z(\lambda) = \int_V e^{\langle \lambda, t \rangle} \, d\mu(t), \tag{1.19}$$

where $\mu = X_*(q)$ is the image of the basic measure q.

To satisfy the second condition of (1.16), we remark that

$$d_\lambda Z = \int_\Omega e^{\langle \lambda, X \rangle} X \, dq = Z(\lambda) \, x.$$

Therefore, if we introduce another function, which we'll call *the Hamiltonian function*,

$$H(\lambda) = -\ln Z(\lambda), \qquad (1.20)$$

then

$$x = -d_\lambda H. \qquad (1.21)$$

We also have

$$I(p, q) = \frac{1}{Z(\lambda)} \int_\Omega e^{\langle \lambda, X \rangle} (H(\lambda) + \langle \lambda, X \rangle) \, dq \qquad (1.22)$$

$$= H(\lambda) + \langle \lambda, x \rangle = H(\lambda) - \langle \lambda, d_\lambda H \rangle. \qquad (1.23)$$

It is possible to get the vector x as a result of using our device X if and only if Eq. (1.21) solvable. In other words, if we consider the Legendre map

$$dH : \lambda \in \boldsymbol{D} \longmapsto d_\lambda H \in \boldsymbol{V},$$

then Eq. (1.21) solvable if and only if the vector x belongs to the image of the map $dH : \quad x \in \widehat{\boldsymbol{D}} = \operatorname{Im} dH \subset \boldsymbol{V}$.

Denote by $\Sigma \subset \boldsymbol{D}$ the set of singular points of the Legendre map $dH : \boldsymbol{D} \to \widehat{\boldsymbol{D}}$, and let $\widehat{\Sigma}$ be the image of Σ. Then $\boldsymbol{D} \smallsetminus \Sigma$ as well as $\widehat{\boldsymbol{D}} \smallsetminus \widehat{\Sigma}$ are unions of "phases" $\boldsymbol{D}_i$ and $\widehat{\boldsymbol{D}}_i$ and the restriction of the Legendre map on them are local diffeomorphisms, and therefore, Eq. (1.21) has a discrete set of solutions $\lambda_{i,j}(x)$ for $x \in \widehat{\boldsymbol{D}}_i$.

Let us first consider vectors $x \in \widehat{\boldsymbol{D}}_i$ such that (1.21) has a unique solution $\lambda(x) \in \boldsymbol{D}_i$ and write (1.22) as a function in x, $I = I(x) = H(\lambda(x)) + \langle \lambda(x), x \rangle$. Taking the differential, we get

$$d_x I = \lambda(x). \qquad (1.24)$$

In the opposite way, given a function $I(x)$ in a domain $\mathbf{D}_i$, then (1.24) and the relation

$$H(\lambda) = I - \langle d_x I, x \rangle$$

define the Hamiltonian and, as a consequence, the partition function Z.

In other words, the Hamiltonian function $H(\lambda)$ and the information gain function $I(x)$, $x \in \widehat{\boldsymbol{D}}_i$, are related by the Legendre transformation (1.22) and our device allows us to measure vectors $x \in \widehat{\boldsymbol{D}}_i$, where densities of the corresponding measures can be found from (1.17).

Geometrically, Eq. (1.21) defines an n-dimensional ($n = \dim V$) manifold L_H in the $2n$-dimensional space $\Phi = V \times V^*$:

$$L_H = \{x = -d_\lambda H, \lambda \in D\}.$$

The manifold Φ is symplectic with respect to the structure 2-form

$$\omega = \sum_{i=1}^{n} d\lambda_i \wedge dx_i,$$

where $\lambda = \sum \lambda_i e_i^*$ and $x = \sum x_i e_i$ in some basis $e_1, \ldots, e_n$ in V, and dual basis $e_1^*, \ldots, e_n^*$ in V^*.

Moreover, the manifold L_H is Lagrangian, i.e., the restriction of ω on L_H equals zero:

$$\omega|_{L_H} = 0.$$

Remark that the functions $(\lambda_1, \ldots, \lambda_n)$ are global coordinates on L_H, but $(x_1, \ldots, x_n)$ are not and the singular set Σ is exactly the set where these functions could not be used as local coordinates.

In general, let $L \subset \Phi$ be such a Lagrangian manifold that functions $(\lambda_1, \ldots, \lambda_n) \in D$ are global coordinates on it.

Then functions H and I, as functions on L, could be found in the following way:

$$I(a) = \int_\gamma \theta, \quad H(a) = \int_\gamma \widehat{\theta}, \tag{1.25}$$

where γ is a path on L connecting the points $o = (0, 0)$ and $a \in L$, and

$$\theta = \sum_{i=1}^{n} \lambda_i dx_i, \quad \widehat{\theta} = -\sum_{i=1}^{n} x_i d\lambda_i. \tag{1.26}$$

The condition that L is Lagrangian, together with the properties $d\theta = d\widehat{\theta} = \omega$, implies that the 1-forms θ and $\widehat{\theta}$ are closed on L. Moreover, the manifold L is diffeomorphic to D and is therefore connected and simply connected.

Therefore, the integrals (1.25) are correctly defined on L and

$$dI = \theta, \quad dH = \widehat{\theta} \text{ on } L, \tag{1.27}$$

which corresponds to relations (1.21) and (1.24).

Moreover,

$$d(I - H) = d(\langle \lambda, x \rangle)$$

on L, and all three functions I, H, $\langle \lambda, x \rangle$ equal zero at point o.

Therefore, we have the relation

$$I = H + \langle \lambda, x \rangle, \tag{1.28}$$

on L.

We shall extend our geometrical picture by adding the function I to the picture. Consider the $(2n + 1)$-dimensional manifold $\widetilde{\Phi} = \mathbb{R} \times \Phi$ and let u be the standard coordinate on the real line $\mathbb{R}$. Then points on $\widetilde{\Phi}$ are triples $(u, x, \lambda) \in \mathbb{R} \times V \times V^*$, or $\widetilde{\Phi} = \mathbb{R}^{2n+1}$, when we choose a basis in the vector space V.

The Lagrangian manifold L extended by function I defines an n-dimensional manifold $\widetilde{L} \subset \widetilde{\Phi}$ in the following way:

$$\widetilde{L} = \{(u = I(a), x = -d_a H, \lambda = d_a I), a \in L\}.$$

This is not an arbitrary manifold. Relations (1.27) mean that the differential 1-form

$$\Theta = du - \theta \tag{1.29}$$

vanishes on $\widetilde{L}$.

In other words, all above formulae (1.21), (1.24), (1.22) could geometrically be reformulated in the following way.

At first, we have the contact manifold $\widetilde{\Phi} = \mathbb{R} \times V \times V^*$ equipped with structure form Θ. Second, the minimal information gain principle leads us to a Legendrian manifold $\widetilde{L} \subset \widetilde{\Phi}$ with the condition that the projection $\pi : \widetilde{\Phi} \to V^*$ induces a diffeomorphism $\pi : \widetilde{L} \to D$. And finally, the Hamiltonian function H, the information gain function I, and the partition function Z, as functions on $\widetilde{L}$, could be found in the following way:

$$I = u|_{\widetilde{L}}, \quad H = I - \langle \lambda, x \rangle|_{\widetilde{L}}, \quad Z = \exp(-H). \tag{1.30}$$

Therefore, the Legendrian manifold $\widetilde{L}$ gives us all necessary information about the measurement of random vectors.

It is important to add that we require that the projection $\widetilde{L}$ on D should be diffeomorphism, but the projection $\widetilde{L}$ on V is not, and the singularities Σ of this projection lead us to a situation where the measurement of some vectors $x \in V$ could be done by different choices of λ's.

Finally, we could remove the condition that the projection $\pi : \widetilde{L} \to D$ is a diffeomorphism by removing from consideration set $\Sigma^* \subset \widetilde{L}$ of singular points of the projection $\widetilde{L} \to V^*$. Then, with possible additional cutting, we represent $\widetilde{L} \setminus \Sigma^*$ as union of Legendrian manifolds $\widetilde{L}_i$, so-called *phases*, for which the projections shall be embeddings. Then we'll assume that formulae (1.30) are still valid for all $\widetilde{L}_i$.

The problem is to find conditions such that the corresponding functions Z_i are the partition functions for some random vectors X_i.

To this end, let's consider the image $X_*\,(q)$ of the basic measure q and assume that this measure on V is absolutely continuous with respect to the Lebesgue's measure dx and let $f\,(x)$ be the density function.

Then integral (1.18) shall take the form

$$Z\,(\lambda) = \int_V e^{\langle \lambda, x \rangle} f\,(x)\ dx, \tag{1.31}$$

i.e., these functions are related by some kind of Laplace transform.

Thus, we arrived at the following problem: for each piece $\widetilde{L}_i$ and the corresponding functions Z_i on them, we are looking for probability density functions f_i such that Eq. (1.31) is satisfied.

These are the technical requirements that should be imposed on Legendrian manifold $\widetilde{L}$.

In this case, we call $\widetilde{L}$ *admissible Legendrian manifold*. In the next sections, we'll find more geometrical conditions for Legendrian manifolds to be admissible.

1.3.4 The Gaussian Distribution

In this section, we consider the case when the measure μ in (1.19) has the normal or Gaussian law of distribution $N\,(m, A)$, i.e., $d\mu = f\,(t)\,dt$, where dt is a Lebesgue measure on V, $t \in V$, and the density function f has the form

$$f\,(t) = \sqrt{\frac{\det A}{(2\pi)^n}}\ \exp\left(-\frac{1}{2}\,\langle A(t - m), t - m \rangle \right),$$

where $A\colon V \to V^*$ is a self-adjoint positive operator and $m \in V$ is the mean of $N\,(m, A)$. The quadratic form $\frac{1}{2}\,\langle At, t \rangle$ is known, in statistics, as the *Mahalanobis metric*.

As we know, the notion of determinant does not exist for operators of the form $A\colon V \to V^*$. In our case the construction of $\det A$ goes in the following way.

At first, we choose an n-form $\tau \in \Lambda^n\,(V^*)$ which generates by translations the Lebesgue measure dt, and we let $\tau^* \in \Lambda^n\,(V)$ be the dual form, i.e., $\langle \tau, \tau^* \rangle = 1$. Then the nth exterior degree of operator A, gives us the operator $\Lambda^n\,(A) : \Lambda^n\,(V) \to \Lambda^n\,(V^*)$, and we define $\det A$ as follows:

$$\Lambda^n\,(A)\,(\tau^*) = \det A \cdot \tau.$$

Said simply, we define $\det A$ as $\det \|a_{ij}\|$, for matrix $\|a_{ij}\|$, where $A\,(e_i) = \sum_j a_{ij} e_j^*$ in a dual bases $\{e_i\}$ and $\{e_j^*\}$, such that $\tau\,(e_1, \ldots, e_n) = 1$.

Then direct computations show that the partition $Z(\lambda)$ function of $N(m, A)$ is defined for all $\lambda \in V^*$ and is equal to

$$Z(\lambda) = \exp\left(\frac{1}{2}\langle\lambda, A^{-1}\lambda\rangle + \langle\lambda, m\rangle\right).$$

Therefore, formulae (1.20)–(1.22) will take the form:

$$H(\lambda) = -\frac{1}{2}\langle\lambda, A^{-1}\lambda\rangle - \langle\lambda, m\rangle,$$

$$x = A^{-1}\lambda + m,$$

$$I = \frac{1}{2}\langle A(x - m), (x - m)\rangle.$$

Moreover, formula (1.17) shows, for this case, that the probability distribution corresponding to the choice of $\lambda \in V^*$ is also the Gaussian distribution $N(x, A)$.

These formulae show that geometrically the normal laws correspond to (positive) Lagrangian affine subspaces $L \subset V \times V^*$, where L is the graph $\lambda = A(x - m)$. It is easy to see that self-adjointness of the operator A corresponds to the condition that the graph L is a Lagrangian manifold.

In the next section, we'll see the geometrical interpretation of this positivity condition.

It is also worth to note that the information gain $I(N(m_1, A_1), N(m_0, A_0))$ is equal to

$$\frac{1}{2}\left(\operatorname{tr}\left(A_1^{-1}A_0\right) + \ln\det A_1 - \ln\det A_0 + \langle A_0(m_1 - m_0), m_1 - m_0\rangle - n\right), \quad (1.32)$$

and the confidence domain, for the given probability ζ, is the ellipsoid

$$\langle A(x - m), (x - m)\rangle \le \chi_n^2(\zeta),$$

where χ_n^2 is the quantile function of the chi-squared distribution.

1.3.5 Central Moments

The kth *moment of a random vector* $X: (\Omega, p) \to V$ is defined as

$$\mu_k(X) = \int_\Omega X(\omega) \otimes \cdots \otimes X(\omega)\, dp \in V^{\otimes k}.$$

It is the integral of a symmetric tensor, and therefore, it is symmetric itself:

$$\mu_k(X) \in S^k(V).$$

The *kth central moment* is the kth moment of centered random vector $X - \mu_1(X)$, i.e.,

$$\sigma_k(X) = \mu_k(X - \mu_1(X)) \in S^k(V).$$

The first central moment equals zero, $\sigma_1(X) = 0$, the second central moment $\sigma_2(X) \in S^2(V)$ is called the *variance*, the third one, $\sigma_3(X) \in S^3(V)$, *skewness*, etc.

It is easy to see that

$$\sigma_2(X) = \mu_2(X) - \mu_1(X) \otimes \mu_1(X),$$

and by the construction $\sigma_2(X)$ is a symmetric positive quadratic form on V^*.

In order to find variation $\sigma_2(X, \lambda)$ of the random vector X with respect to extreme measure $\rho\, dp$ (see (1.17)), we remark that using the affine structure on V^* we are able to identify differentials dH with vectors in V, and using this structure we define the Hessian, Hess $H \in S^2(V)$, as the covariant differential of differential 1-form dH with respect to the affine connection ∇:

$$\text{Hess } H \stackrel{\text{def}}{=} d_\nabla(dH).$$

In coordinates $(\lambda_1, \ldots, \lambda_n)$ on V^* given by the dual basis $\left\{e_j^*\right\}$, we have

$$\text{Hess } H = \sum_{i,j} \frac{\partial^2 H}{\partial \lambda_i \partial \lambda_j} d\lambda_i \otimes d\lambda_j.$$

Now we take formula (1.18) in the form

$$Z = \int_\Omega \exp\left(\sum_i \lambda_i X_i\right) dq,$$

where $X = \sum_i X_i e_i$ and get

$$dZ = \left(\int_\Omega \exp(\langle \lambda, X \rangle)\, X_i dq\right) d\lambda_i.$$

Then

$$\text{Hess } Z = Z \cdot \sum_{i,j} \left(\int_\Omega \frac{\exp(\langle \lambda, X \rangle)}{Z} X_i X_j\, dq\right) d\lambda_i \otimes d\lambda_j = Z \sum_{i,j} \left\langle \mu_2(X), e_i^* \otimes e_j^* \right\rangle d\lambda_i \otimes d\lambda_j.$$

Therefore,

$$\text{Hess } Z = Z \, \mu_2 \, (X, \lambda),$$

and

$$dZ = Z\mu_1 \, (X, \lambda).$$

Finally, the relation $H = -\ln Z$ gives us

$$\text{Hess } H = -\frac{\text{Hess } Z}{Z} + \frac{dZ \otimes dZ}{Z^2} = -\mu_2 \, (X, \lambda) + \mu_1 \, (X, \lambda) \otimes \mu_1 \, (X, \lambda) = -\sigma_2 \, (X, \lambda), \quad (1.33)$$

and we arrive to the following result.

Theorem 1.2 *The variation $\sigma_2 \, (X, \lambda)$ of the random vector X with respect to the extreme measure (1.17) is equal to $-\text{Hess } H$.*

Corollary 1.1 *In the domain $D \subset V^*$, where the partition function Z is defined and smooth, the Hamiltonian function H is strictly concave.*

Example

The variation of the Gaussian distribution $\mathcal{N} \, (m, A)$ considered as a quadratic function on V^* is equal to $-\langle \lambda, A^{-1}\lambda \rangle$.

Consider the Lagrangian manifold $L_H \subset \Phi$, then Hessian Hess H defines a negative quadratic differential form on it.

On the other hand, we define a quadratic differential form $\kappa \in \Sigma^2 \, (\Phi)$ on the entire manifold Φ as follows:

$$\kappa = \frac{1}{2} \sum_{i=1}^{n} (d\lambda_i \otimes dx_i + dx_i \otimes d\lambda_i). \quad (1.34)$$

Or shortly

$$\kappa = d\lambda \cdot dx,$$

where $\cdot$ stands for the symmetric product of differential forms.

It is easy to check that expression does not depend on the choice of dual bases $\{e_i\}$ and $\{e_j^*\}$ in V and V^* and, respectively, on coordinates $\{x_i\}$ and $\{\lambda_j\}$ on Φ.

Moreover, the restriction of κ on the Lagrangian manifold L_H equals

$$\kappa|_{L_H} = -\frac{1}{2} \sum_{i=1}^{n} (d\lambda_i \otimes dH_i + dH_i \otimes d\lambda_i) = -\text{Hess } H,$$

where $H_i = \frac{\partial H}{\partial \lambda_i}$.

Summarizing the above we get the following result.

Theorem 1.3 *The variation $\sigma_2(X, \lambda)$ of the random vector X with respect to the extreme measure is a positive quadratic differential form $\kappa|_{L_H}$ on the Lagrangian manifold L_H.*

Remark 1.6   1. The minimal information gain principle leads us to the following geometrical structures on the space $\Phi = V \times V^*$:

- symplectic structure

$$\omega = d\lambda \wedge dx,$$

  and
- pseudo-Riemannian manifold structure

$$\kappa = d\lambda \cdot dx,$$

  of signature (n, n).

2. The Lagrangian manifold L_H has to be Riemannian with respect to quadratic differential form $\kappa|_{L_H}$. This gives us an extra requirement for selection of phases: one should take only phases where differential form $\kappa|_{L_H}$ is positive.
3. Certainly, we could use all invariants of Riemannian manifolds to study Lagrangian state manifolds L_H but we should also remember that all of them are invariants of the diffeomorphism group and in our case only the group of affine transformations has a sense. Therefore, the measurement problems have much more (and very special) invariants. Thus, central moments $\sigma_k(X, \lambda) \in \Sigma^k(L_H)$ are kth degree differential forms on the Lagrangian manifold and they define invariants which are specific for our case.

1.3.6 Change of Information Gain

In this section, we analyze information gain $I(p_{\lambda+\Delta\lambda}, p_\lambda)$ for extreme measures (1.17):

$$dp_\lambda = \frac{1}{Z(\lambda)} \exp(\langle \lambda, X \rangle) \, dq,$$

for small variation $\Delta\lambda$.

First of all, the density measure $p_{\lambda+\Delta\lambda}$ with respect to p_λ equals

$$\tilde{\rho} = \frac{dp_{\lambda+\Delta\lambda}}{dp_\lambda} = \frac{Z(\lambda)}{Z(\lambda + \Delta\lambda)} \exp(\langle \Delta\lambda, X \rangle).$$

Therefore,

$$
\begin{aligned}
I\left(p_{\lambda+\Delta\lambda}, p_\lambda\right) &= \int_\Omega \widetilde{\rho} \ln \widetilde{\rho} \, dp_\lambda \\
&= \int_\Omega \frac{\exp\left(\langle \lambda + \Delta\lambda, X \rangle\right)}{Z\left(\lambda + \Delta\lambda\right)} \left(H\left(\lambda + \Delta\lambda\right) - H\left(\lambda\right) + \langle \Delta\lambda, X \rangle\right) dq \\
&= H\left(\lambda + \Delta\lambda\right) - H\left(\lambda\right) + \langle \Delta\lambda, x + \Delta x \rangle \\
&= H\left(\lambda + \Delta\lambda\right) - H\left(\lambda\right) - \langle H_\lambda, \Delta\lambda \rangle - \left\langle H_{\lambda\lambda}, \Delta\lambda^2 \right\rangle + o\left(\Delta\lambda^2\right) \\
&= -\frac{1}{2} \left\langle H_{\lambda\lambda}, \Delta\lambda^2 \right\rangle + o\left(\Delta\lambda^2\right).
\end{aligned}
$$

To summarize, we get the following.

Theorem 1.4 *The information gain of extreme measures is equal to*

$$
I\left(p_{\lambda+\Delta\lambda}, p_\lambda\right) = \frac{1}{2}\sigma_2\left(X, \lambda\right) + \mathrm{o}\left(\Delta\lambda^2\right).
$$

Remark 1.7 It is important to note that the first order in terms (in $\Delta\lambda$) of $I\left(p_{\lambda+\Delta\lambda}, p_\lambda\right)$ are trivial.

Example

As we have seen, all extreme distributions for the Gaussian distributions are Gaussian $N\left(x, A\right)$ with the same variation. Due to (1.32) we have

$$
I\left(N\left(x + \Delta x, A\right), N\left(x, A\right)\right) = \frac{1}{2}\langle A\left(\Delta x\right), \Delta x \rangle.
$$

The result of the above theorem could be used in different directions. First, the positivity of $-\mathrm{Hess}\left(H\right)$ at a point $\lambda \in V^*$ shows that means of vector measurement are filled a neighborhood of the vector $x \in V$, corresponding to the choice of $\lambda \in V^*$. Second, if the quadratic form $-\mathrm{Hess}\left(H\right)$ is indefinite then possible directions $\Delta\lambda$, as well as Δx, should be such that $-\mathrm{Hess}\left(H\right)$ is positive. In other words, available deviations $\Delta\lambda$, as well as Δx, belong to the positive cone of $-\mathrm{Hess}\left(H\right)$.

1.3.7 Constraints and Constitutive Relations

By constraints we mean algebraic relations $P\left(X\right) = c$ for random vectors, where P is a polynomial function on vector space V. To be precise, this relation has two meanings:

1. The relation $P(X) = c$ holds for the mean of random vector X, i.e., means x of X belongs to algebraic manifold $P(x) = 0$.
2. The relation $P(X) = c$ holds in mean, i.e., $E(P(X), \lambda) = c$.

In the first case, the Hamiltonian function $H(\lambda)$ satisfies the differential equation

$$P\left(-\frac{\partial H}{\partial \lambda}\right) = c.$$

The second case we'll discuss for quadratic constraints:

$$P = \sum a_{ij} x_i x_j + \sum a_i x_i.$$

Then, due to (1.33), we have

$$E(X_i, \lambda) = -\frac{\partial H}{\partial \lambda_i}, \quad E\left(X_i X_j, \lambda\right) = -\frac{\partial^2 H}{\partial \lambda_i \partial \lambda_j} + \frac{\partial H}{\partial \lambda_i} \frac{\partial H}{\partial \lambda_j},$$

and therefore relation $E(P(X), \lambda) = c$ is equivalent to the second-order differential equation for Hamiltonian function:

$$\sum a_{ij} \frac{\partial^2 H}{\partial \lambda_i \partial \lambda_j} - \sum a_{ij} \frac{\partial H}{\partial \lambda_i} \frac{\partial H}{\partial \lambda_j} + \sum a_i \frac{\partial H}{\partial \lambda_i} + c = 0.$$

In a similar way, we get differential equations of order k if P is a polynomial of degree k.

For the case when constraints constitute an algebraic manifold $M \subset V$, we get correspondingly systems of partial differential equations.

Another way to get a state Lagrangian manifold $L \subset V \times V^*$ consists of adding constitutive relations, or state equations in our picture of measurement.

Namely, the submanifold L could be defined by system equations of the form

$$L = \{f_1(x, \lambda) = 0, \ldots, f_n(x, \lambda) = 0\}, \tag{1.35}$$

with the condition that their Poisson brackets (f_a, f_b) vanish on L, for all $a, b = 1, \ldots, n$.

These equations are called *state or constitutive equations*. As above, they give a system of partial differential equations for the Hamiltonian function:

$$f_1\left(-\frac{\partial H}{\partial \lambda_1}, \lambda\right) = 0, \ldots, f_N\left(-\frac{\partial H}{\partial \lambda_n}, \lambda\right) = 0,$$

and the conditions that Poisson brackets (f_a, f_b) vanish on L are the conditions of solvability of the above system.

1.3.8 Application to Classical Mechanics and Classical Field Theory

1.3.8.1 Mechanics

Let M be a smooth manifold (= configuration space of a mechanical system). We are going to measure velocity vectors $v \in T_a M$, for all $a \in M$, following to the minimal information gain principle.

First of all, we eliminate probability spaces Ω and use the images of the probability measures as measures on the vector spaces $T_a M$.

Second, we consider a contact bundle $\pi : \mathbb{R} \oplus TM \oplus T^* M \to M$ which is the Whitney sum of the trivial, tangent, and cotangent bundles, i.e., fibers $\Phi_a = \pi^{-1}(a)$ of this bundle are contact spaces $\mathbb{R} \oplus T_a M \oplus T_a^* M$.

The measurement process requires an admissible Legendrian manifold $L_a \subset \Phi_a$ at points $a \in M$. We'll assume that there is a smooth subbundle $\pi : \mathcal{L} \to M$ of the bundle π, such that all fibers are admissible Legendrian manifolds $L_a \subset \Phi_a$. We call it an *admissible Legendrian bundle*.

The structures we discussed above on the spaces $\mathbb{R} \oplus V \oplus V^*$ give us fiberwise structures on Φ. Namely, the fiberwise contact form $\theta = du - \lambda dv$, the fiberwise symplectic form $\omega = d\lambda \wedge dv$ on $TM \oplus T^* M$ and fiberwise quadratic form $\kappa = d\lambda \cdot dv$ on $TM \oplus T^* M$.

Let $(q_1, \ldots, q_n)$ be local coordinates on M, $(v_1, \ldots, v_n)$ and $(\lambda_1, \ldots, \lambda_n)$ induced fiberwise coordinates on the tangent and, respectively, cotangent bundles, and let u be the standard coordinate on $\mathbb{R}$.

Then $(q_1, \ldots, q_n, u, v_1, \ldots, v_n, \lambda_1, \ldots, \lambda_n)$ are local coordinates on Φ and the above structures has the following form:

$$\theta = du - \lambda dv \quad - \text{ fiberwise contact form,}$$
$$\omega = d\lambda \wedge dv \quad - \text{ fiberwise symplectic form,}$$
$$\kappa = d\lambda \cdot dv \quad - \text{ fiberwise quadratic form.}$$

For an admissible Legendrian bundle the restriction of θ equals zero, the restriction of κ gives a Riemannian structure, the projection $\mathcal{L} \to T^* M$ is a diffeomorphism and the functions H and I are functions on $T^* M$ and TM, respectively.

Relations (1.22), (1.21), (1.24) shall now take the form

$$v_i = -\frac{\partial H}{\partial \lambda_i}, \ \lambda_i = \frac{\partial I}{\partial v_i}, \ I = H - \sum_i \lambda_i \frac{\partial H}{\partial \lambda_i},$$

which in terms of analytical mechanics correspond to the case when I is a Lagrangian and H is a Hamiltonian of a mechanical system, λ corresponds to momentum and v to velocity.

Moreover, for any trajectory $q = q(t)$, $v = \dot{q}(t)$, the action integral

$$\int L\Big(q\,(t)\,,\dot{q}\,(t)\Big)\,dt = \int I\Big(q\,(t)\,,\dot{q}\,(t)\Big)\,dt$$

corresponds the integral information gain and the principle of least action is a continuation of the principle of minimal information gain.

Example: Maxwell

For the case of Gaussian distribution, we get a Riemannian structure on manifold M.

1.3.8.2   Field Theory

In classical field theory, fields are understood as sections of smooth bundles $\pi : E \rightarrow M$.

We'll expand this notion and, in order to compare the value of sections (i.e., fields) in different points of M, we'll also assume that the bundle π is equipped with a connection ∇.

Geometrically, this means that at any point $s \in F_a = \pi^{-1}\,(a)$, of the fiber F_a, we indicate a subspace $C_s \subset T_s\,(E)$, $\dim C_s = \dim M$, which is transversal to the fiber. In other words, the differential $\pi_{*,s} : T_s\,(E) \rightarrow T_a\,(M)$ gives an isomorphism between C_s and $T_a\,(M)$.

Let's denote by $\pi_k : J^k\,(\pi) \rightarrow M$ the bundle of kth jets of sections of π, and let $\pi_{k,l} : J^k\,(\pi) \rightarrow J^l\,(\pi)$ be projections generated by reductions of k-jets to lth jets of sections, when $k > l$.

If we denote by $[h]_a^k$ the kth jet of the section $h : M \rightarrow E$ at the point $a \in M$, then $\pi_{k,l}\left([h]_a^k\right) = [h]_a^l$.

Let's note that $J^0\,(\pi) = E$ and the connection C gives us a section $C : E \rightarrow J^1\,(\pi)$ of the bundle $\pi_{1,0} : J^1\,(\pi) \rightarrow J^0\,(\pi)$. Sections h of the bundle π we'll call constant at point $a \in M$, if $[h]_a^1 = C_{h(a)}$.

In other words, the introduction of a connection in the bundle π is equivalent to understanding of a special class of constant sections.

Example

In classical mechanics on a manifold M, we work with the trivial bundle $\pi : \mathbb{R} \times M \rightarrow \mathbb{R}$. Then sections of π are just curves $q\,(t)$ on M and the constant sections are constant curves $q\,(t) = a \in M$, for all values of t.

It is known (see, for example, [12, 13]) that the bundle $\pi_{1,0} : J^1\,(\pi) \rightarrow J^0\,(\pi)$ is an affine bundle for any bundle π, i.e., fibers $\pi_{1,0}^{-1}\,(s)$, $s \in E$, are affine spaces and vector spaces, associated with this affine structure, are

$$T_s^v(E) \otimes T_a^*(M),$$

where $T_s^v(E) = T_s(F_a) \subset T_s(E)$ is the vertical tangent space.

The given connection C allows us to identify the affine space $\pi_{1,0}^{-1}(s)$ with vector space $V_s = T_s^v(E) \otimes T_s^*(M)$, (see [13]) by considering the plane C_s as the origin.

Applying now our measurement machinery to this situation, we arrive at a bundle $\Pi: \Phi \to E$, where the fibers $\Pi^{-1}(s)$ are direct sums $\mathbb{R} \oplus V_s \oplus V_s^*$, where as above V_s^* is the dual to V_s:

$$V_s^* = T_s^v(E)^* \otimes T_a(M).$$

As above, Π is a contact and metric bundle, where contact and metric structures are defined as in (1.34) and the measurement procedure in field theory is controlled by an admissible Legendrian bundle $\mathcal{L} \to E$.

Let now $(q_1, \ldots, q_n, u^1, \ldots, u^m)$ be local coordinates in the bundle π, where $(q_1, \ldots, q_n)$ are local coordinates on M and $(u^1, \ldots, u^m)$ are fiberwise coordinates and let $\left(q_1, \ldots, q_n, u^1, \ldots, u^m, \ldots u_i^j \ldots\right)$ be the standard local coordinates in $J^1(\pi)$.

Then the connection C is defined by functions $C_i^j(q_1, \ldots, q_n, u^1, \ldots, u^m)$, where planes of the distribution C are generated by vectors

$$\frac{\partial}{\partial q_i} + \sum_j C_i^j \frac{\partial}{\partial u^j},$$

and $i = 1, \ldots, n$ and $j = 1, \ldots, m$.

Vectors $v \in V$ are elements of the form

$$v = \sum_{i,j} x_i^j \frac{\partial}{\partial u^j} \otimes dq_i,$$

where

$$x_i^j = u_i^j - C_i^j(q, u).$$

Vectors of the dual vector space $\lambda \in V^*$ have the form

$$\lambda = \sum_{i,j} \lambda_i^j du^j \otimes \frac{\partial}{\partial q_i}.$$

For a given Hamiltonian $H(q, \lambda)$, we get

$$u_i^j = C_i^j(q, u) - \frac{\partial H}{\partial \lambda_i^j},$$

$$I(q, u, x) = \sum x_i^j \lambda_i^j - H,$$

$$\lambda_i^j = \frac{\partial I}{\partial x_i^j}, \quad x_i^j = u_i^j - C_i^j(q, u).$$

For a given section $h(q) = \left(h^1(q), \ldots, h^m(q)\right)$, the function of information gain is a function of the form

$$I\left(q, h(q), \ldots, \frac{\partial h^j}{\partial q_i} - C_i^j(q, h), \ldots\right)$$

and, similarly to the mechanical case, this coincides with the field Lagrangian and the principle of least action is a continuation of the principle of minimal information gain.

1.4 Thermodynamics

1.4.1 Laws of Thermodynamics

In the classical interpretation of the thermodynamical laws, they could be presented as follows ([7, 16, 21, 26]).

Zeroth Law   If two systems are in thermodynamical equilibrium with a third system then they are in equilibrium with each other. This law allows us to define the notion of *temperature T of a system.*

First Law   The first law, also known as the *law of conservation of energy*, states that energy cannot be created or destroyed in an isolated system:

$$\Delta E = \Delta Q - \Delta W,$$

where ΔE is a change of *internal energy*, i.e., the energy of the system excluding kinetic and potential energy of external forces, ΔQ is the amount of heat transferred into the system from its surroundings and ΔW is the net work done by the system to its surroundings.

Second Law   This law states that there is an *entropy S*, a physical measure of the lack of information about the microscopic structure of a system, and that

$$\Delta Q = T \Delta S,$$

where T is the temperature of the system. It also states that the entropy of any isolated system always increases.

Third Law   The third law of thermodynamics states that the entropy of a system approaches a constant value as the temperature approaches absolute zero.

Fourth Law   Onsager reciprocal relations ([16, 19]).

1.4.2 Thermodynamics and Measurement

More formally, we reformulate the main thermodynamical laws by saying that a thermodynamical system is characterized by two sets of variables: *extensive ones* such as entropy S, internal energy E, volume V, mass m, number of particles N, etc., and *intensive* ones such as temperature T, pressure p, chemical potentials, etc.

Here extensive means that these variables characterize "extent" or "size" of our system (like volume or mass). The intensive variables do not depend on the "size" of the system (like temperature or pressure). Moreover, if the number of independent extensive variables equals $n + 1$, then the number of independent intensive variables equals n.

In other words, a thermodynamical system is characterized by a point in $\mathbb{R}^{n+1} \times \mathbb{R}^n$. By a *thermodynamical state*, we'll mean a submanifold $L \subset \mathbb{R}^{n+1} \times \mathbb{R}^n$.

Let's denote extensive variables by (S, E, x), where $x \in \mathbb{R}^{n-1}$ and intensive variables by (T, y), where $y \in \mathbb{R}^{n-1}$. Then the second law on the infinitesimal level requires that $\delta Q = T dS$ on L and the work ΔW on the infinitesimal level is presented by the restriction of differential 1-form $\sum_i p_i dq_i$ on L.

Therefore, the first law of conservation of energy on the infinitesimal level says that

$$dE = T dS - \sum_{i=1}^{n-1} p_i dq_i \quad \text{on } L.$$

In other words, if

$$\theta = dE - T dS + \sum_i p_i dq_i$$

is a differential 1-form on $\mathbb{R}^{n+1} \times \mathbb{R}^n$ then the first and second laws mean that $\theta|_L = 0$, or that L is an integral manifold of 1-form θ.

If we now remark that differential 1-form θ defines a contact structure on $\mathbb{R}^{n+1} \times \mathbb{R}^n$ and if we require, in addition, that L is a maximal submanifold, where the law of conservation of energy holds, then we come to the understanding that a thermodynamical state is a Legendrian submanifold of the contact manifold $\left(\mathbb{R}^{n+1} \times \mathbb{R}^n, \theta\right)$.

If we observe now the parallels with the above picture of measurement and that information gain I equals, up to a constant, $-S$, or $dS = -dI$, then we'll write the form θ as proportional to form

$$\tilde{\theta} = -dS + T^{-1}dE + \sum_{i=1}^{n-1} T^{-1} p_i dq_i,$$

which is exactly form of the type (1.29), where

$$u = -S + \text{const}, \ \lambda_1 = -T^{-1},$$
$$\lambda_2 = -T^{-1} p_1, \ldots, \ \lambda_n = -T^{-1} p_{n-1},$$
$$E = x_1, \ q_1 = x_2, \ldots, \ q_{n-1} = x_n.$$

Therefore, any measurement is accompanied by some kind of thermodynamics, and in addition to the definition of thermodynamical state as a Legendrian submanifold, one should add the admissibility of the manifold. In particular, it should be required that the quadratic differential form

$$\kappa = d\lambda \cdot dx = -d\left(T^{-1}\right) \cdot dE - \sum_{i=1}^{n-1} d\left(T^{-1} p_i\right) \cdot d\left(q_i\right) =$$

$$T^{-2} dT \cdot \left(dE + \sum_{i=1}^{n-1} p_i dq_i\right) - T^{-1} \cdot \sum_{i=1}^{n-1} dp_i \cdot dq_i.$$

is positive on L.

In the case when functions $(E, q_1, .., q_{n-1})$ are coordinates on manifold L, we have

$$S = s(E, q),$$
$$T^{-1} = s_E, \ p_1 = \frac{s_{q_1}}{s_E}, \ldots, \ p_{n-1} = \frac{s_{q_{n-1}}}{s_E}.$$

To find a thermodynamical state L, we'll use the constitutive relations (see (1.35)) which, in this case, have the form of physical laws:

$$f_i(T, E, p, q) = 0.$$

As we have seen, finding Legendrian manifold L, which satisfies these laws, is equivalent to finding solutions of the system of first-order differential equations

$$f_i\left(s_E^{-1}, E, \frac{s_q}{s_E}, q\right) = 0,$$

and the condition that the manifold L is Legendrian is equivalent to the requirement that the Poisson brackets $[f_i, f_j]$ vanish on zeroes of $\{f_i\}$ and equivalent to the compatibility of the above system.

1.4.3 Gases

As the first and the simplest example, let's consider the case of gases. They are described by
  extensive variables:

- E—inner energy, S—entropy, V—volume, and m—mass;

  intensive variables:

- T—temperature, p—pressure, and ϕ—chemical potential.

From the above discussion, we see that the thermodynamic state of a gas is defined by a Legendrian 3-dimensional manifold $L \subset \mathbb{R}^7$, in the 7-dimensional contact manifold $\mathbb{R}^7$ equipped with structure form

$$\theta = dE - TdS + pdV - \phi\, dm.$$

Extensivity and intensivity properties mean that in the coordinate description of the system, say

$$E = E\,(S, V, m)\,, \quad T = T\,(S, V, m)\,, \quad p = p\,(S, V, m)\,, \quad \phi = \phi\,(S, V, m)$$

we have invariance under scaling transformations

$$(S, E, V, m) \longmapsto (\lambda S, \lambda E, \lambda V, \lambda m)\,,$$
$$(T, p, \phi) \longmapsto (T, p, \phi)\,,$$

for any positive number λ.

Let's introduce *specific quantities*:

$$\sigma = \frac{S}{m}, \varepsilon = \frac{E}{m}, v = \frac{V}{m},$$

and rewrite differential 1-form θ in new terms. We get

$$\theta = d\,(m\varepsilon) - T d\,(m\sigma) + p d\,(mv) + \phi dm =$$
$$(\varepsilon - T\sigma + pv - \phi)\,dm + m\,(d\varepsilon - T d\sigma + pdv)\,.$$

Therefore, the condition $\theta|_L = 0$, together with scaling invariance of L, is equivalent to the following relations on L:

$$\phi - \varepsilon + T\sigma - pv = 0,$$
$$d\varepsilon - Td\sigma + pdv = 0$$

For practical reasons, it's better to use ε, ρ and m as coordinates on L and write the last differential form as

$$\theta' = d\sigma - T^{-1}d\varepsilon - pT^{-1}dv.$$

Thus, the state manifold L is the direct product of Legendrian manifold $L' \subset \mathbb{R}^5$, $\theta'|_{L'} = 0$ and half line $\mathbb{R}_+$ with coordinate m.

If ε, v are coordinates on L', then L is given by two functions ϕ, σ of the variables ε, v. The first equation shows that

$$\phi = \varepsilon - T\sigma + pv,$$

where $\varepsilon - T\sigma + pv$ is the specific Gibbs free energy, and condition $\theta'|_{L'} = 0$ is equivalent to the relations

$$T = \sigma_\varepsilon^{-1}, \quad p = \sigma_v \sigma_\varepsilon^{-1}.$$

The metric κ' on L' has the form

$$\kappa' = \operatorname{Hess} \sigma = d\left(T^{-1}\right) \cdot d\varepsilon + d\left(pT^{-1}\right) \cdot dv$$

and the condition that κ' is negative is an extra condition for the state equations.

The Poisson bracket on the symplectic manifold $\mathbb{R}^4$ equipped with the structure form

$$d\theta' = -T^{-2}d\varepsilon \wedge dT - pT^{-2}dv \wedge dT - T^{-1}dp \wedge dv$$

equals

$$[f, g] = \frac{p}{T^2}\left(f_p g_\varepsilon - f_\varepsilon g_p\right) + \frac{1}{T}\left(f_T g_\varepsilon - g_T f_\varepsilon\right) + \frac{1}{T^2}\left(f_v g_p - f_p g_v\right).$$

Remark that the only nontrivial brackets between basic variables $(\varepsilon, T, \rho, p)$ are

$$[T, \varepsilon,] = T^{-1}, [p, \varepsilon] = pT^{-2}, [v, p] = T^{-2}.$$

1.4.3.1 Heat Capacities

The notion of "heat capacities" is a contradictory one in thermodynamics. The standard definition as the limit of $\frac{\Delta Q}{\Delta T}$, when $\Delta T \to 0$, is beneath criticism from the mathematical point of view, but completely clear from the physical side.

To understand this notion we recall that a thermodynamical state is defined by a Lagrangian manifold or, in the regular case, by the specific entropy σ as a function of v and ε. Moreover, the conditions $v = \text{const}$ or $p = \text{const}$ we'll consider as extra constraints: $d\rho = 0$ or $dp = 0$.

Let's consider heat capacities by turns.

$\langle dv = 0 \rangle$ In this case, we get

$$dT \bmod \langle dv \rangle = d(\frac{1}{\sigma_\varepsilon}) \bmod \langle dv \rangle = -\frac{\sigma_{\varepsilon\varepsilon}}{\sigma_\varepsilon^2} d\varepsilon \bmod \langle dv \rangle = \frac{1}{T_\varepsilon} d\varepsilon \bmod \langle dv \rangle ,$$

and

$$\delta Q \bmod \langle dv \rangle = T d\sigma \bmod \langle dv \rangle = d\varepsilon \bmod \langle dv \rangle .$$

Therefore, *the heat capacity at constant volume, C_v*, which is usually defined as "$\left(\frac{\partial Q}{\partial T}\right)_v$," equals

$$C_v \overset{\text{def}}{=} \frac{\delta Q \bmod \langle dv \rangle}{dT \bmod \langle dv \rangle} = T_\varepsilon . (\mathrm{C}_v)$$

$\langle dp = 0 \rangle$ First of all,

$$dp = p_v dv + p_\varepsilon d\varepsilon.$$

Therefore,

$$d\varepsilon = -\frac{p_v}{p_\varepsilon} dv \bmod \langle dp \rangle ,$$

and

$$\delta Q \bmod \langle dp \rangle = T d\sigma \bmod \langle dp \rangle = T \left(T^{-1} d\varepsilon + pvT^{-1} dv\right) \bmod \langle dp \rangle =$$
$$\left(pv - \frac{p_v}{p_\varepsilon}\right) dv \bmod \langle dp \rangle ,$$
$$dT \bmod \langle dp \rangle = \left(-T_\varepsilon \frac{p_\rho}{p_\varepsilon} + T_v\right) dv \bmod \langle dp \rangle .$$

Thus, *the heat capacity at constant pressure, C_p*, equals

$$C_p \overset{\text{def}}{=} \frac{\delta Q \bmod \langle dp \rangle}{dT \bmod \langle dp \rangle} = \frac{pv p_\varepsilon - p_v}{-T_\varepsilon p_\rho + T_v p_\varepsilon} . (\mathrm{C}_p)$$

By the *sound speed, C_s*, they usually understand "$\sqrt{\left(\frac{\partial p}{\partial \rho}\right)_\sigma} = v^{-1} \sqrt{-\left(\frac{\partial p}{\partial v}\right)_\sigma}$", or more correctly

$$C_s = v^{-1} \sqrt{\frac{-dp \bmod \langle d\sigma \rangle}{dv \bmod \langle d\sigma \rangle}} .$$

In a similar way as above, we get

$$d\varepsilon = -pdv \bmod \langle d\sigma \rangle,$$

and

$$dp \bmod \langle d\sigma \rangle = (p_v - pp_\varepsilon)\, dv \bmod \langle d\sigma \rangle.$$

Therefore,

$$C_s = v^{-1}\sqrt{pp_\varepsilon - p_v}.(C_s)$$

1.4.3.2 Ideal Gases

The ideal gas is defined by two state equations:

- the Clapeyron–Mendeleev equation

$$pv = RT,$$

where R is the specific gas constant, and
- the equation for internal energy

$$\varepsilon = \frac{n}{2}RT,$$

where n is the degree of freedom.

The corresponding differential equations for the entropy function $\sigma(\varepsilon, v)$ shall take the form

$$v\sigma_v\sigma_\varepsilon^{-1} - R\sigma_\varepsilon^{-1} = 0,$$
$$\frac{n}{2}R(\sigma_\varepsilon)^{-1} - \varepsilon = 0,$$

or

$$v\sigma_v = R, \quad \varepsilon\sigma_\varepsilon = \frac{Rn}{2}.$$

Therefore,

$$\sigma = R\left(\frac{n}{2}\ln\varepsilon + \ln v\right) + \text{const}.$$

and the quadratic form κ' takes the form

$$\kappa' = R\left(-\frac{n}{2}\varepsilon^{-2}d\varepsilon^2 - v^{-2}dv^2\right),$$

which is negative.

Taking the Legendre transformation of the function $I = -\sigma$, we get the Hamiltonian function $H(\lambda_1, \lambda_2)$ in the form

$$H(\lambda_1, \lambda_2) = -\frac{n}{2} \ln\left(\frac{n}{2(-\lambda_1)}\right) + \ln((-\lambda_2)) + \frac{n}{2} + 1.$$

Therefore, the partition function for measuring the volume and energy of ideal gases equals

$$Z(\lambda_1, \lambda_2) = \left(\frac{n}{2}\right)^{\frac{n}{2}} \exp\left(-\frac{n}{2} - 1\right)(-\lambda_1)^{-\frac{n}{2}}(-\lambda_2)^{-1},$$

where $\lambda_1 < 0$ and $\lambda_2 < 0$.

1.4.3.3 van der Waals Gases

The van der Waals state equation is a correction of the Clapeyron–Mendeleev law:

$$\left(p + av^{-2}\right)(v - b) = RT,$$

where constants a, b are positive. The constant a takes into account the intermolecular forces and the constant b, the molecular volume.

In order to find the second equation of state, we assume that the inner energy is linear in temperature, i.e.,

$$\varepsilon = A(v)T + B(v).$$

Then the Poisson bracket between functions $f = \left(p + av^{-2}\right)(v - b) - RT$ and $g = \varepsilon - A(\rho)T - B(\rho)$, restricted on levels $f = g = 0$, is equal to

$$\frac{b - v}{v^2 T^2}\left(TvA' + v^2B' - a\right).$$

Therefore,

$$TvA' + v^2B' - a = 0$$

and we get $A = \text{const}$ and $B = -\frac{a}{v} + \text{const}$ from this equation.

Remark that the case $a = b = 0$ corresponds to the real gas state and therefore $A = \frac{nR}{2}$, similar to the ideal gas.

Finally, we have the second state equation for the van der Waals gases in the form:

$$\varepsilon = \frac{nR}{2}T - \frac{a}{v}.$$

Then the system of differential equations $(f = g = 0)$ on the entropy function $\sigma(\varepsilon, v)$ has the following solution :

$$\sigma = R \ln\left((v - b)\left(\frac{a}{v} + \varepsilon\right)^{n/2}\right) + \sigma_0,$$

where σ_0 is a constant. The corresponding quadratic differential form κ' equals

$$\kappa' = -\frac{nRv^2}{2(\varepsilon v + a)^2} d\varepsilon \cdot d\varepsilon + \frac{nRa}{(a + \varepsilon v)^2} d\rho \cdot d\varepsilon + \left(-\frac{nR\varepsilon^2}{2(\varepsilon v + a)^2} + \frac{Rn}{2v^2} - \frac{R}{(v-b)^2} \right) dv \cdot dv,$$

and condition that κ' is negative is the positivity of the determinant

$$D = 2\frac{RTv^3 - 2a(v-b)^2}{nRT^3v^3(v-b)^2}.$$

Therefore, the conditions of applicability of the van der Waals law are the following:

$$T > \frac{2a}{R}(v-b)^2 v^{-3}.$$

Taking, as above, the Legendre transformation of the function $I = -\sigma$, we get the Hamiltonian function $H(\lambda_1, \lambda_2)$ for van der Waals gases:

$$H(\lambda_1, \lambda_2) = -\frac{4n}{3}\ln\left(\frac{4n}{3(-\lambda_1)}\right) + \frac{3\lambda_1}{\alpha} - \frac{8}{3}\ln(3\alpha - 1) + \frac{4n}{3}$$

The van der Waals partition function is given by

$$Z(\lambda_1, \lambda_2) = (3\alpha - 1)^{\frac{8}{3}}\left(\frac{4n}{3(-\lambda_1)}\right)^{\frac{4n}{3}}\exp\left(\alpha\lambda_2 - \frac{3\lambda_1}{\alpha} - \frac{4n}{3}\right),$$

where, similar to the ideal gases case, $\lambda_1 < 0$ and $\lambda_2 < 0$.

1.4.3.4 Phase Transition for van der Waals Gases

Here we discuss phase transitions by taking the example of van der Waals gases.

Let $L \subset \mathbb{R}^4$ be the Lagrangian manifold defined by the van der Waals state equations

$$(p + av^{-2})(v - b) - RT = 0,$$

$$\varepsilon - \frac{Rn}{2}T + \frac{a}{v} = 0,$$

with the restrictions

$$T > \frac{2a}{R}(v - b)^2 v^{-3}, v > b.$$

Solving these equations with respect to ε and v, gives us

$$\varepsilon = \frac{Rn}{2}T - \frac{a}{\gamma},$$
$$v = \gamma,$$

where γ is a root of the polynomial

$$Q(z) = pz^3 - (RT - bp)z^2 + az - ab.$$

Therefore, the projection on the plane (p, T) of the singular curve $\Sigma \subset L$ is given by zeroes of the discriminant of the polynomial Q:

$$Q_1 = -4ab^4 p^3 - 4ab^2 (3bRT + 2a) p^2 - 4\left(3b^2 R^2 T^2 - 5abRT + a^2\right) p - aR^2 T^2 (4bRT - a).$$

This curve also has singularities where the discriminant Q_1 has zeroes, i.e., at the point

$$p_c = \frac{a}{27b^2}, \ T_c = \frac{8a}{27bR}.$$

The corresponding critical values of v we get from the state equation:

$$v_c = 3b.$$

Let's now change coordinates

$$p = p'p_c, v' = vv_c, T = T'T_c, \varepsilon = \varepsilon'\varepsilon_c, \sigma = \sigma'\sigma_c, \tag{1.36}$$

where constants ε_c and σ_c are taken in such a way that (1.36) is a contact transformation.

Then we get

$$\varepsilon_c = \frac{a}{9b}, \ \sigma_c = \frac{3R}{8}.$$

After this contact transformation the state equations take the reduced form:

$$3pv^3 - (p + 8T)v^2 + 9v - 3 = 0, \tag{1.37}$$
$$\varepsilon - \frac{4n}{3}T + \frac{3}{v} = 0,$$

$$\sigma - \frac{4n}{3}\ln\left(\varepsilon + \frac{3}{v}\right) - \frac{8}{3}\ln(3v - 1) - \sigma'_0 = 0, \tag{1.38}$$

where we continue to use (ε, v, p, T) instead of $(\varepsilon', v', p', T')$.

Fig. 1.1 van der Waals (T, v) area

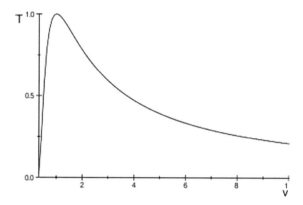

The quadratic differential form κ' in these coordinates takes the following form:

$$\kappa' = -\frac{4nv^2}{3\,(\varepsilon v + 3)^2}\,d\varepsilon\,d\varepsilon + \frac{8n}{(\varepsilon v + 3)^2}\,d\varepsilon\,dv + \left(\frac{4n}{3v^2} - \frac{4n\varepsilon^2}{3\,(\varepsilon v + 3)^2} - \frac{24}{(3v - 1)}\right) dv\,dv$$

with determinant

$$D = \frac{9\left(4Tv^3 - 9v^2 + 6v - 1\right)}{2n\,(3v - 1)^2\,T^3 v^3}.$$

Therefore, in the reduced coordinates, the border of the applicable area for the van der Waals gases on the plane (v, T) is given by equation

$$T = \frac{(3v - 1)^2}{4v^3}, \tag{1.39}$$

and the applicable area is given by inequalities (see Fig. 1.1):

$$T > \frac{(3v - 1)^2}{4v^3}, \quad v > \frac{1}{3}.$$

By eliminating v from the van der Waals state equation and Eq. (1.39), we find the applicable area in (p, T) plane:

$$p^3 + 6\,(4T + 9)\,p^2 + 3\left(64T^2 - 336T + 243\right) p + 16\,(32T - 27)\,T^2 > 0$$

Figure 1.2 shows (in white) applicable area for van der Waals gas on the plane (p, T), and the last picture gives an image of the Lagrangian manifold (Fig. 1.3), corresponding to the van der Waals state equations:

Fig. 1.2 Applicable (p, T)
area

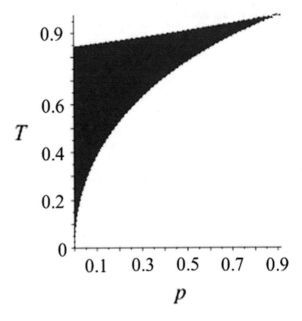

Fig. 1.3 Lagrangian
manifold for van der Waals
gas

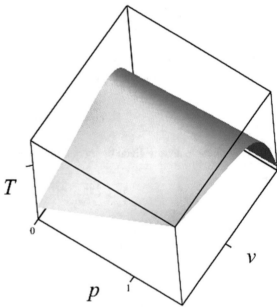

1.4.3.5   Real Gases

A more valuable model for real gases was proposed by Kamerlingh Onnes in 1902. In this model, the first state equation has the form of the *virial equation:*

$$f = p - R\,T\,v^{-1}\,Z(v, T) = 0,$$

where the function $Z(v, T)$, called the *compressibility factor*, takes the form

$$Z\,(v, T) = 1 + \frac{A_1\,(T)}{v} + \frac{A_2\,(T)}{v^2} + \cdots .$$

The functions $A_i\,(T)$ are called *virial coefficients.*

To find the inner energy equation, we'll assume as above that it has the form

$$g = \varepsilon - B\,(v, T)$$

for some function $B\,(v, T)$.

Then, computing the bracket, we get

$$[f, g] = Rv^{-1}Z_T - T^{-2}B_v = 0, \tag{1.40}$$

or

$$B\,(v, T) = RT^2 \int v^{-1}Z_T\,(v, T)\,dv + B_0\,(T). \tag{1.41}$$

The quadratic differential form κ' equals $d\left(T^{-1}\right) \cdot d\varepsilon + d\left(T^{-1}p\right) \cdot dv$, or

$$\kappa' = T^{-1}dp \cdot dv - T^{-2}dT \cdot (d\varepsilon + pdv). \tag{1.42}$$

Computing this form in coordinates v and T gives

$$\kappa' = -T^{-2}dT \cdot dB + Rd\left(v^{-1}Z\right) \cdot dv$$
$$= -T^{-2}B_T dT \cdot dT + \left(-T^{-2}B_v + Rv^{-1}Z_T\right)dT \cdot dv + R\left(-v^{-2}Z + v^{-1}Z_v\right)dv \cdot dv,$$

or, due to (1.40),

$$\kappa' = -T^{-2}B_T dT \cdot dT + R\left(-v^{-2}Z + v^{-1}Z_v\right)dv \cdot dv.$$

Therefore, the domain of applicability of the virial model is given by inequalities

$$B_T > 0, \quad \left(v^{-1}Z\right)_v < 0. \tag{1.43}$$

Example

For the van der Waals gases, we have

$$p = \frac{RTv^2 - av + ab}{v^2(v-b)},$$

and

$$v^{-1}Z(v, T) = \frac{RTv^2 - av + ab}{RT\, v^2(v-b)}.$$

Therefore, the conditions (1.43) in this case take the form

$$RTv^3 - 2av^2 + 4abv - 2ab^2 > 0,$$

and coincide with the previous estimate.

In terms of virial coefficients, relation (1.41) gives us

$$B(v, T) = B_0(T) - RT^2\left(\sum_{i=1}^{\infty} \frac{A_i'(T)}{i v^i}\right), \tag{1.44}$$

and requiring that the state equation coincides with the equation for ideal gases in the case of trivial virial coefficients gives us

$$p = RT\left(\frac{1}{v} + \sum_{i=1}^{\infty} \frac{A_i(T)}{v^{i+1}}\right),$$

$$\varepsilon = \frac{nR}{2}T - RT^2\sum_{i=1}^{\infty} \frac{A_i'(T)}{i v^i},$$

as the general state equations.

Then the conditions of applicability shall take the form

$$1 + \sum_{i=1}^{\infty} \frac{i+1}{i}\frac{A_i(T)}{v^i} > 0, \tag{1.45}$$

$$\frac{n}{2} - \sum_{i=1}^{\infty} \frac{T^2 A_i'' + 2T A_i'}{i v^i} > 0.$$

Example

The case closest to the van der Waals equation is the so-called *cubic virial equation*:

$$Z(v, T) = 1 + \frac{A_1(T)}{v} + \frac{A_2(T)}{v^2}.$$

For this case, we get

$$B(v, T) = \frac{nR}{2}T - RT^2\left(\frac{A_1'(T)}{v} + \frac{A_2'(T)}{2v^2}\right),$$

and the following domain of applicability:

$$\frac{n}{2} - \left(T^2 A_1'(T)\right)_T v^{-1} - \frac{1}{2}\left(T^2 A_2'(T)\right)v^{-2} > 0,$$

$$1 + 2v^{-1}A_1(t) + 3v^{-2}A_2(T) > 0,$$

on the plane (v, T).

1.4.4 Thermodynamic Processes and Contact Transformations

In this last section, we give a very brief exposition of an application of contact geometry to the description of thermodynamic processes. More details shall come later.

By a *thermodynamic process*, we mean a $1-$parameter family of transformations $A_t: \mathbb{R}^{2n+1} \to \mathbb{R}^{2n+1}$ of the thermodynamics phase space $\mathbb{R}^{2n+1}$ which preserves the law of energy conservation, i.e.,

$$A_t^*(\theta) \wedge \theta = 0,$$

or in other words a $1-$parameter family of contact transformations.

From the practical and computational point of view, we'll restrict ourselves by infinitesimal version of this notion and assume that the family A_t is a 1-parameter group of shifts along a contact vector field X :

$$L_X(\theta) \wedge \theta = 0.$$

The contact vector fields (see, for example, [15]) are uniquely defined by *generating functions* f, $X = X_f$, where the vector field X_f is uniquely defined by the relations

$$\theta(X_f) = f,$$
$$(X_f \rfloor \theta + df) \wedge \theta = 0.$$

Example

For the case of gases $n = 2$,

$$\theta = d\sigma - T^{-1} d\varepsilon - pT^{-1} dv,$$

and for a generating function $f = f(\sigma, \varepsilon, T, v, p)$, we have

$$X_f = (f + Tf_T)\,\partial_\sigma + T\left(pf_p + Tf_T\right)\partial_\varepsilon - Tf_p\partial_v \qquad (1.46)$$
$$+T\left(f_v - pf_\varepsilon\right)\partial_p - T\left(f_\sigma + Tf_\varepsilon\right)\partial_T.$$

In many cases, we'll eliminate entropy σ and consider, instead of contact vector fields X_f on the contact manifold $\left(\mathbb{R}^{2n+1}, \theta\right)$, Hamiltonian vector fields X_H on the symplectic manifold $\left(\mathbb{R}^{2n}, d\theta\right)$.

Example

For the case of gases

$$X_H = T\left(pH_p + TH_T\right)\partial_\varepsilon - TH_p\partial_v$$
$$+T\left(H_v - pH_\varepsilon\right)\partial_p - T\left(H_\sigma + TH_\varepsilon\right)\partial_T,$$

for Hamiltonian $H = H(\varepsilon, T, v, p)$.

Consider a thermodynamical system given by a state Legendrian manifold $L \subset \mathbb{R}^{2n+1}$ then by *thermodynamic process* for this system we'll understand a contact flow A_t such that

$$A_t(L) = L.$$

Proposition 1.1 *Assume that a Legendrian manifold* $L \subset \mathbb{R}^{2n+1}$ *is given by state equations*

$$L = \{f_1 = 0, \ldots, f_{n+1} = 0\}$$

and a thermodynamic process for this system is a flow A_t *along contact vector field* X_f *that preserves* L, *i.e.,* $A_t(L) = L$.
Then the restriction of this process on the manifold L *is given by the vector field*

$$G_1 X_{f_1} + \cdots G_{n+1} X_{f_{n+1}}$$

for some functions $G_1, \ldots, G_n$ *on* L.

Proof First of all, let's remark that the restriction of the generating function f on L equals zero, $f|_L = 0$, because X_f is tangent to L, $\theta|_L = 0$ and $f = \theta\left(X_f\right)$. Therefore,

$$f = \lambda_1 f_1 + \cdots + \lambda_{n+1} f_{n+1}$$

for some functions $\lambda_1, \ldots, \lambda_{n+1}$, and $G_i = \lambda_i|_L$.   □

Below we'll consider various thermodynamic processes for ideal and van der Waals gases.

1.4.4.1   Ideal Gases

As we have seen before, the state Legendrian manifold $L \subset \mathbb{R}^5$ for ideal gases is given by equations

$$f_1 = pv - RT,$$
$$f_2 = e - \frac{nR}{2}T,$$
$$f_3 = \sigma - R\left(\frac{n}{2}\ln\varepsilon + \ln v\right).$$

The corresponding contact vector fields are

$$X_{f_1} = (pv - 2RT)\,\partial_\sigma + T\,(pv - RT)\,\partial_\varepsilon + T\,(p\partial_p - v\partial_v),$$
$$X_{f_2} = (\varepsilon - nRT)\,\partial_\sigma - \frac{n}{2}T^2\partial_\varepsilon - T\,(p\partial_p + T\partial_T),$$
$$X_{f_3} = \left(\sigma - \frac{Rn}{2}\ln\varepsilon - R\ln v\right)\partial_\sigma + \frac{T}{\varepsilon}\left(\frac{nR}{2}T - \varepsilon\right)\partial_T + \frac{RT}{v\varepsilon}\left(\varepsilon - \frac{n}{2}pv\right)\partial_p.$$

It is easy to check that the restriction of the vector field X_{f_3} on L equals zero. Therefore, the thermodynamic processes for ideal gases, due to the above proposition, are given by vector fields on L:

$$Y = Av\partial_v + B\varepsilon\partial_\varepsilon,$$

which correspond to contact vector fields X_f with

$$f = Af_1 + Bf_2,$$

for some functions $A(\varepsilon, v)$ and $B(\varepsilon, v)$.

Among these processes, we have

- Adiabatic processes, $X_f(\sigma) = 0$ on L, when

$$2A + nB = 0.$$

- Polytropic processes, $X_f\left(pv^k\right) = 0$ on L, when

$$(k - 1) A + B = 0.$$

Here k is the so-called polytropic index.
- Isobaric processes, $X_f(p) = 0$ on L, when

$$A - B = 0.$$

- Isothermal processes, $X_f(T) = 0$ on L, when

$$B = 0.$$

1.4.4.2 van der Waals Gases

The state Legendrian manifold for van der Waals gases is given by the reduced equations

$$f_1 = 3pv^3 - (p + 8t)v^2 + 9v - 3,$$
$$f_2 = \varepsilon - \frac{4n}{3}T + \frac{3}{v},$$
$$f_3 = \sigma - \frac{4n}{3}\ln\left(\varepsilon + \frac{3}{v}\right) - \frac{8}{3}\ln(3v - 1).$$

The corresponding contact vector fields are

$$X_{f_1} = \left(3pv^3 - (p + 16T)v^2 + 9v - 3\right)\partial_\sigma + (3pTv - T(p + 8T))v^2\partial_\varepsilon - $$
$$Tv^2(3v - 1)\partial_v + T\left(9pv^2 - 2(p + 8T)v + 9\right)\partial_p,$$
$$X_{f_2} = \left(\varepsilon + \frac{3}{v} - \frac{8n}{3}T\right)\partial_\sigma - \frac{4n}{3}T^2\partial_\varepsilon - T^2\partial_T - T\left(p + \frac{3}{v^2}\right)\partial_p,$$
$$X_{f_3} = \left(\frac{4n}{3}\ln\left(\varepsilon + \frac{3}{v}\right) + \frac{8}{3}\ln(3v - 1) - \sigma\right)\partial_\sigma - T\left(1 - \frac{4n}{3}\frac{Tv}{\varepsilon v + 3}\right)\partial_T + $$
$$\frac{4T\left(3npv^3 - (np + 6\varepsilon)v^2 + 9(n - 2)v - 3v\right)}{3v(\varepsilon v + 3)(3v - 1)}\partial_p.$$

As above, the restriction of the vector field X_{f_3} on L equals zero and thermodynamic processes for van der Waals gases, due to the above proposition, are given by the following vector fields on L:

$$Y = \frac{3(\varepsilon v + 3)\left(9Av^2 + B\varepsilon v - 3Av + 3B\right)}{4nv^2}\partial_v + \frac{3v(\varepsilon v + 3)(3v - 1)A}{4n}\varepsilon\partial_\varepsilon,$$

for some functions $A(\varepsilon, v)$ and $B(\varepsilon, v)$.

As above, we have the following realizations for thermodynamic processes:

- Adiabatic processes, when $X_f(\sigma) = 0$ on L and

$$6v^2 A + nB = 0.$$

- Polytropic processes, $X_f\left(pv^k\right) = 0$ on L and

$$A\left(3(k-1)\varepsilon v^3 + \left(9k - k\varepsilon - \frac{9n}{2}(k-2)\right) - 9)v^2 + 3((k-2)n - k)v + \frac{n(k-2)}{2}\right) +$$
$$(\varepsilon v + 3) B = 0.$$

- Isobaric processes, $X_f(p) = 0$ on L and

$$\left(3\varepsilon v^3 - 9(n-1)v^2 + 6nv - n\right) A - (\varepsilon v + 3) B = 0.$$

- Isothermal processes, $X_f(T) = 0$ on L and $B = 0$.

Acknowledgements This work is partially supported by Russian Foundation for Basic Research (project 18-29-10013).

References

1. Alekseevsky D.V., Lychagin V.V. and Vinogradov A.M. Basic ideas and concepts of differential geometry, Springer – Verlag, 1991.
2. Ash, Robert B. Basic probability theory. Dover publications. NY. 1970.
3. Clausius, R. The Mechanical Theory of Heat – with its Applications to the Steam Engine and to Physical Properties of Bodies. London: John van Voorst, (1867).
4. Cohn, Donald. Measure Theory, Springer, (2013).
5. Dymond J.D., Wilhoit R.C., Virial coefficients of pure gases and mixtures, Springer (2003).
6. Jaynes, E.T. "Information theory and statistical mechanics". Physical Review. 106 (4): 620–630,(1957) and "Information theory and statistical mechanics II". Physical Review. 108 (2): 171–190, (1957).
7. Guggenheim, E.A. Thermodynamics. An Advanced Treatment for Chemists and Physicists, North Holland, Amsterdam, (1985).
8. Gibbs, Josiah Willard. Elementary Principles in Statistical Mechanics, developed with especial reference to the rational foundation of thermodynamics. New York: Charles Scribner's Sons, (1902).
9. Hirschfelder J.O., Curtiss Ch.F., Bird R.B. Molecular theory of gases and liquids, NY, (1954).
10. Kallenberg, Olav. Foundations of Modern Probability. 2. edition. Springer, New York (2002).
11. Kolmogorov, A.N.: Foundations of the Theory of Probability (2nd ed.). New York: Chelsea, (1956).
12. Krasilshchik I.S., Lychagin V.V. and Vinogradov A.M. Geometry of jet spaces and non-linear differential equations, Gordon and Breach, 1986.
13. Kruglikov Boris, Lychagin Valentin. Geometry of differential equations, in Handbook of global analysis, 725–771, Elsevier, Amsterdam, 2008.
14. Kullback, S. Information Theory and Statistics, John Wiley & Sons. (1959), Dover Publications (1968).

15. Kushner A.G. , Lychagin V.V. and Roubtsov V.N. Contact Geometry and Nonlinear Differential Equations, Cambridge University Press, 2007.
16. Landau, L.D.; Lifshitz, E.M. Statistical Physics. Oxford, (1975).
17. Maxwel, James. Theory of heat. Dover Publications. (2001).
18. Mac Lane, Saunders. Categories for the working mathematician. Graduate texts in mathematics Springer-Verlag, (1971).
19. Onsager, L., Reciprocal Relations in Irreversible Processes. I., Phys. Rev. 37, 405–426 (1931).
20. Kamerlingh Onnes H., Expression of state of gases and liquids by means of series, KNAW Proceedings, 4, 1901–1902, 125-147 (1902).
21. Pippard, A.B. Elements of Classical Thermodynamics for Advanced Students of Physics, Cambridge University Press,(1966).
22. Planck, Max. The Theory of Heat Radiation. Dover (1991).
23. Ruppeiner, George. "Riemannian geometry in thermodynamic fluctuation theory", Reviews of Modern Physics, 67 (3): 605–659, (1995).
24. Shannon Claude E. A Mathematical Theory of Communication, Bell System Technical Journal, Vol. 27, pp. 379–423, 623–656, 1948.
25. Shilov, G.E., Gurevich B.L., Integral, Measure, and Derivative: A Unified Approach. Dover Publications, (1978).
26. Sommerfeld, Arnold. Thermodynamics and statistical mechanics, Academic Press INC, (1956).

Chapter 2
Lectures on Geometry of Monge–Ampère Equations with Maple

Alexei Kushner, Valentin V. Lychagin and Jan Slovák

2.1 Introduction

The main goal of these lectures is to give a brief introduction to application of contact geometry to Monge–Ampère equations. These equations have the form

$$Av_{xx} + 2Bv_{xy} + Cv_{yy} + D(v_{xx}v_{yy} - v_{xy}^2) + E = 0, \qquad (2.1)$$

where A, B, C, D and E are functions on independent variables x, y, unknown function $v = v(x, y)$ and its first derivatives v_x, v_y.

Equations of this type arise in various fields. For example, G. Monge considered such equations in connection with the problem of the optimal transportation of sand or soil. This problem was of great importance for the construction of fortifications. A modern modification of this problem has the applications to mathematical economics, especially in taxations problem (Kantorovich–Monge problem [7]).

A. Kushner (✉)
Lomonosov Moscow State University, GSP-2, Leninskie Gory, Moscow 119991, Russia
e-mail: kushner@physics.msu.ru

Moscow Pedagogical State University, 1/1 M. Pirogovskaya Str., Moscow, Russia

A. Kushner · V. V. Lychagin
V. A. Trapeznikov Institute of Control Sciences of Russian Academy of Sciences, 65
Profsoyuznaya Str., Moscow 117997, Russia
e-mail: valentin.lychagin@uit.no

V. V. Lychagin
UiT Norges Arktiske Universitet, Postboks 6050, Langnes, 9037 Tromso, Norway

J. Slovák
Department of Mathematics and Statistics, Masaryk University, Kotlářská 2,
61137 Brno, Czech Republic
e-mail: slovak@muni.cz

© Springer Nature Switzerland AG 2019
R. A. Kycia et al. (eds.), *Nonlinear PDEs, Their Geometry, and Applications*,
Tutorials, Schools, and Workshops in the Mathematical Sciences,
https://doi.org/10.1007/978-3-030-17031-8_2

J.G. Darboux studied and applied such equations in his lectures on general theory of surfaces [3–5]. At that time, geometry was a source of various types of equations. For example, the problem of reconstructing a surface with a given Gaussian curvature $K(x, y)$ is equivalent to solving the following equation:

$$v_{xx}v_{yy} - v_{xy}^2 = K(x, y)\left(1 + v_x^2 + v_y^2\right)^2.$$ (2.2)

Nowadays, the number of sources of Monge–Ampère equations has increased. Equations arise in physics, aerodynamics, hydrodynamics, filtration theory, in models of the development of oil and gas fields, in meteorology and so on. Some of these applications will be discussed. On the other hand, as we shall see, the Monge–Ampère equations themselves generate geometric structures. For instance, some hyperbolic equations can be considered as almost product structures, and elliptic ones as almost complex structures.

The class of equations is rather wide and contains all linear and quasi-linear equations as we can see. On the other hand, it is the minimal class that contains quasi-linear equations and that is closed with respect to contact transformations.

This fact was known to Sophus Lie, who applied contact geometry methods to this kind of equations. In this paper, S. Lie posed some classification problems for equations with respect to contact pseudogroup. In particular, he posed the problem of equivalence of equations to the quasi-linear and linear forms. This problem was solved by V.V. Lychagin and V.N. Rubtsov [20] (see also [21]) in symplectic case and by A.G. Kushner [12] in contact case. Conditions when equations can be transformed to equations with constant coefficients by contact transformations were found by D.V. Tunitskii [23]. The problem of classification for mixed type equations was solved by A.G. Kushner [9–11].

In 1978, V.V. Lychagin noted that the classical Monge–Ampère equations and its multi-dimensional analogues admit effective description in terms of differential forms on the space of 1-jets of smooth functions [16]. His idea was fruitful, and it generated a new approach to Monge–Ampère equations.

The lectures has the following structure.

The first lecture is an introduction to geometry of 1-jets space. We define 1-jets of scalar functions, Cartan distribution, contact transformations and contact vector fields on the 1-jets space [8, 15].

In the second lecture, we describe V.V. Lychagin approach and an introduction to geometry of the Monge–Ampère equations. We follow papers [16, 17] and books [15, 18].

The third lecture is devoted to contact transformations of the Monge–Ampère equations. We consider examples of such transformations and apply them to construct multivalued solutions. We illustrate this on the example of equation arising in filtration theory of two immiscible fluids (oil and water, for example) in porous media [1].

In the fourth lecture, we study geometrical structures associated with non-degenerated (i.e. hyperbolic and elliptic) equations. We consider also the class of so-called symplectic equations and give a criterion of their linearization by symplectic transformation [18, 19].

The last, fifth lecture is devoted to tensor invariants of the Monge–Ampère equations. We construct here differential 2-forms that generalize the well-known Laplace invariants. We follow the papers [12, 14].

All calculations in these lectures are illustrated in the program Maple. The Maple files can be found on the website d-omega.org.

2.2 Lecture 1. Introduction to Contact Geometry

2.2.1 Bundle of 1-Jets

Let M be an n-dimensional smooth manifold, $C^\infty(M)$ be the ring of smooth functions on M and $T_a^* M$ be the cotangent space at the point $a \in M$.

Definition 2.1 A 1-*jet* $[f]_a^1$ *of a function* $f \in C^\infty(M)$ *at the point* a is a pair

$$(f(a), df|_a) \in \mathbb{R} \times T^* M.$$

The set of 1-jets at the point $a \in M$ of all functions

$$J_a^1 M := \left\{ [f]_a^1 \mid f \in C^\infty(M) \right\}$$

is a vector space with respect to operations of addition and multiplication by real numbers which are pointwise is defined as

$$[f]_a^1 + [g]_a^1 := [f + g]_a^1, \qquad k[f]_a^1 := [kf]_a^1.$$

Denote by

$$J^1 M := \mathbb{R} \times T^* M$$

the set of 1-jets of all smooth functions $f \in C^\infty(M)$ at all points $a \in M$.

This is a smooth manifold of dimension $2 \dim M + 1$ with local coordinates $x_1, \ldots, x_n u, p_1, \ldots, p_n$, where $x_1, \ldots, x_n$ are local coordinates on M, $p_1, \ldots, p_n$ are the induced coordinates on the cotangent bundle and u is the standard coordinate on $\mathbb{R}$. In other words, the values of these functions at point $[f]_k^1 \in J^1 M$ are the following:

$$x_i([f]_a^1) = x_i(a), \quad u([f]_a^1) = f(a), \quad p_i([f]_a^1) = f_{x_i}(a), \quad i = 1, \ldots, n. \quad (2.3)$$

These coordinates are called *canonical*.

In what follows we'll call $J^1 M$ the *manifold of 1-jets*, and the projection

$$\pi_1 : J^1 M \longrightarrow M, \quad \text{where} \quad \pi_1 : [f]_a^1 \longmapsto a$$

the *1-jet bundle*.

Any function $f \in C^\infty(M)$ defines the following map:

$$j_1(f): M \longrightarrow J^1M, \tag{2.4}$$

where

$$j_1(f): M \ni a \longmapsto [f]^1_a \in J^1_a M \subset J^1M.$$

The image

$$\Gamma^1_f := j_1(f)(M) \subset J^1M,$$

which is a smooth submanifold of J^1M, is called the 1-*graph* of the function f.

Consider the following differential 1-form

$$\varkappa := du - p_1dx_1 - \cdots - p_ndx_n$$

on the 1-jet space J^1M which we'll call *Cartan form*.

It is easy to check that this form does not depend on a choice of canonical coordinates in J^1M.

This form allows us to separate submanifolds of the form $\Gamma^1_f \subset J^1M$ from arbitrary submanifolds of dimension n by observation that

$$\varkappa|_{\Gamma^1_f} = 0,$$

for any $f \in C^\infty(M)$. Indeed,

$$\varkappa|_{\Gamma^1_f} = df - f_{x_1}dx_1 - \cdots - f_{x_i}dx_i = 0.$$

On the other hand, if a submanifold $N \subset J^1M$ is a graph of section $s: M \longrightarrow J^1M$, i.e. $\pi_1: N \longrightarrow M$ is a diffeomorphism, and

$$\varkappa|_N = 0,$$

then one can easily check that $N = \Gamma^1_f$ for some smooth function $f \in C^\infty(M)$.

This observation shows that zeroes of the Cartan form (but not the form itself) is important to distinguish 1-graphs from arbitrary submanifolds in J^1M.

Denote by C the $2n$-dimensional distribution (Cartan distribution) on J^1M given by zeroes of the Cartan form:

$$C: J^1M \ni \theta \longmapsto C(\theta) := \ker \varkappa_\theta \subset T_\theta(J^1M).$$

In the dual way, the Cartan distribution can be defined by vector fields tangent to this distribution. Namely, vector fields

$$\partial_{x_1} + p_1 \partial_u, \ldots, \partial_{x_n} + p_n \partial_u, \partial_{p_1}, \ldots, \partial_{p_n}$$

give us a local basis in the module of vector fields tangent to C. This module will be denoted by $D(C)$.

Then a submanifold $N \subset J^1 M$ is a graph of a smooth function if and only if

1. N is an integral submanifold of the Cartan distribution and
2. $\pi_1 \colon N \to M$ is a diffeomorphism.

Remind that a contact structure on an odd-dimensional manifold K, $\dim K = 2k + 1$, consists of $2k$-dimensional distribution P on K such that

$$\lambda \wedge (d\lambda)^k \neq 0$$

for any differential 1-form λ, such that locally $P = \ker \lambda$.

In our case, we have

$$\varkappa \wedge (d\varkappa)^n \neq 0$$

and therefore the Cartan distribution defines the contact structure on the manifold of 1-jets $J^1 M$.

2.2.2 Contact Transformations

A transformation Φ of the space $J^1 M$ is called *contact*, if it preserves the Cartan distribution, i.e.

$$\Phi_*(C) = C.$$

In terms of the Cartan form, a transformation Φ is contact if

$$\Phi^*(\varkappa) = h_\Phi \varkappa \tag{2.5}$$

for some function h_Φ, or equivalently

$$\Phi^*(\varkappa) \wedge \varkappa = 0.$$

Examples of Contact Transformations

1. Translations:

$$(x_1, x_2, u, p_1, p_2) \longmapsto (x_1 + \alpha_1, x_2 + \alpha_2, u + \beta, p_1, p_2),$$

where α_1, α_2 and β are constants.
2. The Legendre transformation:

$$(x_1, x_2, u, p_1, p_2) \longmapsto (p_1, p_2, u - x_1 p_1 - x_2 p_2, -x_1, -x_2).$$

3. Partial Legendre's transformation:

$$(x_1, x_2, u, p_1, p_2) \longmapsto (p_1, x_2, u - p_1 x_1, -x_1, p_2).$$

Infinitesimal versions of contact transformations are contact vector fields.

A vector field X on $J^1 M$ is called *contact* if its local translation group consists of contact transformations.

It means that

$$\Phi_t^*(\varkappa) = \lambda_t \varkappa \tag{2.6}$$

for some function λ_t on $J^1 M$. Here, Φ_t are shifts along vector field X.

After differentiating both parts of (2.6) by t at $t = 0$, we get:

$$\frac{d}{dt}\Big|_{t=0} (\Phi_t^*(\varkappa)) = \left(\frac{d\lambda}{dt}\Big|_{t=0}\right)\varkappa.$$

The left-hand side of the equation is the Lie derivative $L_X(\varkappa)$ of the Cartan form in the direction of the vector field X and therefore, we get

$$L_X(\varkappa) = h\varkappa,$$

where h is a function on $J^1 M$.

Multiplying both parts of the last equation by $\varkappa$, we get:

$$L_X(\varkappa) \wedge \varkappa = 0. \tag{2.7}$$

In canonical coordinates, each contact vector field has the form

$$X_f = -\sum_{i=1}^n \frac{\partial f}{\partial p_i} \frac{\partial}{\partial x_i} + \left(f - \sum_{i=1}^n p_i \frac{\partial f}{\partial p_i}\right)\frac{\partial}{\partial u} + \sum_{i=1}^n \left(\frac{\partial f}{\partial x_i} + p_i \frac{\partial f}{\partial u}\right)\frac{\partial}{\partial p_i}$$

for some function f which is called *generating function* of the contact vector field. Note that

$$\varkappa(X_f) = f.$$

Maple Code: Main Operation on $J^1\mathbb{R}^2$

1. Load libraries:

```
with(DifferentialGeometry): with(JetCalculus):
```

2. Set jet notation, declare coordinates on the manifold M and generate coordinates on the 1-jet space:

```
Preferences("JetNotation", "JetNotation2"):
DGsetup( [x1,x2],[u], M, 1, verbose);
```

3. Generate the Cartan form:

```
kappa:= convert(Cu[0,0],DGform);
```

4. Define partial Legendre transformation:

```
PartLegendre:=Transformation(M,M,[x1=-u[1,0],x2=x2,
u[0,0]=u[0,0]-u[1,0]*x1, u[1,0]=x1, u[0,1]=u[0,1]]);
```

5. Apply this transformation to the Cartan form:

```
Pullback(PartLegendre,kappa);
```

6. Prolongation of transformations from J^0M to J^1M:

```
Phi:=Transformation(M,M,
[x1=x2,x2=x1+x2,u[0,0]=-u[0,0]]);
Prolong(Phi,1);
```

7. Define the contact vector field X_f with generating function $f = p_2$:

```
X:=GeneratingFunctionToContactVector(u[0, 1]);
```

8. Prolongation of vector fields from the plane $M = \mathbb{R}^2$ to J^1M:

```
Y:=evalDG(-x2*D_x1+x_1*D_x2);
Prolong(Y,1);
```

2.3 Lecture 2. Geometrical Approach to Monge–Ampère Equations

2.3.1 Non-linear Second-Order Differential Operators

Following [16], any differential n-form ω on $J^1 M$ is associated with the differential operator

$$\Delta_\omega : C^\infty(M) \longrightarrow \Omega^n(M),$$

which acts in the following way:

$$\Delta_\omega(v) := j_1(v)^*(\omega), \qquad (2.8)$$

where (see formula (2.4))

$$j_1(v)^* : \Omega^n(J^1 M) \longrightarrow \Omega^n(M).$$

This construction does not cover all non-linear second-order differential operators, but only a certain subclass of them.

Examples

1. The differential 1-form on $J^1\mathbb{R}$

$$\omega = (1 - x^2)dp + (\lambda u - xp)\,dx,$$

   where

$$\lambda = \frac{a^2}{b^2},$$

   generates the Lissajou differential operator

$$\Delta_\omega(y) = \left((1 - x^2)y'' - xy' + \frac{a^2}{b^2}y \right) dx. \qquad (2.9)$$

   Indeed,

$$\Delta_\omega(v) = (1 - x^2)d\left(y'\right) + \left(-xy' + \frac{a^2}{b^2}y\right) dx$$

$$= \left((1 - x^2)y'' - xy' + \frac{a^2}{b^2}y \right) dx.$$

2. The differential 2-form on $J^1\mathbb{R}^2$

$$\omega = dp_1 \wedge dp_2$$

generates the Hesse operator

$$\Delta_\omega(v) = (\det \text{Hess } v)\, dx_1 \wedge dx_2. \tag{2.10}$$

Indeed,

$$\begin{aligned}
\Delta_\omega(v) &= d\left(v_{x_1}\right) \wedge d\left(v_{x_2}\right) \\
&= \left(v_{x_1 x_1} dx_1 + v_{x_1 x_2} dx_2\right) \wedge \left(v_{x_2 x_1} dx_1 + v_{x_2 x_2} dx_2\right) \\
&= \left(v_{x_1 x_1} v_{x_2 x_2} - v_{x_1 x_2}^2\right) dx_1 \wedge dx_2 \\
&= (\det \text{Hess } v)\, dx_1 \wedge dx_2,
\end{aligned}$$

where Hess v is the Hessian of the function v.

3. The differential 3-form

$$\omega = p_1 dp_1 \wedge dx_2 \wedge dx_3 - dx_1 \wedge dp_2 \wedge dx_3 - dx_1 \wedge dx_2 \wedge dp_3 \tag{2.11}$$

on $J^1\mathbb{R}^3$ produces the von Karman differential operator

$$\left(v_x v_{xx} - v_{yy} - v_{zz}\right) dx \wedge dy \wedge dz,$$

where $x = x_1$, $y = x_2$, $z = x_3$.

4. The differential 2-form

$$\omega = dp_1 \wedge dx_2 - dp_2 \wedge dx_1$$

on $J^1\mathbb{R}^2$ represents the *two*-dimensional Laplace operator

$$\Delta_\omega(v) = \left(v_{xx} + v_{yy}\right) dx \wedge dy,$$

where $x = x_1$, $y = x_2$.

5. Two differential 2-forms

$$\omega = dx_1 \wedge du \qquad \text{and} \qquad \varpi = p_2 dx_1 \wedge dx_2 \tag{2.12}$$

on $J^1\mathbb{R}^2$ generate the same operator:

$$\begin{aligned}
\Delta_\omega(v) &= dx_1 \wedge \left(v_{x_1} dx_1 + v_{x_2} dx_2\right) = v_{x_2}\, dx_1 \wedge dx_2, \\
\Delta_\varpi(v) &= v_{x_2}\, dx_1 \wedge dx_2.
\end{aligned}$$

6. Any differential n-form

$$\omega = \varkappa \wedge \alpha + d\varkappa \wedge \beta \tag{2.13}$$

on $J^1 M$, where $\alpha \in \Omega^{n-1}\left(J^1 M\right)$, $\beta \in \Omega^{n-2}\left(J^1 M\right)$ and $\varkappa$ is the Cartan form, gives the zero operator.

All differential operators Δ_ω generate differential equations of second order:

$$\Delta_\omega(v) = 0. \tag{2.14}$$

For example, operator (2.9) generates Lissajou equation

$$(1 - x^2)y'' - xy' + \frac{a^2}{b^2}y = 0. \tag{2.15}$$

Note that the differential operators Δ_ω and $\Delta_{h\omega}$ generate the same equation for each non-zero function h.

Equation (2.14) are called *Monge–Ampère equations* [16].

The following observation justifies this definition: being written in local canonical contact coordinates on $J^1 M$, the operators Δ_ω have the same type of non-linearity as the Monge–Ampère equations.

Namely, the non-linearity involves the determinant of the Hesse matrix and its minors. For instance, in the case $n = 2$, for

$$\omega = E dx_1 \wedge dx_2 + B\left(dx_1 \wedge dp_1 - dx_2 \wedge dp_2\right) + \tag{2.16}$$
$$C dx_1 \wedge dp_2 - A dx_2 \wedge dp_1 + D dp_1 \wedge dp_2.$$

we get classical Monge–Ampère equations

$$A v_{xx} + 2B v_{xy} + C v_{yy} + D(v_{xx}v_{yy} - v_{xy}^2) + E = 0. \tag{2.17}$$

An advantage of this approach is the reduction of the order of the jet space: we use the simpler space $J^1 M$ instead of the space $J^2 M$ where Monge–Ampère equations should be ad hoc as second-order partial differential equations [8].

The differential equation which is associated with a differential n-form ω will be denote by $\mathcal{E}_\omega$:

$$\mathcal{E}_\omega := \{\Delta_\omega(v) = 0\}.$$

The following Maple code generates the corresponding differential operator Δ_ω for a differential 2-form ω on $J^1 \mathbb{R}^2$.

Maple Code: $\omega \longmapsto \Delta_\omega$

```
with(DifferentialGeometry): with(JetCalculus):
Preferences("JetNotation", "JetNotation2"):
DGsetup( [x1,x2],[u], M, 1);
DGsetup( [x,y], N, verbose);
```

Construct the differential operator Δ:

```
Delta := proc(z, h)
    Pullback(Prolong(Transformation(N,M,
    [x1=x,x2=y,u[0,0]=h]),2),z);
end proc;
```

Define a differential 2-form:

```
omega:=evalDG(dx1 &w du[1,0]-dx2 &w du[0,1]);
```

Apply the differential operator to this differential form $\omega = dx_1 \wedge dp_1 - dx_2 \wedge dp_2$:

```
simplify(Delta(omega,v(x,y)),size);
```

As a result, we get the differential operator

$$2\frac{\partial^2}{\partial y \partial x}dx \wedge dy.$$

2.3.2 Multivalued Solutions of Monge–Ampère Equations

Let v be a classical solution of the Monge–Ampère equation $\mathcal{E}_\omega$, i.e. $\Delta_\omega(v) = 0$. Then

$$j_1(v)^*(\omega) = 0.$$

It means that the restriction of the differential form ω to 1-graph of the function v is zero:

$$\omega \mid_{\Gamma_v^1} = 0.$$

An n-dimensional submanifold $L \subset J^1 M$ is called a *multivalued solution* of Monge–Ampère equation if

1. L is an integral manifold of the Cartan distribution, i.e. the restriction of the Cartan form to L is zero: $\varkappa |_L = 0$;
2. the restriction of the differential n-form ω to L is zero, too: $\omega |_L = 0$.

Examples: Multivalued Solutions

1. Parameterized curves

$$L = \left\{ x = \sin bt, \ y = \cos at, \ p = -\frac{a \sin at}{b \cos bt} \right\}$$

in the space $J^1 \mathbb{R}$ are multivalued solutions of the Lissajou equation

$$(1 - x^2)y'' - xy' + \frac{a^2}{b^2} y = 0. \tag{2.18}$$

Indeed, the restriction of the differential 1-form

$$\omega = (1 - x^2)dp + \left(\frac{a^2}{b^2} y - xp \right) dx$$

on the curve L is zero. The projections of these curves on the plane (x, y) are well-known Lissajou curves (see Figs. 2.1, 2.2).
2. Projections of multivalued solutions of the Monge equation

$$v_{xx}v_{yy} - v_{xy}^2 = \left(1 + v_x^2 + v_y^2\right)^2$$

to the space $\mathbb{R}^3$ with coordinates x, y, v are spheres with radius 1 (see Eq. (2.2).
3. Projections of multivalued solutions of the equation

$$v_{xx}v_{yy} - v_{xy}^2 = 0 \tag{2.19}$$

to the space $\mathbb{R}^3$ with coordinates x, y, v are deployable surfaces.

2.3.3 Effective Forms

Last two examples (2.12) and (2.13) show that the constructed map

$$\text{"differential } n\text{-forms"} \rightarrow \text{"differential operators"}$$

has a huge kernel.

Fig. 2.1 Multivalued
solutions of the Lissajou
equation for $a = 3, b = 2$

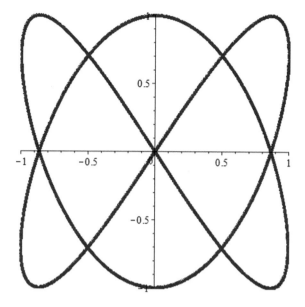

Fig. 2.2 Multivalued
solutions of the Lissajou
equation for $a = 1, b = \sqrt{2}$
is a curve, everywhere dense
in the square (Lissajou's
Black Square)

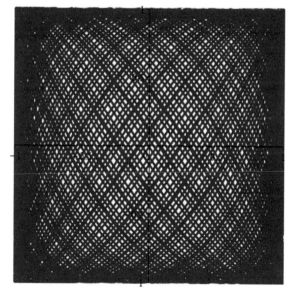

This kernel consists of differential forms that vanish on any integral manifold of the Cartan distribution. All such forms have form (2.13) (see [15]).

Let's find a submodule of the module $\Omega^2(J^1M)$ of differential 2-forms such that the map is bijective (dim $M = 2$).

Differential 2-form $\omega \in \Omega^2(J^1M)$ is called *effective* if

1. $X_1 \lrcorner \omega = 0$;
2. $\omega \wedge d\varkappa = 0$.

Here, X_1 is the contact vector field with generating function 1. In canonical coordinates (2.3)

$$X_1 = \partial_u.$$

The first condition means that coordinate representation of ω does not contain terms $du \wedge *$, and therefore $\omega \neq \varkappa \wedge \alpha$ for some differential 1-form α. Second condition means that $\omega \neq \beta d\varkappa$, for a function β.

The module of effective differential 2-forms will be denoted by $\Omega_\epsilon^2(J^1M)$.

There is the projection p which maps module $\Omega^2(J^1M)$ to the module $\Omega^2(C)$ of "differential forms" on the Cartan distribution.

Namely, define

$$p : \Omega^2(J^1M) \longrightarrow \Omega^2(C)$$

as follows:

$$p(\omega) := \omega - \varkappa \wedge (X_1 \lrcorner \omega).$$

Here, $\Omega^2(J^1M)$ and $\Omega^2(C)$ are modules of 2-forms on the 1-jet manifold J^1M and on the Cartan distribution C respectively. Remark that

$$X_1 \lrcorner p(\omega) = 0,$$

i.e. 2-form $p(\omega) \in \Omega^2(C)$.

Theorem 2.1 *Any differential 2-form $\omega \in \Omega^2(C)$ has the unique representation*

$$\omega = \omega_\epsilon + \beta d\varkappa, \qquad (2.20)$$

where $\omega_\epsilon \in \Omega_\epsilon^2(J^1M)$ is an effective 2-form and β is a function.

Proof In our case, the Cartan distribution C is four-dimensional. The exterior differential of the Cartan form is non-degenerated 2-form on each Cartan subspace, i.e. $d\varkappa_\theta$ is a symplectic structure on $C(\theta)$ for any $\theta \in J^1M$. Therefore, formula

$$\omega \wedge d\varkappa = \beta d\varkappa \wedge d\varkappa$$

uniquely defines a function β. Define now differential form

$$\omega_\epsilon = \omega - \beta d\varkappa.$$

Since $\omega_\epsilon \wedge d\varkappa = 0$, the form ω_ϵ is effective. $\square$

The constructed differential form ω_ϵ is called the *effective part* of the differential form ω.

Define the operator

$$\mathrm{Eff} : \Omega^2(J^1 M) \longrightarrow \Omega^2_\epsilon(J^1 M), \quad \mathrm{Eff}(\omega) := (p(\omega))_\epsilon,$$

which for any differential 2-form ω on the space $J^1 M$ gives its effective part.

It is obvious that differential 2-forms ω and $\mathrm{Eff}(\omega)$ generate the same Monge–Ampère equations.

In canonical coordinates

$$d\varkappa = dx_1 \wedge dp_1 + dx_2 \wedge dp_2$$

and any effective differential 2-form has the following representation:

$$\omega = E dx_1 \wedge dx_2 + B(dx_1 \wedge dp_1 - dx_2 \wedge dp_2) + \qquad (2.21)$$
$$C dx_1 \wedge dp_2 - A dx_2 \wedge dp_1 + D dp_1 \wedge dp_2,$$

where A, B, C, D and E are smooth functions on $J^1 M$. This form corresponds to Eq. (2.17).

The following Maple code contains two procedures which generate effective parts of differential 2-forms.

Maple Code: $\omega \longmapsto \omega_\epsilon$

1. Projection of a 2-form to the Cartan distribution:

```
ProjC:=proc (omega)
    GeneratingFunctionToContactVector(1);
    evalDG(omega-kappa &w Hook(evalDG(D_u[0,0]),omega));
end proc:
```

2. Calculation of effective parts of a differential 2-forms:

```
Eff:=proc (omega)
    evalDG(evalDG(omega-kappa &w Hook(evalDG(D_u[0,0]),
    omega))-(solve(op(Tools:-DGinfo(evalDG(g*Omega&w Omega-
    (evalDG(omega-kappa &w Hook(evalDG(D_u[0,0]),omega)))
    &w Omega),"CoefficientSet")),g))*Omega);
end proc:
```

2.4   Lecture 3. Contact Transformations of Monge–Ampère Equations

By the definition, contact transformations preserve the Cartan distribution and multiply the Cartan form $\varkappa$ by a function (see formula (2.5)).

Therefore, contact transformations do not preserve the contact vector field X_1 in general. Because of this, the image of an effective differential form can be not effective.

Let $\Phi : J^1 M \to J^1 M$ be a contact transformation and ω be an effective differential 2-form. Then by the image of differential 2-form ω, we shall understand the effective differential form $\mathrm{Eff}(\Phi^*(\omega))$.

Two Monge–Ampère equations $\mathcal{E}_\omega$ and $\mathcal{E}_\varpi$ are *contact equivalent* if there exist a contact transformation Φ such that $\varpi = h\mathrm{Eff}(\Phi^*(\omega))$ for some function h.

Theorem 2.2 *If two equations $\mathcal{E}_\omega$ and $\mathcal{E}_\varpi$ are contact equivalent, then their contact transformation maps multivalued solutions of one to multivalued solutions of the other.*

Note that, in general, contact transformations do not preserve the class of classical solutions: classical solutions can transform to multivalued solutions and vice versa.

Examples of Linearization of Equations by Contact Transformations

1. The von Karman equation

$$v_{x_1} v_{x_1 x_1} - v_{x_2 x_2} = 0 \tag{2.22}$$

becomes the linear equation

$$x_1 v_{x_2 x_2} + v_{x_1 x_1} = 0 \tag{2.23}$$

after Legendre transformation (2.24).
The last equation is known as the *Triccomi* equation.
2. Equation

$$\det \mathrm{Hess}\, v = 1$$

is generated by the effective differential 2-form

$$\omega = dp_1 \wedge dp_2 - dx_1 \wedge dx_2.$$

After the partial Legendre transformation

$$\Phi : (x_1,\ x_2,\ u,\ p_1,\ p_2) \mapsto (p_1,\ x_2,\ u - p_1 x_1,\ -x_1,\ p_2)$$

this form becomes

$$\omega = dx_2 \wedge dp_1 - dx_1 \wedge dp_2,$$

and corresponds to the Laplace equation

$$v_{x_1 x_1} + v_{x_2 x_2} = 0.$$

3. Quasi-linear equation:

$$A\left(v_x, v_y\right) v_{xx} + 2B\left(v_x, v_y\right) v_{xy} + C\left(v_x, v_y\right) v_{yy} = 0.$$

This equation is represented by the following effective form:

$$\omega = B\,(p_1, p_2)\,(dx_1 \wedge dp_1 - dx_2 \wedge dp_2) + C\,(p_1, p_2)\,dx_1 \wedge dp_2 - A\,(p_1, p_2)\,dx_2 \wedge dp_1.$$

After the Legendre transformation

$$\Phi : (x_1,\ x_2,\ u,\ p_1,\ p_2) \mapsto (p_1,\ p_2,\ u - p_1 x_1 - p_2 x_2,\ -x_1,\ -x_2,)\quad (2.24)$$

we get the following effective form

$$\varphi^*(\omega) = B\,(-x_1, -x_2)\,(dx_1 \wedge dp_1 - dx_2 \wedge dp_2) + \\ - A\,(-x_1, -x_2)\,dx_1 \wedge dp_2 + C\,(-x_1, -x_2)\,dx_2 \wedge dp_1,$$

which corresponds to the linear equation:

$$-A\,(-x_1, -x_2)\,v_{x_2 x_2} + 2B\,(-x_1, -x_2)\,v_{x_1 x_2} - C\,(-x_1, -x_2)\,v_{x_1 x_1} = 0.$$

Example

The following equation arises in filtration theory of two immiscible fluids in porous media [1]:

$$u_{xy} - u_x u_{yy} = 0. \tag{2.25}$$

It is used for finding a strategy to control wavefronts in the development of oil fields.

The corresponding differential 2-form is

$$\omega = 2p_1 dp_2 \wedge dx_1 + dx_1 \wedge dp_1 - dx_2 \wedge dp_2,$$

where $x_1 = x, x_2 = y$. Applying the Legendre transformation

$$\Phi : (x_1, x_2, u, p_1, p_2) \mapsto (p_1, p_2, u - x_1 p_1 - x_2 p_2, -x_1, -x_2)$$

we get the following differential 2-form:

$$\Phi^*(\omega) = 2x_1 dx_2 \wedge dp_1 + dx_1 \wedge dp_1 - dx_2 \wedge dp_2.$$

This form corresponds to the linear equation

$$u_{x_1 x_2} - x_1 u_{x_1 x_1} = 0. \tag{2.26}$$

The general solution of the last equation is

$$u(x_1, x_2) = e^{-x_2} F_1(x_1 e^{x_2}) + F_2(x_2), \tag{2.27}$$

where F_1 and F_2 are arbitrary functions. Differentiating both sides of (2.27), we get

$$u_{x_1} = F_1'(x_1 e^{x_2}),$$
$$u_{x_2} = -e^{-x_2} F_1(x_1 e^{x_2}) - F_1'(x_1 e^{x_2})x_1 + F_2'(x_2).$$

Thus, solution (2.27) generate a surface $L \subset J^1 M$:

$$L : \begin{cases} u - e^{-x_2} F_1(x_1 e^{x_2}) + F_2(x_2) = 0, \\ p_1 - F_1'(x_1 e^{x_2}) = 0, \\ p_2 + e^{-x_2} F_1(x_1 e^{x_2}) + F_1'(x_1 e^{x_2})x_1 - F_2'(x_2) = 0. \end{cases}$$

Applying the inverse Legendre transformation

$$\Phi^{-1} : (x_1, x_2, u, p_1, p_2) \longmapsto (-p_1, -p_2, u - x_1 p_1 - x_2 p_2, x_1, x_2)$$

to L, we get multivalued solutions of equation (2.25) in parametric form (Fig. 2.3):

$$\Phi^{-1}(L) : \begin{cases} u - x_1 p_1 - x_2 p_2 - e^{p_2} F_1(-p_1 e^{-p_2}) + F_2(-p_2) = 0, \\ x_1 - F_1'(-p_1 e^{-p_2}) = 0, \\ x_2 + e^{p_2} F_1(-p_1 e^{-p_2}) + p_1 F_1'(-p_1 e^{-p_2}) + F_2'(-p_2) = 0. \end{cases} \tag{2.28}$$

In order to simplify the last formula, we introduce new parameters

$$a = -p_1 e^{-p_2}, \qquad b = -p_2,$$

and new functions

$$k(a) = F_1(a), \qquad r(b) = F_2(b).$$

In these notation, multivalued solutions of equation (2.25) takes the form:

Fig. 2.3 Projection of the multivalued solution $\mathcal{L}$ to the space x, y, u for $k(a) = a^9 - 20a^5$ and $r(b) = b^{0.01}$

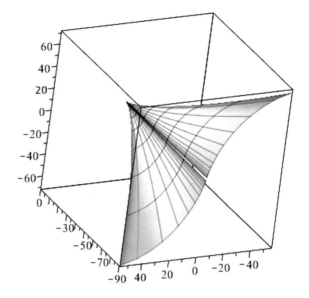

$$\mathcal{L}: \begin{cases} x = k'(a), \\ y = e^{-b}(ak'(a) - k(a)) - r'(b), \\ u = (b+1)e^{-b}(k(a) - ak'(a)) + br'(b) - r(b), \\ p_1 = -ae^{-b}, \\ p_2 = -b, \end{cases}$$

where $k(a)$ and $r(b)$ are arbitrary functions.

Maple Code: Equation $u_{xy} - u_x u_{yy} = 0$

Define coordinates on M:

```
with(DifferentialGeometry): with(JetCalculus):
Preferences("JetNotation", "JetNotation2"):
DGsetup( [x1,x2],[u], M, 1);
DGsetup( [x,y],N,1);
```

Construct the differential operator Δ:

```
Delta := proc(z, h)
    description "M-A operator";
    Pullback(Prolong(Transformation
```

```
    (N,M,[x1=x,x2=y,u[0,0]=h]),2),z);
end proc;
```

Define the differential 2-form ω:

```
omega:=evalDG(2*u[1,0]*du[0,1] &w dx1 +
dx1 &w du[1,0]-dx2 &w du[0,1]);
```

$$\omega = 2p_1 dp_2 \wedge dx_1 + dx_1 \wedge dp_1 - dx_2 \wedge dp_2,$$

The Legendre transformation:

```
Legendre:=Transformation(M,M,[x1=u[1,0],x2=u[0,1],
u[0,0]=u[0,0]-x1*u[1,0]-u[0,1]*x2, u[1,0]=-x1, u[0,1]=-x2]):
```

Apply the Legendre transformation to ω:

```
omega1:=Pullback(Legendre,omega);
```

Construct the differential operator Δ_{ω_1}:

```
Delta(omega1,u(x,y));
```

$$\left(2 \left(\frac{\partial^2}{\partial y \partial x} u(x, y) \right) - 2x \left(\frac{\partial^2}{\partial x^2} u(x, y) \right) \right) dx \wedge dy$$

Check solution:

```
sub:={u(x,y)=exp(-y)*F1(x*exp(y))+F2(y)};
eval(diff(u(x, y), x, y)-x*diff(u(x, y), x, x), sub);
```
$$0$$

Inverse Legendre transformation:

```
InvLegendre:=InverseTransformation(Legendre):
```

Apply this transformation to the surface L:

```
z1:=convert(u(x1,x2)-exp(-x2)*F1(x1*exp(x2))+F2(x2),DGjet):
```

```
z2:=convert(diff(u(x1,x2)-exp(-x2)*
F1(x1*exp(x2))+F2(x2),x1),DGjet):
```

```
z3:=convert(diff(u(x1,x2)-exp(-x2)*
```

```
F1(x1*exp(x2))+F2(x2),x2),DGjet):

u1:=Pullback(InvLegendre,z1):

u2:=Pullback(InvLegendre,z2):

u3:=Pullback(InvLegendre,z3):
```

As a result, we get formula (2.28).
Check that $\mathcal{L}$ is a multivalued solution of equation (2.25), i.e. $\omega \mid \mathcal{L} = 0$:

```
DGsetup( [x1,x2,u,p1,p2], M);
DGsetup( [a,b], N);

omega:=evalDG(2*p1*dp2 &w dx1 + dx1 &w dp1-dx2 &w dp2):

NtoM:=Transformation(N,M,[x1=diff(k(a),a),
x2=exp(-b)*(a*diff(k(a),a)-k(a))-diff(r(b),b),
u=(b+1)*exp(-b)*(-a*diff(k(a),a)+k(a))+b*diff(r(b),b)-r(b),
p1=-a*exp(-b), p2=-b]):

Pullback(NtoM,omega);
```
$$0$$

Visualization of the multivalued solution $\mathcal{L}$:

```
plot3d(eval([diff(k(a),a),exp(-b)*(a*diff(k(a),a)-k(a))
-diff(r(b),b), (b+1)*exp(-b)*(-a*diff(k(a),a)+k(a))+
b*diff(r(b),b)-r(b)], {k(a)=a^9-20*a^5,r(b)=b^0.01}),
a = -1 .. 1, b = -6 .. 6);
```

2.5   Lecture 4. Geometrical Structures

2.5.1   Pfaffians

First of all, we remark that the restriction of the differential 2-form $d\varkappa$ on the Cartan distribution

$$\Omega = d\varkappa \mid_C$$

defines a symplectic structure on Cartan space $\mathcal{C}(\theta) \subset T_\theta(J^1 M)$.

Using this structure and an effective 2-form $\omega \in \Omega_\epsilon^2(J^1M)$ we define function $\mathrm{Pf}(\omega)$, called *Pfaffian*, in the following way [20]:

$$\mathrm{Pf}(\omega)\, \Omega \wedge \Omega = \omega \wedge \omega. \tag{2.29}$$

This is a correct construction because $\omega \wedge \omega$ and $\Omega \wedge \Omega$ are 4-forms on the four-dimensional Cartan distribution.

In the case when

$$\omega = E dx_1 \wedge dx_2 + B\,(dx_1 \wedge dp_1 - dx_2 \wedge dp_2) + \tag{2.30}$$
$$C dx_1 \wedge dp_2 - A dx_2 \wedge dp_1 + D dp_1 \wedge dp_2,$$

we get

$$\mathrm{Pf}(\omega) = B^2 + DE - AC.$$

We say that the Monge–Ampère equation $\mathcal{E}_\omega$ is *hyperbolic, elliptic* or *parabolic* at a domain $\mathcal{D} \subset J^1M$ if the function $\mathrm{Pf}(\omega)$ is negative, positive or zero at each point of $\mathcal{D}$, respectively.

If the Pfaffian changes the sign in some points of $\mathcal{D}$, then the equation $\mathcal{E}_\omega$ is called a *mixed type* equation (see [10]).

The hyperbolic and elliptic equations are called *non-degenerate*.

Maple Code: Pfaffian

```
kappa:=convert(Cu[0,0],DGform):
Omega:=ExteriorDerivative(kappa):
omega:=evalDG(dq1 &w du[1,0]+ du[0,0] &w du[0,1]):

Pf:=proc (omega)
solve(op(DGinfo(evalDG(z*Omega &w Omega-omega &w omega),
"CoefficientSet")),z)
end proc:
```

For example, the Pfaffian of the differential 2-form

$$\omega = dx_1 \wedge dp_1 - dx_2 \wedge dp_2$$

which corresponds to wave equation $u_{xy} = 0$ is equal to -1, and as we know this equation is hyperbolic.

The Pfaffian of the differential 2-form

$$\omega = dx_1 \wedge dp_2 - dx_2 \wedge dp_1$$

which corresponds to Laplace equation $u_{xx} + u_{yy} = 0$ is equal to 1. Indeed,

```
omega:=evalDG(dx1 &w du[1,0]-dx2 &w du[0,1]):
Pf(omega);
```

$$-1$$

```
omega:=evalDG(dx1 &w du[0,1]-dx2 &w du[1,0]):
Pf(omega);
```

$$1$$

2.5.2 Fields of Endomorphisms

The standard linear algebra allows us to construct a field of endomorphisms

$$A_\omega : D(C) \longrightarrow D(C)$$

which is associated with an effective 2-form ω. Here $D(C)$ is the module of vector fields tangent to C.

Namely, the 2-form Ω is non-degenerated on C and the operator A_ω is uniquely determined by the following formula [19]:

$$A_\omega X \rfloor \, \Omega = X \rfloor \, \omega \tag{2.31}$$

for all vector fields X tangent to C.

Proposition 2.1 *Operators A_ω satisfy the following properties:*

1. $\Omega(A_\omega X, X) = 0$.
2. $\Omega(A_\omega X, Y) = \Omega(X, A_\omega Y)$.

Proof 1. $\Omega(A_\omega X, X) = \omega(X, X) = 0$.
2. $\Omega(A_\omega X, Y) = \omega(X, Y) = -\omega(Y, X) = -\Omega(A_\omega Y, X) = \Omega(X, A_\omega Y)$. $\square$

Proposition 2.2 *The squares of operators A_ω are scalar and*

$$A_\omega^2 + \mathrm{Pf}(\omega) = 0. \tag{2.32}$$

Proof First of all

$$A_\omega X \rfloor (\omega \wedge \Omega) = (A_\omega X \rfloor \omega) \wedge \Omega + \omega \wedge (A_\omega X \rfloor \Omega).$$

Using Proposition 2.1,

$$\begin{aligned}
X \lrcorner (A_\omega X \lrcorner (\omega \wedge \Omega)) &= \omega (A_\omega X, X)\Omega - (A_\omega X \lrcorner \omega) \wedge (X \lrcorner \Omega) \\
&\quad + (X \lrcorner \omega) \wedge (A_\omega X \lrcorner \Omega) + \Omega (A_\omega X, X)\omega \\
&= \Omega (A_\omega^2 X, X)\Omega - (A_\omega^2 X \lrcorner \Omega) \wedge (X \lrcorner \Omega) \\
&\quad + (A_\omega X \lrcorner \Omega) \wedge (A_\omega X \lrcorner \Omega) + \Omega (A_\omega X, X)\omega \\
&= -(A_\omega^2 X \lrcorner \Omega) \wedge (X \lrcorner \Omega).
\end{aligned}$$

Since ω is effective, $\omega \wedge \Omega = 0$. Then

$$(A_\omega^2 X \lrcorner \Omega) \wedge (X \lrcorner \Omega) = 0,$$

i.e. differential 1-forms $A_\omega^2 X \lrcorner \Omega$ and $X \lrcorner \Omega$ are linearly dependent. Therefore the square of the operator A_ω is a scalar: $A_\omega^2 = \alpha$.

Let $X \in D(C)$ be an arbitrary vector field. Applying the operators $A_\omega X \lrcorner$ and $X \lrcorner$ to both parts of formula (2.29) we get

$$Pf(\omega)(A_\omega X \lrcorner \Omega) \wedge (X \lrcorner \Omega) = (A_\omega X \lrcorner \omega) \wedge (X \lrcorner \omega) = (\alpha X \lrcorner \Omega) \wedge (A_\omega X \lrcorner \Omega).$$

Then

$$(Pf(\omega) + \alpha)(A_\omega X \lrcorner \Omega) \wedge (X \lrcorner \Omega) = 0. \tag{2.33}$$

Suppose that $(A_\omega X \lrcorner \Omega) \wedge (X \lrcorner \Omega) = 0$. Then the vector fields X and $A_\omega X$ are linearly dependent. Since X is an arbitrary vector field we see that the operator A_ω is scalar, i.e. $A_\omega X = \lambda X$ for any X. Then

$$X \lrcorner \omega = A_\omega X \lrcorner \Omega = \lambda X \lrcorner \Omega.$$

Therefore $\omega = \lambda \Omega$, which is impossible. So from (2.33), it follows that $Pf(\omega) + \alpha = 0$, i.e. $A_\omega^2 + Pf(\omega) = 0$. $\qquad\square$

Let's find a coordinate representation of the operator A_ω. Let

$$\frac{\partial}{\partial x_1} + p_1 \frac{\partial}{\partial u}, \quad \frac{\partial}{\partial x_2} + p_2 \frac{\partial}{\partial u}, \quad \frac{\partial}{\partial p_1}, \quad \frac{\partial}{\partial p_2} \tag{2.34}$$

be a local basis of the module $D(C)$. Then formula (2.31) gives:

$$A_\omega = \begin{Vmatrix} B & -A & 0 & -D \\ C & -B & D & 0 \\ 0 & E & B & C \\ -E & 0 & -A & -B \end{Vmatrix} \tag{2.35}$$

in this basis.

Maple Code: Operator A_ω

```
with(DifferentialGeometry): with(LinearAlgebra): with(Tensor):
```

Coordinates on the 1-jet space:

```
DGsetup( [x1,x2,u,p1,p2], J):
```

Cartan's form and its exterior differential:

```
kappa:=evalDG(du-p1*dx1-p2*dx2):
Omega:=ExteriorDerivative(kappa):
```

Define 2-form ω:

```
omega:=evalDG(2*p1*dp2 &w dx1+ dx1 &w dp1-dx2 &w dp2);
```

Vector fields and 1-forms on Cartan's distribution:

```
VectCartan:=evalDG([D_x1+p1*D_u,D_x2+p2*D_u,D_p1,D_p2]):
CovectCartan:=evalDG([dx1,dx2,dp1,dp2]):
```

Checking duality:

```
m := proc (i, j) options operator, arrow;
    Hook(VectCartan[i],CovectCartan[j])
end proc:
Matrix(4,m):
```

Construct an arbitrary vector field on Cartan's distribution:

```
V:=DGzip([a, b, c,d], VectCartan, "plus"):
```

$$V = a\frac{\partial}{\partial x_1} + b\frac{\partial}{\partial x_2} + (bp_2 + ap_1)\frac{\partial}{\partial u} + c\frac{\partial}{\partial p_1} + d\frac{\partial}{\partial p_2}$$

General form of $A = A_\omega$. Here $a_{i,j}$ are arbitrary functions:

```
A:=evalDG(sum(sum(a[i,j]*VectCartan[i] &t
CovectCartan[j],i=1..4),j=1..4)):
```

Action of A_ω on vector fields:

```
Act:=Z->convert(ContractIndices(evalDG(A &tensor Z),
[[2,3]]), DGvector):
```

Equations with respect to $a_{i,j}$:

```
for i from 1 to 4 do
e[i]:=evalDG(Hook(Act(evalDG(VectCartan[i])),Omega)-
```

```
Hook(VectCartan[i], omega));
end do:

AEq:=[]:
for i from 1 by 1 to 4 do
AEq:=[op(AEq),op(GetComponents(e[i],CovectCartan))]
end do:
AEq;
```

$$
\begin{array}{llll}
-a_{3,1}, & -a_{4,1}, & -1+a_{1,1}, & 2p_1+a_{2,1}, \\
-a_{3,2}, & -a_{4,2}, & a_{1,2}, & 1+a_{2,2}, \\
1-a_{3,3}, & -a_{4,3}, & a_{1,3}, & a_{2,3}, \\
-2p_1-a_{3,4}, & -1-a_{4,4}, & a_{1,4}, & a_{2,4}
\end{array}
$$

```
sol:=solve(AEq,[a[1,1],a[1,2],a[1,3],a[1,4],
a[2,1],a[2,2],a[2,3],a[2,4],
a[3,1],a[3,2],a[3,3],a[3,4],
a[4,1],a[4,2],a[4,3],a[4,4]]);

assign(sol);

m := proc (i, j) options operator, arrow; a[i,j] end proc;

Am:=Matrix(4,4,m);
```

$$
A_\omega = \begin{Vmatrix}
1 & 0 & 0 & 0 \\
-2p_1 & -1 & 0 & 0 \\
0 & 0 & 1 & -2p_1 \\
0 & 0 & 0 & -1
\end{Vmatrix} \tag{2.36}
$$

```
Determinant(Am);
```

$$
1
$$

```
Am.Am;
```

$$
\begin{Vmatrix}
1 & 0 & 0 & 0 \\
0 & 1 & 0 & 0 \\
0 & 0 & 1 & 0 \\
0 & 0 & 0 & 1
\end{Vmatrix}
$$

2.5.3 Characteristic Distributions

Effective forms ω and $h\omega$, where h is any non-vanishing function, define the same Monge–Ampère equation. Therefore, for a non-degenerated equation $\mathcal{E}_\omega$ the form ω can be normed in such a way that $|\operatorname{Pf}(\omega)| = 1$. It is sufficient to replace ω by

$$\frac{\omega}{\sqrt{|\operatorname{Pf}(\omega)|}}.$$ (2.37)

By (2.32), the hyperbolic equations generate a product structure

$$A_{\omega,a}^2 = 1$$

and elliptic equations generate a complex structure

$$A_{\omega,a}^2 = -1$$

on the Cartan space $C(a)$ [18].

Therefore, a non-degenerated Monge–Ampère equation generates two two-dimensional (complex—for elliptic case) distributions on $J^1 M$, which are eigenspaces of the operator A_ω.

These distributions $C_+(a)$ and $C_-(a)$ correspond to the eigenvalues 1 and -1 for the hyperbolic equations or to ι and $-\iota$ for the elliptic ones, respectively. Here $\iota = \sqrt{-1}$.

The distributions C_+ and C_- are called *characteristic*.

The characteristic distributions are real for the hyperbolic equations and complex for the elliptic ones. They are complex conjugate for the elliptic equations.

Proposition 2.3 ([18]) *1. The characteristic distributions C_+ and C_- are skew orthogonal with respect to the symplectic structure Ω, i.e. $\Omega(X_+, X_-) = 0$ for $X_\pm \in D(C_\pm)$.*
2. On each of them, the 2-form Ω is non-degenerate.

On the other hand, any pair of arbitrary real distributions $C_{1,0}$ and $C_{0,1}$ on $J^1 M$ such that

1. $\dim C_{1,0} = \dim C_{0,1} = 2$;
2. $C = C_{1,0} \oplus C_{0,1}$;
3. $C_{1,0}$ and $C_{0,1}$ are skew-orthogonal with respect to the symplectic structure Ω

determines the operator A. Therefore, a hyperbolic Monge–Ampère equation can be regarded as such pair $\{C_{1,0}, C_{0,1}\}$ of distributions.

Maple Code: Characteristic Distributions

Calculation of eigenvalues and eigenvectors of the operator A_ω:

```
EV,e:=Eigenvectors(Am):
```

Find the vector fields from the Cartan distribution

```
Cp:=[]:Cm:=[]:

for i from 1 to 4 do
if EV[i]=EV[1] then Cp:=[op(Cp),
   (convert((Transpose(e[1..-1,i])),list))]
else
   Cm:=[op(Cm),(convert((Transpose(e[1..-1,i])),list))]
end if
end do:

Vp1:=DGzip(Cp[1], VectCartan, "plus");
Vp2:=DGzip(Cp[2], VectCartan, "plus");
Vm1:=DGzip(Cm[1], VectCartan, "plus");
Vm2:=DGzip(Cm[2], VectCartan, "plus");
```

For example, the characteristic distribution C_+ and C_- of operator (2.36) are generated by the following vector fields:

$$C_+ = \left\langle p_1 \frac{\partial}{\partial p_1} + \frac{\partial}{\partial p_2}, \quad \frac{\partial}{\partial x_2} + p_2 \frac{\partial}{\partial u} \right\rangle$$

and

$$C_- = \left\langle \frac{\partial}{\partial p_1}, \quad p_1 \frac{\partial}{\partial x_2} + p_1(p_2 - 1)\frac{\partial}{\partial u} - \frac{\partial}{\partial x_1} \right\rangle.$$

2.5.4 Symplectic Monge–Ampère Equations

Monge–Ampère equation (2.17) is called *symplectic* if its coefficients A, B, C, D, E do not depend on v.

In this case, the structures described above (effective differential forms, the differential operator Δ_ω, field of endomorphisms A_ω) can be considered on the four-dimensional cotangent bundle T^*M instead of the five-dimensional jet bundle J^1M.

Below, we repeat main constructions for the symplectic case.

A smooth function $f \in C^\infty(M)$ defines a section $s_f : M \longrightarrow T^*M$ of the cotangent bundle

$$\pi : T^*M \longrightarrow M$$

by the following formula:

$$s_f : a \longmapsto df_a.$$

Let ω be a differential 2-form on T^*M. Define a differential operator

$$\Delta_\omega : C^\infty(M) \longrightarrow \Omega^2(M), \quad \Delta_\omega(v) := (s_v)^*(\omega).$$

Then equation $\Delta_\omega(v) = 0$ is a symplectic Monge–Ampère equation.

Let Ω be the symplectic structure on T^*M. In canonical coordinates x_1, x_2, p_1, p_2 on T^*M

$$\Omega = dx_1 \wedge dp_1 + dx_2 \wedge dp_2.$$

The differential form ω is said to be *effective* if

$$\omega \wedge \Omega = 0.$$

Pfaffian $\mathrm{Pf}(\omega)$ of the differential 2-form ω is defined by the following equality:

$$\mathrm{Pf}(\omega)\, \Omega \wedge \Omega = \omega \wedge \omega,$$

and formula

$$A_\omega X \rfloor \, \Omega = X \rfloor \, \omega$$

defines the field of endomorphisms A_ω on T^*M.

The square of operator A_ω is scalar:

$$A_\omega^2 + \mathrm{Pf}(\omega) = 0.$$

Consider now the case when equation is non-degenerated, i.e. $\mathrm{Pf}(\omega) \neq 0$ on T^*M. Then, the operator A_ω can be normed (see formula (2.37).

For hyperbolic equations we get almost product structure: $A_\omega^2 = 1$, and for elliptic ones we get almost complex structure: $A_\omega^2 = -1$.

We say that two symplectic equation $\mathcal{E}_\omega$ and $\mathcal{E}_\varpi$ are *symplectically equivalent* if there exist a symplectic transformation Φ such that

$$\Phi^*(\omega) = h\varpi$$

for some function h.

The following theorem gives a criterion of symplectic equivalence of non-degenerated Monge–Ampère equation to linear equations with constant coefficients.

Theorem 2.3 ([19]) *Non-degenerated symplectic Monge–Ampère equation $\mathcal{E}_\omega$ is symplectically equivalent to wave equation*

$$v_{xx} - v_{yy} = 0 \tag{2.38}$$

(in hyperbolic case), or to Laplace equation

$$v_{xx} + v_{yy} = 0$$

(in elliptic case) if and only if the Nijenhuis tensor

$$N_{A_\omega} = 0, \tag{2.39}$$

where A_ω is the normed operator.

Recall that the Nijenhuis tensor N_A of an operator A is a tensor field of rank (1, 2) given by

$$N_A(X, Y) := -A^2[X, Y] + A[AX, Y] + A[X, AY] - [AX, AY]$$

for vector fields X and Y.

Condition (2.39) can be written in the following equivalent form [20]:

$$d\omega = \frac{1}{2} d \left(\ln | \mathrm{Pf}(\omega)| \right) \wedge \omega.$$

Maple Code: Symplectic Equation and Nijenhuis Tensor

Below we construct the operator A_ω for non-linear wave equation

$$v_{xy} = f(x, y, v_x, v_y). \tag{2.40}$$

Then we calculate the Nijenhuis tensor N_{A_ω} and find conditions under which is this equation symplectically equivalent to the linear wave equation with constant coefficients.

```
with(DifferentialGeometry): with(Tools):
with(PDETools): with(Tensor):with(LinearAlgebra):
DGsetup( [x1,x2,p1,p2], M):

Omega:=evalDG(dx1 &w dp1+dx2 &w dp2):
omega:=evalDG(-2*f(x1,x2,p1,p2)*dx1 &w dx2+
dx1 &w dp1-dx2 &w dp2);

Vect:=evalDG([D_x1,D_x2,D_p1,D_p2]):
```

```
Covect:=evalDG([dx1,dx2,dp1,dp2]):

V:=DGzip([a, b, c,d], Vect, "plus"):

A:=evalDG(sum(sum(a[i,j]*(Vect[i] &t
Covect[j]),i=1..4),j=1..4)):

Act:=Z->convert(ContractIndices
(evalDG(A &tensor Z),[[2,3]]),DGvector):

for i from 1 to 4 do
e[i]:=evalDG(Hook(Act(evalDG(Vect[i])),Omega)-
Hook(Vect[i],omega));
end do:

AEq:=[]:

for i from 1 by 1 to 4 do AEq:=
[op(AEq),op(GetComponents(e[i], Covect))] end do:

sol:=solve(AEq,[a[1,1],a[1,2],a[1,3],a[1,4],
a[2,1],a[2,2],a[2,3],a[2,4],
a[3,1],a[3,2],a[3,3],a[3,4],
a[4,1],a[4,2],a[4,3],a[4,4]]);

assign(sol):
A:=DGsimplify(convert(A, DGtensor)):

N := TensorBrackets(A, A, "Frolicher--Nijenhuis"):

eq:=Tools:-DGinfo(N, "CoefficientSet");

pdsolve(eq);
```

As a result we get
$$f = F1(x1, x2),$$

where F is an arbitrary function.

So, Eq. (2.40) is symplectically equivalent to wave equation (2.38) if and only if f is a function in x_1 and x_2 only.

Fig. 2.4 Splitting of the
tangent space $T_a(J^1M)$

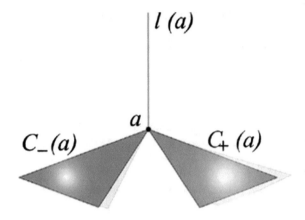

2.5.5 Splitting of Tangent Spaces

Let us return to the space J^1M.

A non-degenerate equation is called *regular* if the derivatives $C_\pm^{(k)}$ ($k = 1, 2, 3$) of the characteristic distributions are constant rank distributions, too.

Below we consider regular equations only. Then, the first derivatives of the characteristic distributions

$$C_\pm^{(1)} := C_\pm + [C_\pm, C_\pm]$$

are three-dimensional. Their intersection

$$l := C_+^{(1)} \cap C_-^{(1)}$$

is a one-dimensional distribution, which is transversal to Cartan distribution.

Therefore, for hyperbolic equations, the tangent space $T_a(J^1M)$ splits into the direct sum (see Fig. 2.4)

$$T_a(J^1M) = C_+(a) \oplus l(a) \oplus C_-(a) \tag{2.41}$$

at each point $a \in J^1M$ [18].

For elliptic equations, we get a similar decomposition of the complexification of $T_a(J^1M)$. In this case, the distribution l is real, too.

2.6 Lecture 5. Tensor Invariants of Monge–Ampère Equations

2.6.1 Decomposition of de Rham Complex

Let us construct the decomposition of the de Rham complex, which is generated by the splitting of tangent spaces.

Decomposition (2.41) generates a decomposition of the module of exterior s-forms (or its complexification for elliptic equations). Denote the distributions C_+, l, and C_- by P_1, P_2, and P_3, respectively.

Let $D(J^1 M)$ be the module of vector fields on $J^1 M$, and let D_j be the module of vector fields tangent to distribution P_j.

Define the following submodules of modules of differential s-forms $\Omega^s(J^1 M)$:

$$\Omega_i^s := \{\alpha \in \Omega^s(J^1 M) \mid X \lrcorner \alpha = 0 \ \forall \ X \in D_j, \ j \neq i\} \quad (i = 1, 2, 3).$$

Then we get the following decomposition of the module of differential s-forms on $J^1 M$:

$$\Omega^s(J^1 M) = \bigoplus_{|\mathbf{k}|=s} \Omega^{\mathbf{k}}, \tag{2.42}$$

where $\mathbf{k} = (k_1, k_2, k_3)$ is a multi-index, $k_i \in \{0, 1, \ldots, \dim P_i\}$,

$$|\mathbf{k}| = k_1 + k_2 + k_3,$$

and

$$\Omega^{\mathbf{k}} := \left\{ \sum_{j_1+j_2+j_3=|\mathbf{k}|} \alpha_{j_1} \wedge \alpha_{j_2} \wedge \alpha_{j_3}, \text{ where } \alpha_{j_i} \in \Omega_i^{k_i} \right\} \subset \bigotimes_{i=1}^{3} \Omega_i^{k_i}.$$

Three first terms of the decomposition are presented in the diagram (see Fig. 2.5).

The exterior differential also splits into the direct sum

$$d = \bigoplus_{|\mathbf{t}|=1} d_{\mathbf{t}},$$

where

$$d_{\mathbf{t}} : \Omega^{\mathbf{k}} \to \Omega^{\mathbf{k}+\mathbf{t}}.$$

Theorem 2.4 ([12]) *If the multi-index* $\mathbf{t}$ *contains one negative component and this component is* -1, *then the operator* $d_{\mathbf{t}}$ *is a* $C^\infty(J^1 M)$-*homomorphism, i.e.,*

$$d_{\mathbf{t}}(f\alpha) = f d_{\mathbf{t}} \alpha \tag{2.43}$$

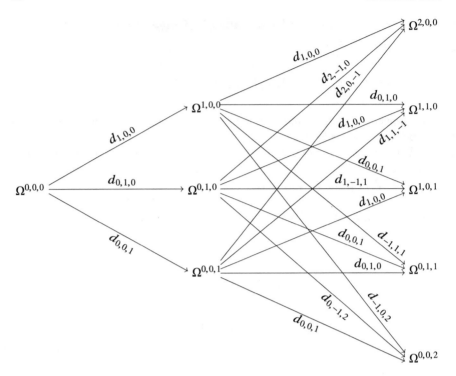

Fig. 2.5 Decomposition of de Rham complex

for any function f and any differential form $\alpha \in \Omega^{\mathbf{k}}$.

Due to this theorem, we have the seven homomorphisms, and three of them are zeroes. The non-trivial homomorphisms are the following:

$$d_{2,-1,0}, \quad d_{0,-1,2}, \quad d_{-1,1,1} \quad \text{and} \quad d_{1,1,-1}.$$

2.6.2 Tensor Invariants

Consider a case

$$\mathbf{t} = \mathbf{1}_j + \mathbf{1}_k - \mathbf{1}_s.$$

Then the differential $d_{\mathbf{t}}$ is a $C^\infty(J^1 M)$-homomorphism. Note that

$$d_{\mathbf{1}_j + \mathbf{1}_k - \mathbf{1}_s} : \Omega^{\mathbf{1}_q} \to 0,$$

if $q \neq s$. Then, the only non-trivial of $d_{\mathbf{1}_j + \mathbf{1}_k - \mathbf{1}_s}$ is the restriction to the module $\Omega^{\mathbf{1}_s}$:

$$d_{1_j+1_k-1_s} : \Omega^{1_s} \to \Omega^{1_j} \wedge \Omega^{1_k}.$$

Therefore, the homomorphism $d_{1_j+1_k-1_s}$ defines a tensor field of the type $(2,1)$. This tensor field we denote by $\tau_{1_j+1_k-1_s}$:

$$\tau_{1_j+1_k-1_s} \in \Omega^{1_j} \wedge \Omega^{1_k} \otimes D_s.$$

A unique non-trivial component of this tensor field is its restriction to Ω^{1_s}. Note that

$$\tau_{1_j+1_k-1_s} : \Omega^{1_s} \to \Omega^{1_j} \wedge \Omega^{1_k}$$

coincides with $d_{1_j+1_k-1_s}$.

Tensor fields $\tau_{1_j+1_k-1_s}$ are differential invariants of Monge–Ampère equations. So, we get four tensors of $(2,1)$-type [12]:

$$\tau_{2,-1,0}, \quad \tau_{0,-1,2}, \quad \tau_{-1,1,1} \quad \text{and} \quad \tau_{1,1,-1}. \tag{2.44}$$

Maple Code: Tensor Invariants

Below, we present a program for calculating the tensor $\tau_{-1,1,1}$. The remaining tensors can be found similarly after a small adjustment of the program. In this program, we omit the calculation of the characteristic distributions. They must be calculated in advance (see "Maple Code: Operator A_ω" and "Maple Code: Characteristic distributions").

```
with(DifferentialGeometry): with(LinearAlgebra):
with(Tensor):with(Tools): with(PDETools):

DGsetup( [x1,x2,u,p1,p2], J):

kappa:=evalDG(du-p1*dx1-p2*dx2):
Omega:=ExteriorDerivative(kappa):

omega:=evalDG(2*u*dx2 &w dp1+ dx1 &w dp1-
dx2 &w dp2-2*k*p1^2*dx1 &w dx2):
```

Construct the distribution l (transversal to the Cartan distribution). We are looking for l as an intersection of derivatives of the characteristic distributions $C_-^{(1)}$ and $C_+^{(1)}$. This intersection is one-dimensional and it is generated by the vector field Z which we are looking for.

```
S:=evalDG(a1*Vp1+a2*Vp2+a3*LieBracket(Vp1,Vp2)-
(b1*Vm1+b2*Vm2+b3*LieBracket(Vm1,Vm2))):
```

```
sol:=solve(Tools:-DGinfo(S, "CoefficientSet"),
[a1,a2,a3,b1,b2,b3]):
```

```
assign(sol):
```

```
Z:=evalDG(a1*Vp1+a2*Vp2+a3*LieBracket(Vp1,Vp2)):
```

Basis of the module of vector fields on J^1M and dual basis:

```
BV:=[Vm1,Vm2,Vp1,Vp2,Z]:
BC:=evalDG(DualBasis(BV)):
```

Decomposition of de Rham complex. Bases of $\Omega^1(J^1M)$ and $\Omega^2(J^1M)$:

```
Lambda[1,0,0]:=evalDG([BC[1], BC[2]]);
```

```
Lambda[0,1,0]:=evalDG([BC[5]]);
```

```
Lambda[0,0,1]:=evalDG([BC[3], BC[4]]);
```

```
Lambda[2,0,0]:=evalDG([BC[1] &w BC[2]]); #1
```

```
Lambda[1,1,0]:=evalDG([BC[1] &w BC[5], BC[2] &w BC[5]]); #2,3
```

```
Lambda[1,0,1]:=evalDG([BC[1] &w BC[3], BC[1] &w BC[4],
BC[2] &w BC[3], BC[2] &w BC[4]]); #4,5,6,7
```

```
Lambda[0,1,1]:=evalDG([BC[3] &w BC[5], BC[4] &w BC[5]]); #8,9
```

```
Lambda[0,0,2]:=evalDG([BC[3] &w BC[4]]); #10
```

List of elements of the basis of Ω^2:

```
Lambda2:=[op(Lambda[2,0,0]),op(Lambda[1,1,0]),
op(Lambda[1,0,1]), op(Lambda[0,1,1]),op(Lambda[0,0,2])];
```

Construct the tensor $\tau_{-1,1,1}$:

```
unassign('z1','z2','z3','z4','z5','z6','z7','z8','z9','z10');
```

Arbitrary differential 2-form:

```
V:=evalDG(DGzip([z1,z2,z3,z4,z5,z6,z7,z8,z9,z10],
Lambda2, "plus")):
```

Arbitrary 2-form from $\Omega^{1,0,0}$:

```
S:=evalDG(ExteriorDerivative(C1*Lambda[1,0,0][1]+
C2*Lambda[1,0,0][2])-V):

S_coeff:=Tools:-DGinfo(S, "CoefficientSet"):

sol:=solve(S_coeff,{z1,z2,z3,z4,z5,z6,z7,z8,z9,z10});

assign(sol);
```

Projection of a differential 2-form to $\Omega^{0,1,1}$:

```
Pr_011:=evalDG(DGzip([z8,z9],
[Lambda2[8],Lambda2[9]],"plus")):

Pr_011:=convert(Pr_011, DGtensor):

unassign('a','b','c','d'):

Tau:=evalDG(a*Lambda[0,1,1][1] &t BV[1]+
b*Lambda[0,1,1][2] &t BV[1]+
c*Lambda[0,1,1][1] &t BV[2]+
d*Lambda[0,1,1][2] &t BV[2]):

aTau:=ContractIndices(evalDG(Tau &t
(C1*Lambda[1,0,0][1]+C2*Lambda[1,0,0][2])),[[3,4]]):

eq0:=DGsimplify(evalDG(aTau-Pr_011)):
eq:=Tools:-DGinfo(eq0, "CoefficientSet"):
e1:=op(eval(eq,{C1=1,C2=0})):
e2:=op(eval(eq,{C1=0,C2=1})):

sol:=solve([e1,e2],[a,b,c,d]):
assign(sol):

Tau1:=DGsimplify(Tau):

tau[-1, 1, 1]:=DGsimplify(Tau);
```

Example: Hunter–Saxton Equation

Consider the Hunter–Saxton equation

$$v_{tx} = v v_{xx} + \kappa v_x^2, \tag{2.45}$$

where κ is a constant. This equation is hyperbolic, and it has applications in the theory of liquid crystals [6].

The corresponding effective differential 2-form and the operator A_ω are the following:

$$\omega = 2u dq_2 \wedge dp_1 + dq_1 \wedge dp_1 - dq_2 \wedge dp_2 - 2\kappa p_1^2 dq_1 \wedge dq_2$$

and

$$A_\omega = \begin{Vmatrix} 1 & 2u & 0 & 0 \\ 0 & -1 & 0 & 0 \\ 0 & -2\kappa p_1^2 & 1 & 0 \\ 2\kappa p_1^2 & 0 & 2u & -1 \end{Vmatrix}.$$

Let's take the following base in the module of vector fields on $J^1 M$:

$$X_1 = \frac{\partial}{\partial q_1} + p_1 \frac{\partial}{\partial u} + \kappa p_1^2 \frac{\partial}{\partial p_2},$$

$$X_2 = \frac{\partial}{\partial p_1} + u \frac{\partial}{\partial p_2},$$

$$Z = \frac{\partial}{\partial u} + (2\kappa - 1) p_1 \frac{\partial}{\partial p_2},$$

$$Y_1 = \frac{\partial}{\partial q_2} + \kappa p_1^2 \frac{\partial}{\partial p_1} - u \frac{\partial}{\partial q_1} + (p_2 - u p_1) \frac{\partial}{\partial u},$$

$$Y_2 = \frac{\partial}{\partial p_2}.$$

The dual basis of the module of differential 1-forms is

$$\alpha_1 = dq_1 + u dq_2,$$
$$\alpha_2 = dp_1 - \kappa p_1^2 dq_2,$$
$$\theta = du - p_1 dq_1 - p_2 dq_2,$$
$$\beta_1 = dq_2,$$
$$\beta_2 = dp_2 + (1 - 2\kappa) p_1 du + (\kappa - 1) p_1^2 dq_1 + (2\kappa - 1) p_1 p_2 dq_2 - u dp_1.$$

The vector fields X_1, X_2 and Y_1, Y_2 form bases in the modules $D(C_+)$ and $D(C_-)$ respectively. Tensor invariants of Eq. (2.45) have the form

$$\tau_{-1,1,1} = -\left(p_1 dq_1 \wedge dq_2 + dq_2 \wedge du\right) \otimes \left(\frac{\partial}{\partial q_1} + p_1 \frac{\partial}{\partial u} + \kappa p_1^2 \frac{\partial}{\partial p_2}\right),$$

$$\tau_{1,1,-1} = 2(\kappa - 1)\left(\kappa p_1^3 dq_1 \wedge dq_2 + \kappa p_1^2 dq_2 \wedge du - \right.$$

$$\left. dp_1 \wedge du - p_1 dq_1 \wedge dp_1 - p_2 dq_2 \wedge dp_1\right) \otimes \frac{\partial}{\partial p_2},$$

$$\tau_{2,-1,0} = \left(dq_1 \wedge dp_1 - \kappa p_1^2 dq_1 \wedge dq_2 + u dq_2 \wedge dp_1\right) \otimes$$

$$\left(\frac{\partial}{\partial u} + (2\kappa - 1) p_1 \frac{\partial}{\partial p_2}\right),$$

$$\tau_{0,-1,2} = \left(dq_2 \wedge dp_2 + (1 - 2\kappa) p_1 dq_2 \wedge du + (1 - \kappa) p_1^2 dq_1 \wedge dq_2 - u dq_2 \wedge dp_1\right) \otimes$$

$$\left(\frac{\partial}{\partial u} + (2\kappa - 1) p_1 \frac{\partial}{\partial p_2}\right).$$

2.6.3   The Laplace Forms

Define bracket $\langle \alpha \otimes X, \beta \otimes Y \rangle$ for decomposable tensors $\alpha \otimes X$ and $\beta \otimes Y$ of types $(2,1)$ as follows [12]:

$$\langle \alpha \otimes X, \beta \otimes Y \rangle = (Y \rfloor \alpha) \wedge (X \rfloor \beta).$$

For non-decomposable tensors the bracket is defined by linearity.

Define two differential 2-forms λ_- and λ_+ from the module $\Omega^{1,0,1}$ as "wedge contractions" of the tensor fields:

$$\lambda_+ := \langle \tau_{0,-1,2}, \tau_{1,1,-1} \rangle, \qquad \lambda_- := \langle \tau_{2,-1,0}, \tau_{-1,1,1} \rangle. \tag{2.46}$$

Then tensors (2.46) are called *Laplace forms* of Monge–Ampère equations $\mathcal{E}_\omega$.

Example: Laplace Form for Linear Equations

For linear hyperbolic equation

$$v_{xy} = a(x, y)v_x + b(x, y)v_y + c(x, y)v + g(x, y), \tag{2.47}$$

the Laplace forms are

$$\lambda_- = k dx \wedge dy \quad \text{and} \quad \lambda_+ = -h dx \wedge dy, \tag{2.48}$$

where

$$k = ab + c - b_y \qquad h = ab + c - a_x \tag{2.49}$$

are the classical Laplace invariants. This observation justifies our definition.

For linear elliptic equations

$$v_{xx} + v_{yy} = a(x, y)v_x + b(x, y)v_y + c(x, y)v + g(x, y), \qquad (2.50)$$

Laplace forms generalize Cotton invariants [2].

We emphasize that the classical Laplace invariants (2.49) of Eq. (2.50) are not absolute invariants even with respect to transformations

$$\phi : (x, y, v) \mapsto (X(x), Y(y), A(x, y)v), \qquad A(x, y) \neq 0 \qquad (2.51)$$

in contrast to forms $\lambda_{\pm}$, which are contact invariants.

Example: Laplace Forms for Hunter–Saxton Equation

The Laplace forms for the Hunter–Saxton equation (2.45) are

$$\lambda_- = -dq_2 \wedge dp_1, \quad \lambda_+ = 2(1 - \kappa) dq_2 \wedge dp_1.$$

2.6.4   Contact Linearization of the Monge–Ampère Equations

It is well known that if the classical Lagrange invariants h and k of a linear hyperbolic equation is zero, then the equation can be reduced to the wave equation (see [22], for example).

Similar statement is true for the Monge–Ampère equations [14]:

Theorem 2.5 *A hyperbolic Monge–Ampère equation is locally contact equivalent to the wave equation*

$$v_{xy} = 0$$

if and only if its Laplace invariants are zero: $\lambda_+ = \lambda_- = 0$.

Corollary 2.1 *The equation*

$$v_{xy} = f(x, y, v, v_x, v_y)$$

is locally contact equivalent to the wave equation $v_{xy} = 0$ *if and only if the function* f *has the following form:*

$$f = \varphi_y v_x + \varphi_x v_y + (\varphi_v + \Phi_v)v_x v_y + R,$$

where the function $R = R(x, y, v)$ satisfies to the following ordinary linear differential equation:

$$R_v = (\varphi_v + \Phi_v)R + \varphi_{xy} - \varphi_x\varphi_y.$$

Solving this equation we get

$$R = e^{\varphi + \Phi}\left(\int(\varphi_{xy} - \varphi_x\varphi_y)e^{-\varphi - \Phi}dv + g\right),$$

where $\varphi = \varphi(x, y, v)$, $\Phi = \Phi(v)$, and $g = g(x, y)$ are arbitrary functions.

The general problem of linearization of non-degenerated Monge–Ampère equations with respect to the contact transformations was solved in [13].

Acknowledgements This work is partially supported by the Russian Foundation for Basic Research (project 18-29-10013). The third author was also supported by the Czech Grant Agency, grant no. GA17-01171S.

References

1. Akhmetzianov, A.V., Kushner, A.G., Lychagin, V.V. Integrability of Buckley–Leverett's filtration model. In: IFAC Proceedings Volumes (IFAC-PapersOnline). **49**(12), pp. 1251–1254 (2016)
2. Cotton, E. Sur les invariants différentiels de quelques équations lineaires aux dérivées partielles du second ordre. In: Ann. Sci. Ecole Norm. Sup. **17**, pp. 211–244 (1900)
3. Darboux, G. Leçons sur la théorie générale des surfaces. Vol. I. Paris, Gauthier-Villars, Imprimeur–Libraire, 1887, vi+514 pp.
4. Darboux, G. Leçons sur la théorie générale des surfaces. Vol. II. Paris, Gauthier-Villars, 1915, 579 pp.
5. Darboux, G. Leçons sur la théorie générale des surfaces. Vol. III. Paris, Gauthier-Villars at fils, Imprimeur– Libraire, 1894, viii+512 pp.
6. Hunter, J.K., Saxton, R. Dynamics of director fields. In: SIAM J. Appl. Math. **51**(6), pp. 1498–1521 (1991)
7. Kantorovich, L. On the translocation of masses. // C.R. (Doklady) Acad. Sci. URSS (N.S.), **37**, 199–201 (1942)
8. Krasil'shchik, I.S., Lychagin, V.V., Vinogradov, A.M. Geometry of jet spaces and nonlinear partial differential equations. Advanced Studies in Contemporary Mathematics, **1**, Gordon and Breach Science Publishers, New York, 1986, xx+441 pp.
9. Kushner, A.G. Classification of mixed type Monge-Ampère equations. In: Pràstaro, A., Rassias, Th.M. (eds) Geometry in Partial Differential Equations, pp. 173–188. World Scientific (1993)
10. Kushner, A.G. Symplectic geometry of mixed type equations. In: Lychagin, V.V. (ed) The Interplay between Differential Geometry and Differential Equations. Amer. Math. Soc. Transl. Ser. 2, **167**, pp. 131–142 (1995)
11. Kushner, A.G. Monge–Ampère equations and e-structures. In: Dokl. Akad. Nauk **361**(5), pp. 595–596 (1998) (Russian). English translation in Doklady Mathematics, **58**(1), pp. 103–104 (1998)
12. Kushner, A.G. A contact linearization problem for Monge–Ampère equations and Laplace invariants. In: Acta Appl. Math. **101**(1–3), pp. 177–189 (2008)

13. Kushner, A.G. Classification of Monge-Ampère equations. In: "Differential Equations: Geometry, Symmetries and Integrability". Proceedings of the Fifth Abel Symposium, Tromso, Norway, June 17–22, 2008 (Editors: B. Kruglikov, V. Lychagin, E. Straume) pp. 223–256.
14. Kushner, A.G. On contact equivalence of Monge–Ampère equations to linear equations with constant coefficients. Acta Appl. Math. **109**(1), pp. 197–210 (2010)
15. Kushner, A.G., Lychagin, V.V., Rubtsov, V.N. Contact geometry and nonlinear differential equations. Encyclopedia of Mathematics and Its Applications **101**, Cambridge University Press, Cambridge, 2007, xxii+496 pp.
16. Lychagin, V.V. Contact geometry and nonlinear second-order partial differential equations. In: Dokl. Akad. Nauk SSSR **238**(5), pp. 273–276 (1978). English translation in Soviet Math. Dokl. **19**(5), pp. 34–38 (1978)
17. Lychagin, V.V. Contact geometry and nonlinear second-order differential equations. In: Uspekhi Mat. Nauk **34**(1 (205)), pp. 137–165 (1979). English translation in Russian Math. Surveys **34**(1), pp. 149–180 (1979)
18. Lychagin, V.V. Lectures on geometry of differential equations. Vol. 1,2. "La Sapienza", Rome, 1993.
19. Lychagin, V.V., Rubtsov, V.N. The theorems of Sophus Lie for the Monge–Ampère equations (Russian). In: Dokl. Akad. Nauk BSSR **27**(5), pp. 396–398 (1983)
20. Lychagin, V.V., Rubtsov, V.N. Local classification of Monge–Ampère differential equations. In: Dokl. Akad. Nauk SSSR **272**(1), pp. 34–38 (1983)
21. Lychagin, V.V., Rubtsov, V.N., Chekalov, I.V. A classification of Monge–Ampère equations. In: Ann. Sci. Ecole Norm. Sup. (4) **26**(3), pp. 281–308 (1993)
22. Ovsyannikov, L.V. Group Analysis of Differential Equations. New York, Academic Press, 1982, 416 pp.
23. Tunitskii, D.V. On the contact linearization of Monge-Ampère equations. In: Izv. Ross. Akad. Nauk, Ser. Matem. **60**(2), pp. 195–220 (1996)

Chapter 3
Geometry of Monge–Ampère Structures

Volodya Rubtsov

3.1 About These Lectures

These lectures were designed for the *Summer school Wisła -18 'Nonlinear PDEs, their geometry, and applications'* of Bałtycki Instytut Matematyki, in Wisła, Poland, 20–30 August 2018.

The intended audience is graduate students with some minimal background in differential geometry and in geometry of non-linear PDE. I was encouraged to produce lecture notes by Organisers, since not so much of the literature in the subject exists on the corresponding level. So that notes could serve some purpose.

There are (almost) no new results in my notes, though some presentation is a new or a nonstandard. I have used the results obtained during long-term collaboration with V.V. Lychagin, whose ideas mainly gave the basis and the foundation of all further approach. The lectures are following the five oral presentations during the Wisla-2018 School and carry therefore some features of these expositions. Part of subjects is presented in more details while other has some punctual line presentation. We assume that the interested reader is able to find the omitted details in the original source literature in the references.

Some part of material was taken from our book with A.Kushner and V.V. Lychagin (CUP, 2007) and from Angers PhD Thesis of B. Banos (Angers, 2002). Generalities about Nijenhuis or recursion operators and their basic properties are taken from

---

To Sir Michael Atiyah.

V. Rubtsov (✉)

Maths Department, University of Angers, 2, Lavoisier Boulevard, 49045
CEDEX 01, Angers, France
e-mail: volodya@univ-angers.fr

Theory Division, Mathematical Physics Laboratory, ITEP, 25, Bol. Tcheremushkinskaya,
117218 Moscow, Russia

© Springer Nature Switzerland AG 2019

R. A. Kycia et al. (eds.), *Nonlinear PDEs, Their Geometry, and Applications*,
Tutorials, Schools, and Workshops in the Mathematical Sciences,
https://doi.org/10.1007/978-3-030-17031-8_3

published papers of G. Bande and D. Kotschick. New presentational part of Chap. 4 is based on the joint review in progress with I. Roulstone, M. Wolf and J. McYorist.

I am deeply grateful to all of them for their input and collaboration on various subjects of these lectures, discussions and improvements. My deep thanks also to Y. Kosmann-Schwarzbach and to I. Mencattini for collaboration and many useful discussions of Monge–Ampère operators and higher structures and to Misha Verbitsky for his various lectures and very inspiring discussions. The last but not the least my deep thanks to Jerzy Szmit for organising a very stimulating summer school and for a personal help.

During preparation of the material for this minicourse, I was partly supported by support of the project IPaDEGAN (H2020-MSCA-RISE-2017), Grant Number 778010 and of the Russian Foundation for Basic Research under the Grants RFBR 18-01-00461 and 19-51-53014 GFEN. Part of this work was carried out within framework of the State Programme of the Ministry of Education and Science of the Russian Federation, Project N 1.2873.2018/12.1.

When these lectures were prepared and printed, I have got a sad news that the great mathematician Sir Michael Atiyah had passed away. His ideas and work had a huge influence on me and my mathematical exercises since my studentship and up to now. I have met him the first time in the Newton Institute in Cambridge in the fall of 1996 (it was his last fall of directorship). I cannot forget his sincere and deep satisfaction and positive emotional reaction when some of participants of two initially different institute programmes—4-dimensional Geometry and Quantum Field Theory and Mathematics of Atmosphere and Ocean Dynamics—had successfully interacted and started to discuss and to cooperate.

I would like with pain and sorrow to dedicate these lectures to his memory…

3.2 Lecture One: What Is It All About?

In this lecture, we begin with a very brief introduction to the basic of various geometric structures used in my course. I remind the basic definitions and (in some cases) elementary examples. This chapter cannot be used for a detailed study of differential geometry and basic structures. I have collected here the definitions and terminology for reader's convenience.

3.2.1 Basic Geometric Structures

3.2.1.1 (Almost) Complex Structures

Let M be a smooth manifold.

Definition 3.1 An **almost-complex structure** is an operator $I : TM \longrightarrow TM$ which satisfies $I^2 = -\operatorname{Id}_{TM}$.

In other words, this is a smooth section I of the vector bundle $TM \otimes T^*M \longrightarrow M$ such that $\forall x \in M \; I_x^2 = \mathrm{Id}$.

The almost-complex structure permits to enable each tangent space T_xM with a structure of $\mathbb{C}$-vector space defining $i.v = I_x(v)$ for all tangent vectors $v \in T_xM$.

Example

Let M^n be a holomorphic variety of complex dimension n and let $U_j \overset{\phi_j}{\to} \mathbb{C}^n$ be a holomorphic atlas $\{U_j, j \in J\}$ on this variety. One can enable the real variety M with an almost-complex structure I defining for $x \in U_j$ and $U \subset T_xM$:

$$I_x(U) = (d_x\phi_j)^{-1}(i.d_x\phi_j(U))$$

(This definition is independent on the choice of $j \in J$ because the changing of charts $d_{\phi_j(x)}\phi_k \circ \phi_j^{-1}$ is $\mathbb{C}$-linear).

The eigenvalues of the operator I are $\pm\sqrt{-1}$, and we shall denote the corresponding eigenvalue decomposition iby $TM \otimes \mathbb{C} = T^{0,1}M \oplus T^{1,0}(M)$. Let us describe it in details.

Let I be an almost-complex structure on a variety M. Denote by D_+ and by D_- the distributions on $TM \otimes \mathbb{C}$ defined by the sub-eigenspaces for I :

$$\begin{cases} D_+(x) := \{V \in T_xM \otimes \mathbb{C} \mid I(V) = i.V\}, \\ D_-(x) := \{V \in T_xM \otimes \mathbb{C} \mid I(V) = -i.V\}. \end{cases}$$

We shall denote as usual the conjugation on T_xM by $V \longrightarrow \bar{V}$. It defines a real isomorphism $D_+ \simeq D_-$ and the map

$$T_xM \longrightarrow T_xM \otimes \mathbb{C}, \quad V \longrightarrow V - i.I_x(V)$$

induces a complex isomorphism $(T_xM, I_x) \simeq (D_+, i)$.

We denote by $\mathcal{D}_+$ (respectively by $\mathcal{D}_-$) the set of smooth sections of the fibre bundle $TX \otimes \mathbb{C} \longrightarrow X$ with values in D_+ (respectively in D_-).

Definition 3.2 An almost-complex structure is **integrable in the sense of Frobenius** if $\mathcal{D}_+$ and $\mathcal{D}_-$ are Lie subalgebras of the vector field Lie algebra $\mathcal{D}(M) \otimes \mathbb{C}$.

One can verify that I is integrable in sense of Frobenius if and only if the Nijenhuis tensor $\mathcal{N}_I = 0$ where

$$\mathcal{N}_I(U; V) := [U; V] - [IU; IV] + I([IU; V] + [U; IV]).$$

Definition 3.3 (*Integrability-1*) An almost-complex structure I on a manifold M is **integrable** if there exist a holomorphic atlas on M such that I is the associated almost-complex structure. We shall say then that I **is a complex structure** on M. In this case, I is called **a complex structure operator**. A manifold with an integrable almost complex structure is called **a complex manifold**.

The difficult Newlander–Nirenberg theorem says that an almost-complex structure I is integrable if and only if it is integrable in the sense of Frobenius:

Theorem 3.1 (Newlander–Nirenberg-1) *An almost-complex structure I is inte-grable iff $\mathcal{N}_I = 0$.*

Definition 3.4 (*Integrability-2*) An almost-complex structure is **integrable** if $\forall X, Y \in T^{1,0}M$, one has $[X, Y] \in T^{1,0}M$.

Theorem 3.2 (Newlander–Nirenberg-2) *The definitions -1 and -2 are equivalent.*

3.2.1.2 (Almost) Product and Para-Complex Structures

Let M be a smooth manifold.

Definition 3.5 An **almost-product structure** is an operator $J : TM \longrightarrow TM$ if $J^2 = \mathrm{Id}_{TM}$.

It is called an **almost para-complex structure** if it is an almost product and $\mathrm{tr}J = 0$. In this case, M is called an **almost para-complex manifold**.

In the almost-product case, we have the (real) decomposition $TM \cong T^{1,0}M \oplus T^{0,1}M$, where

$$
\begin{cases}
T^{1,0}M := \{X + JX \mid X \in \Gamma(TM)\}, \\
T^{0,1}M := \{X - JX \mid X \in \Gamma(TM)\},
\end{cases}
\tag{3.1}
$$

and which are the ± 1 eigenspaces of J.

In the almost para-complex case, the traceless condition yields that $\mathrm{rk}(T^{1,0}M) = \mathrm{rk}(T^{0,1}M)$. In addition, we have $\Omega^p(M) \cong \bigoplus_{r+s=p} \Omega^{r,s}(M)$. The analogue of the Nijenhuis tensor is then

$$
\mathcal{N}_J(X, Y) := -[X, Y] + J[X, JY] + J[JX, Y] - [JX, JY].
\tag{3.2}
$$

I leave to readers proofs of the following two propositions:

Proposition 3.1 *The distribution $T^{1,0}M$ (respectively, $T^{0,1}M$) is integrable in the sense of Frobenius whenever $\mathcal{N}_J|_{T^{0,1}M}$ (respectively, $\mathcal{N}_J|_{T^{1,0}M}$) vanishes.*

Proposition 3.2 *The almost para-complex structure J is integrable if and only if the Nijenhuis tensor vanishes, that is, whenever both distributions $T^{1,0}M$ and $T^{0,1}M$ are integrable.*

Remark 3.1 Note that these are two independent conditions as the integrability of either of those distributions does not imply the integrability of the other.

3.2.2 Kähler, Special and Other Related Structures

3.2.2.1 Kähler Manifolds

Definition 3.6 A Riemannian metric g on a complex manifold (M, I) is called **Hermitian** if $g(Ix, Iy) = g(x, y)$.

In this case, $g(x, Iy) = g(Ix, I^2 y) = -g(y, Ix)$; hence, $\omega(x, y) := g(x, Iy)$ is skew-symmetric.

Definition 3.7 The differential form $\omega \in \Lambda^{1,1}(M)$ is called **the Hermitian form** of (M, I, g).

Definition 3.8 A complex Hermitian manifold (M, I, ω) is called **Kähler** if $d\omega = 0$. The cohomology class $[\omega] \in H^2(M)$ of a form ω is called **the Kähler class** of M, and ω **the Kähler form**.

Remark 3.2 This condition is equivalently read as $\nabla\omega = 0$, where ∇ is the Levi-Civita connection.

3.2.2.2 Special Complex Manifolds

Definition 3.9 A complex manifold (M, I) is called **special complex** if there is a flat, torsion-free (linear) connection ∇ such that

$$d_\nabla I = 0.$$

Definition 3.10 If there is a covariantly constant symplectic form Ω on M (i.e. $\nabla\Omega = 0$), then the triple (M, I, Ω) is called **special symplectic.**

3.2.2.3 Special Kähler Manifolds

Definition 3.11 (*Special Kähler-1*) A special symplectic manifold (M, Ω, I) is a **special Kähler** if the symplectic form Ω is I-invariant:

$$I^*\Omega(X, Y) := \Omega(IX, IY) = \Omega(X, Y), \quad \forall X, Y \in \Gamma(TM).$$

Notice that the 2-tensor $g(\cdot, \cdot) = \Omega(\cdot, I\cdot)$, while it is always symmetric, in general, not positive definite. So, the name of **special pseudo-Kähler** would be more appropriate.

The notion of special Kähler structure was refined by D. Freed ([25]).

Definition 3.12 (*Special Kähler-2*) Special Kähler variety is a Kähler variety (M, g, I, ω) together with a flat zero torsion connection ∇ such that $\nabla\omega = 0$ and $d_\nabla\omega = 0$.

One can characterise the special Kähler varieties as **bi-Lagrangian** subvarieties of some complex symplectic space. More precisely, let (V^{2n}, Ω) be a real symplectic vector space. Hitchin ([35]) has supplied the space $V \times V$ with two symplectic forms ω_1 and ω_2 and with a pseudo-metric g which are defined as

$$\begin{cases} \omega_1((X, Y), (X', Y')) = \Omega(X, X') - \Omega(Y, Y'), \\ \omega_2((X, Y), (X', Y')) = \Omega(X, Y') + \Omega(Y, X'), \\ g((X, Y), (X', Y')) = \frac{1}{4}[\Omega(X, Y') - \Omega(X', Y)] \end{cases}$$

He has proven that if a bi-Lagrangian subvariety $M \subset (V \times V, \omega_{1,2})$ is transversal to both projections onto V, the g induces a special (pseudo) Kähler metric on M and vice versa, and any special Kähler metric on M is induced locally by a such embedding $M \hookrightarrow V \times V$.

3.2.3 Holomorphic Symplectic Structures

We suppose $\mathbb{C}$ to be the base field throughout this section.

3.2.3.1 Holomorphically Symplectic Structures

Definition 3.13 A **holomorphically symplectic manifold** is a complex manifold equipped with non-degenerate, holomorphic $(2, 0)$-form.

Remark 3.3 In these lectures, all holomorphically symplectic manifolds are assumed to be Kähler but not always compact.

Remark 3.4 A hyperkähler manifold has three symplectic forms:

$$\omega_I := g(I\cdot, \cdot), \quad \omega_J := g(J\cdot, \cdot), \quad \omega_K := g(K\cdot, \cdot).$$

Claim In these assumptions, $\omega_J + \sqrt{-1}\,\omega_K$ is holomorphic symplectic on (M, I).

Theorem 3.3 (Calabi–Yau) *A compact, Kähler, holomorphically symplectic manifold admits a unique hyperkähler metric in any Kähler class.*

3.2.3.2 Calabi–Yau Manifolds

Definition 3.14 A Kähler manifold M of complex dimension n enabled with a covariantly constant holomorphic $n-$ form $\Omega \in \Lambda^{n,0}(M)$ is called a **Calabi–Yau manifold**.

The corresponding geometry is completely described by a set of few closed real forms: by the Kähler form ω and by two $n-$forms $\omega_1 := \operatorname{Re} \Omega$ and $\omega_2 := \operatorname{Im} \Omega$ which are defined up to a non-zero constant.

Proposition 3.3 *Let M be a Calabi–Yau manifold. Then, the triple $\omega, \omega_1, \omega_2$ satisfies the following properties:*

1. ω *is a symplectic form on M;*
2. ω_1 *and ω_2 are* **primitive (effective)** *with respect to ω : $\omega \wedge \omega_1 = \omega \wedge \omega_2 = 0$;*
3. $d\omega_1 = d\omega_2 = 0$;
4. $\Omega \wedge \bar{\Omega} = c\omega^n, \quad c \in \mathbb{C}.$

3.2.4 Lagrangian, Special Lagrangian and Complex Lagrangian Submanifolds

3.2.4.1 Lagrangian Subvarieties

Let $(X; \Omega)$ be a symplectic variety, as it is well known. X should be an even dimensional, say, dim $X = 2n$.

Definition 3.15 Any $n-$dimensional subvariety $L \subset X$ such that $\Omega|_L = 0$ is a **Lagrangian submanifold** in X.

In our main examples, let X be often the cotangent bundle $X = T^*M$ to a variety X and $\Omega = d\rho$ be the canonical structure with the Liouville 1-form ρ. If $(\bar{q}, \bar{p})$ the local canonical coordinates on T^*M, then $\rho = \bar{p}d\bar{q}$.

3.2.4.2 Complex Lagrangian Subvarieties

In fact we are more interested in the bi-Lagrangian property than the complex structure but these two notions are equivalent like it was proved by Hitchin [35].

Proposition 3.4 (Hitchin) *Let $(N; \Omega = \omega_1 + i\omega_2)$ be a complex symplectic variety of complex dimension $2n$ and $L \hookrightarrow M$ be a real subvariety or real dimension $2n$. The subvariety L is a complex Lagrangian (i.e. $\omega_{|L} = 0$ et L is a complex subvariety) iff L is a real bi-Lagrangian, i.e.*

$$\omega_{1|L} = \omega_{2|L} = 0.$$

The complex symplectic structure we are interested in is the following which one can define on the real cotangent bundle T^*M of a complex variety M: let $z = (z_1, \ldots z_n)$ be a system of complex coordinates on M. The coordinate system

$(x; y) = (x_1, \ldots, x_n; y_1, \ldots, y_n)$ is the associated real coordinates. The real cotangent bundle T^*M naturally identifies with the holomorphic cotangent bundle $T^*M_\mathbb{C}$ by the isomorphism

$$\theta(x; y; p; q) = (x + iy; p - iq).$$

The isomorphism θ permits to assign to T^*M:

1. a complex structure I:

$$\begin{cases} I(\frac{\partial}{\partial x_j}) = \frac{\partial}{\partial y_j}, \\ I(\frac{\partial}{\partial p_j}) = -\frac{\partial}{\partial q_j}. \end{cases}$$

2. two real symplectic forms ω_1 and $\omega_2 = -\omega_1(I.; .)$:

$$\begin{cases} \omega_1 = \sum_{j=1}^{n} dx_j \wedge dp_j + dy_j \wedge dq_j, \\ \omega_2 = \sum_{j=1}^{n} dy_j \wedge dp_j - dx_j \wedge dq_j. \end{cases}$$

3. a holomorphic symplectic form $\Omega = \omega_1 + i\omega_2$:

$$\Omega = \sum_{j=1}^{n} dz_j \wedge dw_j,$$

where $dz_j = dx_j + idy_j$ and $dw_j = dp_j - idq_j$.

Definition 3.16 A smooth function $f : M \to \mathbb{R}$ is called a **pluriharmonic** if $(df)^*(\Omega) = 0$ where $df : M \to T^*M$ is the section of the cotangent bundle associated with f.

Denote by $H(M)$ the set of pluriharmonic functions on M. 2-form ω_1 is the natural symplectic on T^*M; therefore, $(df)^*\omega_1 = 0$ for all smooth f. Hence, a function f is pluriharmonic iff $(df)^*\omega_2 = 0$, i.e.

$$\begin{cases} \frac{\partial^2 f}{\partial x_j \partial y_k} = \frac{\partial^2 f}{\partial x_k \partial y_j}, \\ \frac{\partial^2 f}{\partial x_j \partial x_k} = -\frac{\partial^2 f}{\partial y_k \partial y_j} \end{cases}$$

for $j; k = 1, \ldots, n$. Any pluriharmonic function is the real part of a holomorphic function on M.

The natural objects associated with pluriharmonic operators are the complex Lagrangian subvarieties.

If $f : M \to \mathbb{R}$ is a pluriharmonic then the image of the graph Graph(df) in $T^*M_\mathbb{C}$ under θ is the graph de graph(dF) of a holomorphic function $F : M \to \mathbb{C}$ such that $f = 2\operatorname{Re}(F)$. The graph of dF is a complex Lagrangian subvariety in $(T^*M_\mathbb{C}; \omega_\mathbb{C})$, and hence the graph of df is a complex Lagrangian subvariety in $(T^*M; \Omega)$. Reciprocally, any complex Lagrangian subvariety of $(T^*M; \Omega)$ whose projection is non-singular locally is the graph of df for a certain function f.

3.2.4.3 Special Lagrangian Subvarieties

Let $(X; \omega, \omega_1, \omega_2)$ be a Calabi–Yau variety.

Definition 3.17 Any n-dimensional subvariety $L \subset X$ such that $\omega|_L = \omega_2|_L = 0$ is a **special Lagrangian submanifold** in X.

Remark 3.5 The real part $\operatorname{Re} \Omega = \omega_1$ is (after the restriction and under appropriate choice of c) the volume form on L with respect to the induced metric.

To find an explicite example of special Lagrangian manifolds is a difficult task. The most popular are complex Lagrangian submanifolds in **hyperkähler** manifolds and special Lagrangian subvarieties in noncompact Calabi–Yau.

One of the most famous examples comes from solutions of differential equations (Special Lagrangian Differential equation) related to examples of Calabi–Yau metrics.

Example

There are very few explicit examples of Calabi–Yau metrics. One of these is the Stenzel metric on T^*S^n (see for instance [3, 52]). This metric is not flat; therefore, the special Lagrangian equation associated with is not the classical one.
$T^*S^n = \left\{ (u, v) \in \mathbb{R}^{n+1} \times \mathbb{R}^{n+1} : \|u\| = 1, <u, v> = 0 \right\}$ can be seen as the complex manifold $Q^n = \left\{ z \in \mathbb{C}^{n+1} : z_1^2 + \ldots + z_{n+1}^2 = 1 \right\}$ using the isomorphism

$$\xi(x + iy) = (\frac{x}{\sqrt{1 + \|y\|^2}}, y).$$

The holomorphic form is then

$$\alpha_z(Z_1, \ldots, Z_n) = det_{\mathbb{C}}(z, Z_1, \ldots, Z_n).$$

and the Kähler form is $\Omega = i\partial\bar{\partial}\phi$ with $\phi = f(\tau)$ where τ is the restriction to Q^n of $|z_1|^2 + \ldots + |z_n|^2$ and f is a solution of the ordinary differential equation

$$x(f')^n + f''(f')^{n-1}(x^2 - 1) = c > 0.$$

To write the special Lagrangian equation, we have to find some Darboux coordinates. Consider the case when $n = 3$. Using the relations

$$\begin{cases} \sum_{k=1}^{4} u_k du_k + v_k dv_k = 0 \\ \sum_{k=1}^{4} u_k dv_k + v_k du_k = 0 \end{cases}$$

on T^*S^3, we see that on the chart $u_4 \neq 0$,

$$\Omega = \sum_{k=1}^{3} dw_k \wedge du_k$$

with

$$w_k = 4 \frac{f'(1 + \|v\|^2)\sqrt{1 + \|v\|^2}}{u_4}(u_k v_4 - v_k u_4).$$

Denote by ψ the map $(u, w) \mapsto (x + iy)$. The special Lagrangian equation on T^*S^3 is then

$$(\psi \circ df)^*(Im(\alpha)) = 0.$$

Note that it is difficult to explicit this equation and it does not seem possible to write it in a simple way.

3.2.5 Hyperkähler Manifolds

Definition 3.18 A **hyperkähler structure** on a manifold M is a Riemannian structure g and a triple of complex structures I, J, K, satisfying quaternionic relations $I \circ J = -J \circ I = K$, such that g is Kähler for I, J, K.

Remark 3.6 This is equivalent to $\nabla I = \nabla J = \nabla K = 0$: the parallel translation along the connection preserves I, J, K.

Definition 3.19 Let M be a Riemannian manifold, $x \in M$ a point. The subgroup of $GL(T_xM)$ generated by parallel translations (along all paths) is called **the holonomy group** of M.

Remark 3.7 A hyperkähler manifold can be defined as a manifold which has holonomy in $Sp(n)$ (the group of all endomorphisms preserving I, J, K).

Claim A compact hyperkähler manifold M **has maximal holonomy of Levi-Civita connection** $Sp(n)$ if and only if $\pi_1(M) = 0$, $h^{2,0}(M) = 1$.

3.2.5.1 Almost Hyper-complex and Hyper-para-complex Structures

Next, let us recall the definition of almost hyper-complex and hyper-para-complex structures.

Definition 3.20 Let M be a manifold equipped with three endomorphisms $J_1, J_2, J_3 : TM \to TM$. Then, the triple (J_1, J_2, J_3) is called an

- **almost hyper-complex structure** if and only if the J_is are almost-complex structures and $J_1 J_2 J_3 = -\,\mathrm{Id}$. In this case, M is called an **almost hyper-complex manifold**.
- **almost hyper-para-complex structure** if and only if (up to permutation) the J_1 is an almost-complex structure, J_2 and J_3 are almost para-complex structures and $J_1 J_2 J_3 = \mathrm{Id}$. In this case, M is called an **almost hyper-para-complex manifold**.

Note that such manifolds must be $4m$-dimensional.

Remark 3.8 In this case, we may define

$$J := a J_1 + b J_2 + c J_3, \quad \text{for a, b, c} \in \mathbb{R}. \tag{3.3}$$

Using the fact that $J_1^2 = -\,\mathrm{Id}$ and $J_2^2 = -\epsilon\,\mathrm{Id} = J_3^2$ and $J_1 J_2 J_3 = -\epsilon\,\mathrm{Id}$ with $\epsilon = 1$ in the hyper-complex case and $\epsilon = -1$ in the almost hyper-para-complex case, we obtain

$$J^2 = -(a^2 + \epsilon b^2 + \epsilon c^2)\,\mathrm{Id} \tag{3.4}$$

implying that, in fact, we have a two-sphere worth of such structures for $\epsilon = 1$ while a hyperboloid worth of such structures for $\epsilon = -1$, respectively.

Furthermore, we shall need metrics that are compatible with almost hyper-complex and hyper-para-complex structures in the following sense.

Definition 3.21 Let M be a manifold with an almost-complex (respectively, para-complex) structure J and let g be a non-degenerate metric on M. The triple (M, g, J) is called an **almost Hermitian** (respectively, **para-Hermitian**) manifold if and only if $g(JX, Y) = -g(X, JY)$ for all $X, Y \in \Gamma(TM)$. In this case, g is called an **almost Hermitian** (respectively, **para-Hermitian**) **metric**.

Note that any almost-complex (respectively, para-complex) manifold admits such a metric. Indeed, if h is any non-degenerate metric, then with $J^2 = -\epsilon\,\mathrm{Id}$, the metric

$$g(X, Y) := h(X, Y) + \epsilon h(JX, JY), \quad \forall X, Y \in \Gamma(TM) \tag{3.5}$$

has the desired properties. Next, we introduce a non-degenerate differential two-form ω, called the **almost Kähler form**, by means of

$$\omega(X, Y) := g(JX, Y), \quad \forall X, Y \in \Gamma(TM). \tag{3.6}$$

Likewise, any almost hyper-complex (respectively, hyper-para-complex) manifold admits a metric that almost Hermitian (respectively, para-Hermitian) with respect to each of the J_i. Indeed, as before, let h be any non-degenerate metric, then with $J_i^2 = -\epsilon_i\,\mathrm{Id}, \quad \epsilon_i = \pm 1$, the metric

$$g(X, Y) := h(X, Y) + \sum_{i=1}^{3} \epsilon_i h(J_i X, J_i Y), \quad \forall X, Y \in \Gamma(TM) \tag{3.7}$$

has the desired properties. Consequently, we may introduce three almost Kähler forms by

$$\omega_i(X, Y) := g(J_i X, Y), \quad \forall X, Y \in \Gamma(TM). \tag{3.8}$$

Since $J_1 J_2 J_3 = -\epsilon$ Id we have that (J_i, ω_j) are compatible for $i \neq j$ in the sense of Definition 3.21. Furthermore,

$$\iota_{J_i} \omega_j(.,.) := \omega_j(J_i.,.) = c_{ij}{}^k \omega_k, \tag{3.9}$$

where the $c_{ij}{}^k$s are the structure constants of $\mathfrak{su}(2)$ (respectively, $\mathfrak{sl}(2, \mathbb{R})$).

Definition 3.22 Let (M, g, ω_i, J_i) be an almost hyper-complex (respectively, hyper-para-complex) manifold. Then, (M, g, ω_i, J_i) is called

1. **hyper-complex** (respectively, **hyper-para-complex**) if and only if the J_is are integrable;
2. **hyper-Kähler** (respectively, **hyper-para-Kähler**) if and only if the ω_is are closed.

Remark 3.9 As shown in [33], any hyper-Kähler (respectively, hyper-para-Kähler) manifold is hyper-complex (respectively, hyper-para-complex). Indeed, using (3.9) and the closedness of the ω_is, we can use Proposition 3.7 (the so-called 'Hitchin lemma') to conclude immediately that the J_is are integrable. Note, however, the converse is not necessarily true, i.e. not every hyper-complex (respectively, hyper-para-complex) manifold is hyper-Kähler (respectively, hyper-para-Kähler).

In four dimensions, we may define a hyper-complex (respectively, hyper-para-complex) structure differently. Indeed, suppose we are given three non-degenerate differential two-forms $\omega_i \in \Omega^2(M)$ for $i, j, \ldots = 1, 2, 3$ with $\omega_i \wedge \omega_j = 0$ for $i \neq j$ and $\omega_i \wedge \omega_i \neq 0$ such that (up to permutation) $\omega_2 \wedge \omega_2 = \epsilon \omega_1 \wedge \omega_1 = \omega_3 \wedge \omega_3$, where $\epsilon = 1$ (respectively, $\epsilon = -1$). Then, these forms determine a non-degenerate metric

$$g(X, Y) = \iota_{\varpi} \tfrac{1}{3!} \varpi^{ijk}(\iota_X \omega_i) \wedge (\iota_Y \omega_j) \wedge \omega_k, \quad \forall X, Y \in \Gamma(TM), \tag{3.10}$$

where $\varpi \in \Gamma(\bigwedge^4 TM)$ is a volume form with the normalisation $\tfrac{1}{2!} \iota_{\varpi}(\omega_1 \wedge \omega_1) = -\epsilon$. This metric is Riemannian in the hyper-complex case and Kleinian (means pseudo-Riemannian of neutral signature, $(2, 2)$) in the hyper-para-complex case, respectively. Using this metric, we can define endomorphisms $J_i : TM \to TM$ by means of $\omega_i(X, Y) = g(J_i X, Y)$, and it is easy to check that they obey the conditions listed in Definition 3.20. We summarise as follows.

Proposition 3.5 *An **almost hyper-complex structure** (respectively, **almost hyper-para-complex structure**) on a 4-dimensional manifold M is a set of three differential two-forms $\omega_i \in \Omega^2(M)$ for $i, j, \ldots = 1, 2, 3$ with $\omega_i \wedge \omega_j = 0$ for $i \neq j$ and $\omega_i \wedge \omega_i \neq 0$ such that (up to permutation) $\omega_2 \wedge \omega_2 = \epsilon \omega_1 \wedge \omega_1 = \omega_3 \wedge \omega_3$ where $\epsilon = 1$ (respectively, $\epsilon = -1$).*

If the almost-complex structures are instead not assumed to be integrable, the manifold M is called **quaternionic, or almost hyper-complex**. Every hyperkähler manifold is also hyper-complex but not vice versa. A typical counterexample is given by the **Hopf surface**:

Example

The surface $\{\mathbb{H}\backslash 0\}/\mathbb{Z}$ (with $\mathbb{Z}-$action by a quaternion q, $|q| > 1$) is hyper-complex, but not Kähler, hence not hyperkähler either.

3.2.6 Generalised Complex Structure

3.2.6.1 General Complex Structure on a Vector Space

Let V be a finite dimensional vector space over $\mathbb{R}$.

Recall that a **complex structure** on V is a linear map $I : V \longrightarrow V$ such that $I^2 = -\mathrm{Id}_V$.

A **symplectic structure** on V is equivalently a linear isomorphism $\omega : V \longrightarrow V^*$ such that $\omega^* = -\omega$, where V^* denotes the dual vector space and ω^* the dual linear map.

The following definition may be thought of as combining these two concepts.

Definition 3.23 A **generalised complex structure** on V is a linear map

$$\mathbb{J} : V \oplus V^* \longrightarrow V \oplus V^*$$

(an endomorphism of the direct sum of V with its dual vector space) such that it is both a complex structure on $V \oplus V^*$ in that $\mathbb{J}^2 = -\mathrm{Id}_{V \oplus V^*}$; and a symplectic structure on $V \oplus V^*$ in that $\mathbb{J}^* = -\mathbb{J}$.

The following shows that this is indeed a joint generalisation of complex and symplectic structures.

Example

Let $I : V \longrightarrow V$ be an ordinary complex structure on V. Then the linear endomorphism of $V \oplus V^*$ defined by the matrix as

$$\mathbb{J}_I = \begin{pmatrix} I & 0 \\ 0 & -I^* \end{pmatrix} \tag{3.11}$$

is a generalised complex structure on V.

Example

Similarly, let $\omega : V \longrightarrow V^*$ be an ordinary symplectic structure on V. Then the endomorphism

$$\mathbb{J}_\omega = \begin{pmatrix} 0 & -\omega^{-1} \\ \omega & 0 \end{pmatrix} \tag{3.12}$$

is a generalised complex structure on V.

3.2.6.2 Hitchin Pairs

Let us denote by T the tangent bundle of M and by T^* its cotangent bundle. The natural indefinite interior product on $T \oplus T^*$ is

$$(X + \xi, Y + \eta) = \frac{1}{2}(\xi(Y) + \eta(X)),$$

and the **Courant bracket** on sections of $T \oplus T^*$ is defined by

$$[X + \xi, Y + \eta] = [X, Y] + L_X \eta - L_Y \xi - \frac{1}{2} d(\iota_X \eta - \iota_Y \xi).$$

Definition 3.24 (*Gualtieri* [29]) An **almost generalised complex structure** is a bundle map $\mathbb{J} : T \oplus T^* \to T \oplus T^*$ satisfying

$$\mathbb{J}^2 = -1,$$

and

$$(\mathbb{J}\cdot, \cdot) = -(\cdot, \mathbb{J}\cdot).$$

We can consider an integrability condition for an almost generalised complex structure in the same line like we have introduced the similar notion for integrability of a usual almost-complex structure (see Definitions 3.2 and 3.4).

Such an almost generalised complex structure is said to be **integrable** if the spaces of sections of its two eigenspaces are closed under the Courant bracket.

The standard examples above are reads now as

$$\mathbb{J}_1 = \begin{pmatrix} J & 0 \\ 0 & -J^* \end{pmatrix}$$

and

$$\mathbb{J}_2 = \begin{pmatrix} 0 & \Omega^{-1} \\ -\Omega & 0 \end{pmatrix}$$

with J a complex structure and Ω a symplectic form.

Lemma 3.1 (Crainic [17]) *Let Ω be a symplectic form and ω any 2-form. Define the tensor R by $\omega = \Omega(R\cdot, \cdot)$ and the form $\tilde\omega$ by $\tilde\omega = -\Omega(1 + R^2\cdot, \cdot)$.*
The almost generalised complex structure

$$\mathbb{J} = \begin{pmatrix} R & \Omega^{-1} \\ \tilde\omega & -R^* \end{pmatrix} \tag{3.13}$$

is integrable if and only if ω is closed.

Such a pair (ω, Ω) with $d\omega = 0$ is called a **Hitchin pair of 2-forms**.

Definition 3.25 A **Hitchin pair of bivectors** is a pair consisting of two bivectors π and Π, Π being non-degenerate, and satisfying

$$\begin{cases} [\Pi, \Pi] = [\pi, \pi] \\ [\Pi, \pi] = 0. \end{cases} \tag{3.14}$$

Proposition 3.6 *There is a 1–1 correspondence between generalised complex structure*

$$\mathbb{J} = \begin{pmatrix} R & \pi_R \\ \tilde\omega & -R^* \end{pmatrix}$$

with $\tilde\omega$ non-degenerate and Hitchin pairs of bivector (π, Π). In this correspondence, we have

$$\begin{cases} \tilde\omega = \Pi^{-1} \\ R = \pi \circ \Pi^{-1} \\ \pi_R = -(1 + R^2)\Pi. \end{cases}$$

Example

If $\pi + i\Pi$ is non-degenerate, it defines a 2-form $\omega + i\Omega$ which is necessarily closed (this is the complex version of the classical result which says that a non-degenerate Poisson bivector is actually symplectic). We find again a Hitchin pair. So new examples occur only in the degenerate case. Note that $\pi + i\Pi = (R + i)\Pi$, so $\det(\pi + i\Pi) = 0$ if and only if $-i$ is an eigenvalue for R. In dimension 4, this implies that $R^2 = -1$ but this is no more true in higher ($n \geq 3$) dimensions (see for example the classification of pair of 3-forms on 6-dimensional manifolds in [43]). Nevertheless, the case $R^2 = -1$ is interesting by itself. It corresponds to generalised complex structure of the form

$$\mathbb{J} = \begin{pmatrix} J & 0 \\ \tilde\omega & -J^* \end{pmatrix}$$

with J an integrable complex structure and $\tilde{\omega}$ a 2-form satisfying $J^*\tilde{\omega} = -\tilde{\omega}$ and

$$d\tilde{\omega}_J = d\tilde{\omega}(J\cdot, \cdot, \cdot) + d\tilde{\omega}(\cdot, J\cdot, \cdot) + d\tilde{\omega}(\cdot, \cdot, J\cdot).$$

where $\tilde{\omega}_J = \tilde{\omega}(J\cdot, \cdot)$ (see [17]). Or equivalently $\tilde{\omega} + i\tilde{\omega}_J$ is a $(2, 0)$-form satisfying

$$\partial(\tilde{\omega} + i\tilde{\omega}_J) = 0.$$

One typical example of such geometry is the **hyperkähler geometry with torsion** which is an elegant generalisation of hyperKähler geometry [27]. Unlike the hyperkäler case, such geometry is always generated by potentials [10].

3.2.7 Notes and Further Reading

The original paper by Gualtieri in which he defines the notion of generalised complex structure and establishes the basic properties is [29]. Various 'para'-generalisations of (almost) complex structure were defined in so many papers [42, 51, 53] that I am unable to list all of them. We will say more about their combinations and origins in small dimensions in the next chapter. Lecture notes and expositions of Misha Verbitsky [54] and S. Salamon [50] are extremely readable and have also been very influential.

3.3 Lecture Two: Recursion (Nijenuijs) Operators and Some Related Algebraic Constructions

The second chapter is based mainly on the material of two papers [4, 5] which (in spite of its elementary nature) gives rise to some unexpected and non-trivial geometric and topological examples which are resulted from a fact on existence of few recursion or Nijenhuis operators and some compatibility conditions.

3.3.1 Recursion Operators and Its Properties

3.3.1.1 Recursion (Nijenuijs) Operators

Let M be a manifold enabled with given two non-degenerate 2-forms ω and η in $\Omega^2(M)$.

Then there exists (a unique!) field of non-degenerated endomorphisms $R \in \text{End}(TM)$ of the tangent bundle to M which is defined by the equation

$$\iota_X \omega = \iota_{RX} \eta , \tag{3.15}$$

where $X \in \Gamma(TM)$—a vector field on M.

The case when the two 2-forms ω and η are closed ($d\omega = d\eta = 0$), and therefore symplectic is very important in the perspectives of integrable systems, where it arises in the context of bi-Hamiltonian systems, but it is important also from a geometric point of view.

Definition 3.26 The field of endomorphisms R is called a field of **recursion or Nijenuijs operator**.

We shall study the relation of geometry and topology of the manifold M endowed with two (or more) symplectic (or even more general) forms with the associated Nijenhuis operators R.

3.3.1.2 Symplectic Pairs-1 $R^2 = \mathrm{Id}$

Let M be a manifold with two symplectic forms ω, η and their recursion operator R.

Consider first the case $R^2 = \mathrm{Id}$, but $R \neq \pm Id$. Then the eigenvalues of R are ± 1, and $\forall X \in \Gamma(TM)$ admits the following decomposition:

$$X = \frac{1}{2}(X + RX) + \frac{1}{2}(X - RX).$$

This is the unique splitting of a tangent vector X into a sum of eigenvectors of R. Thus the eigenspaces $D_\pm$ of R give a decomposition of tangent bundle $TM = D_+ \oplus D_-$.

Lemma 3.2 *The kernels of 2-forms $\Omega^\mp := \omega \mp \eta$ are precisely the eigenspaces $D_\pm$ for the eigenvalues ± 1 .*

Proof We shall use the arguments given in the paper [5].

Let X be an arbitrary tangent vector. Then taking the substitution or the inner product of X and both two-forms $\Omega_\pm$ one have

$$\iota_X \Omega^\mp = \iota_X \omega \mp \iota_X \eta$$

and we obtain

$$\iota_X \Omega^\mp = \iota_{RX} \eta \mp \iota_X \eta = \iota_{RX \mp X} \eta .$$

The form η is non-degenerate; hence, the condition $\iota_X \Omega^\mp = 0$ implies $RX = \mp X$ and vice versa, if $RX = \pm X$ then $X \in \mathrm{Ker}\, \Omega^\mp$. □

Lemma 3.3 *The forms $\Omega^\mp$ have constant ranks.*

Proof We use the observation made in [5] that the dimensions of kernels of $\Omega^\mp$ are semi-continuous functions on a closed submanifold and that such function can only be increased. The previous lemma (3.2) shows that if the dimension of the kernel of one of the two-forms $\Omega^{\mp}$ jumps up, then the dimension of the kernel of the other one has to decrease. Therefore, the dimensions of the kernels are actually constant on a connected manifold M, hence the result.   □

? Exercises

Check that the closedeness of $\Omega^{\mp}$ implies the Frobenius integrability of their kernel distributions.

Definition 3.27 (*see* [5]) A **symplectic pair** on a smooth manifold M is a pair of non-trivial closed two-forms ω_1, ω_2 of constant and complementary ranks, for which ω_1 restricts as a symplectic form to the leaves of the kernel foliation of ω_2, and vice versa.

Thus, the forms $\Omega^{\mp}$ in 3.2 form a symplectic pair in the sense of this definition.

Conversely, suppose that we have a symplectic pair $\Omega^{\pm}$ on M, that is, a pair of closed 2-forms of constant ranks, whose kernel foliations $\mathcal{F}_{\mp}$ are complementary and integrable.

? Exercises

Check that the forms $\omega = \frac{1}{2}(\Omega^+ + \Omega^-)$ and $\eta = \frac{1}{2}(\Omega^+ - \Omega^-)$ are symplectic.

Now the corresponding recursion operator is $R = \mathrm{Id}_{T\mathcal{F}_+} - \mathrm{Id}_{T\mathcal{F}_-}$. Thus $R^2 = \mathrm{Id}_{TM}$.

Summing up we have the following.

Theorem 3.4 ([5]) *Two symplectic forms ω and η on a connected manifold M whose recursion operator R satisfies $R^2 = Id$ and $R \neq \pm Id$ give rise to a symplectic pair $\Omega^{\pm}$, and every symplectic pair $\Omega^{\pm}$ arises in this way.*

Remark 3.10 The condition $R^2 = Id$ implies that the Nijenhuis tensor of R vanishes identically. Therefore, in this case, ω and η are compatible in the sense of Poisson geometry.

The latter means that the Poisson structures defined by $\omega^{-1}, \eta^{-1} \in \Gamma(\Lambda^2(TM))$ form a **Poisson pencil** of antisymmetric bivector fields $\pi_\lambda \in \Gamma(\Lambda^2(TM))$ such that

$$\pi_\lambda = \omega^{-1} + \lambda \eta^{-1}$$

and the following operation (**Poisson brackets**)

$$\{f, g\} := \langle \pi_\lambda, df \wedge dg \rangle$$

defines a Lie algebra structure on $C^\infty(M)$ and $X_f := \iota_{df}\, \pi_\lambda$ is a vector field on M (a **Hamiltonian vector field**) or in other terms, the derivation $X_f := \{f, -\}$ of the algebra $C^\infty(M)$.

3.3.1.3   Symplectic Pairs-2 $R^2 = -\mathrm{Id}$

Throughout this subsection, we assume that we have two symplectic forms ω and η on a manifold M of dimension $2n$, such that the recursion operator defined by $\iota_X \omega = \iota_{RX} \eta$ satisfies $R^2 = -\mathrm{Id}_{TM}$. This implies $\iota_{RX} \omega = -\iota_X \eta$.
We shall prove the following.

Theorem 3.5 *If the recursion operator R satisfies $R^2 = -\mathrm{Id}_{TM}$, then it defines an integrable complex structure with a holomorphic symplectic form (i.e. closed non-degenerate, holomorphic $(2, 0)$-form) whose real and imaginary parts are ω and η. Every holomorphic symplectic form arises in this way.*

Proof In this case, R defines an almost-complex structure on M. We extend R complex linearly to the complexified tangent bundle $T_\mathbb{C} M = TM \otimes_\mathbb{R} \mathbb{C}$. The eigenvalues of R are $\pm i$, and

$$X = \frac{1}{2}(X - iRX) + \frac{1}{2}(X + iRX)$$

is the unique decomposition of a complex tangent vector X into a sum of eigenvectors of R. As usual, the eigenspaces of R give a splitting $T_\mathbb{C} M = T^{1,0} \oplus T^{0,1}$, where $T^{1,0}$ is the $+i$ eigenspace, and $T^{0,1}$ is the $-i$ eigenspace. The two are complex conjugates of each other.   □

Lemma 3.4 *The eigenspaces $T^{0,1}$ and $T^{1,0}$ are precisely the kernels of $\Omega = \omega + i\eta$ and of its complex conjugate $\bar\Omega = \omega - i\eta$.*

Proof It suffices to prove the statement for the $-i$ eigenspace $T^{0,1}$. The other case then follows by complex conjugation.
Let $X = Y + iZ$ be a complex tangent vector. Then

$$\iota_X \Omega = \iota_Y \omega - \iota_Z \eta + i(\iota_Y \eta + \iota_Z \omega).$$

The real part of the equation $\iota_X \Omega = 0$ is equivalent to its imaginary part, and each is equivalent to $RY = Z$, which is obviously equivalent to $X \in T^{0,1}$.   □

Proposition 3.7 *The almost-complex structure R is in fact integrable.*

Proof By the Newlander–Nirenberg theorem, it suffices to check that one, and hence both eigendistributions of R are closed under commutation. To do this, suppose X and Y are complex vector fields in $T^{1,0}$, so that $RX = iX, RY = iY$. Then, extending the Lie derivative $L_X = \iota_X \circ d + d \circ \iota_X$ complex linearly to complex tangent vectors, and using that ω and η are closed, we find

$$\iota_{R[X,Y]} \eta = \iota_{[X,Y]} \omega = L_X \iota_Y \omega - \iota_Y L_X \omega = L_X \iota_{RY} \eta - \iota_Y L_{RX} \eta = i(L_X \iota_Y \eta - \iota_Y L_X \eta) = \iota_{i[X,Y]} \eta \,.$$

The non-degeneracy of η now implies that $R[X, Y] = i[X, Y]$, so that in $T^{1,0}$ is closed under commutation.

Thus, we have seen that two symplectic forms ω and η whose recursion operator satisfies $R^2 = -Id$ give rise to an integrable complex structure, for which $T^{0,1}$ is precisely the kernel of $\Omega = \omega + i\eta$. Thus Ω is a closed form of type $(2, 0)$ and rank n, where n is the complex dimension of M.

Conversely, if a manifold is complex and carries a holomorphic symplectic form, then the real and imaginary parts of this form are real symplectic forms whose recursion operator is just the complex structure.

This completes the proof of Theorem 3.5.   □

Remark 3.11 The above proof of the integrability of the almost-complex structure defined by the recursion operator is the same as that in Lemma (6.8) of Hitchin's paper [31] (the 'Hitchin lemma'). However, unlike this reference, we do not assume the symplectic forms to be compatible with any metric and our proof mimics the classification result in Theorem 1.5 of [43].

3.3.2 Triples of Symplectic Forms

We now want to discuss the geometries defined by a triple of symplectic forms ω_1, ω_2, ω_3 whose recursion operators R_i defined by

$$\iota_X \omega_i = \iota_{R_{i+2}X} \omega_{i+1} \tag{3.16}$$

satisfy $R_i^2 = \pm Id$ and $R_i \neq \pm Id$. Here and in the sequel all indices are taken modulo 3. Note that by the definition all cyclic compositions $R_{i+2} \circ R_{i+1} \circ R_i = Id$.

There are four different cases depending on a sign in front of the squares of the R_i. In the two cases when there is an odd number of R_i with square $-Id$, there are natural pseudo-Riemannian metrics defined by the triple of two-forms. If there is exactly one R_i with the square $-Id$, one can get a notion of a hypersymplectic structure considered in Chap. 1. When all three R_i have square $-Id$, we find a hyper-complex structure for which all complex structures admit holomorphic symplectic forms. Examples for this geometric structure, which one can naturally call a hyper-complex symplectic structure, are provided by hyper-Kähler structures.

3.3.2.1 Hypercomplex Symplectic Structures

Remind from the previous chapter (Sects. 5 and 5.1) that a hyper-complex structure on a manifold is a triple of integrable complex structures satisfying the quaternion relations; see for example [37, 51] for more details.

The first example of a structure given by a triple of symplectic forms is as follows.

Definition 3.28 A triple of symplectic forms ω_i whose pairwise recursion operators satisfy $R_i^2 = -Id$ for all $i = 1, 2, 3$ is called a hyper-complex symplectic structure.

In this case $R_{i+2} \circ R_{i+1} \circ R_i = Id$ implies that the R_i anti-commute and satisfy the quaternion relations. By Theorem 3.5 each R_i is an integrable complex structure, and so the R_i together form a hyper-complex structure. Furthermore, each R_i admits a holomorphic symplectic form, justifying the name hyper-complex symplectic structure for such a triple.[1]

Hypercomplex symplectic structures are much more restrictive than hyper-complex ones, but every hyper-complex structure on M does give rise to a natural hyper-complex symplectic structure on T^*M.

Example

(see [5]) Let M be a manifold with an integrable complex structure J. Then lifting J to T^*M, the total space of the cotangent bundle is also a complex manifold. It is also holomorphic symplectic, because if ω is the exact symplectic form given by the exterior derivative of the Liouville 1-form, then $\Omega(X, Y) = \omega(X, Y) + i\omega(JX, Y)$ is holomorphic symplectic for the lifted J. If M has a hyper-complex structure, then the lifts of the three complex structures to T^*M still satisfy the quaternion relations, and are the recursion operators for the triple of symplectic forms given by the imaginary parts of the three holomorphic symplectic forms.

Now we show that hyper-complex symplectic structures have natural metrics associated with them.

Proposition 3.8 *Let M be a manifold with a hyper-complex symplectic structure. Then the bilinear form on TM defined by*

$$g(X, Y) = \omega_i(X, R_i Y)$$

is independent of $i = 1, 2, 3$. It is non-degenerate and symmetric, and invariant under all R_i.

---

[1] This is different from the hypersymplectic structures discussed in Sect. 3.3.2.2 below.

Proof We first prove independence of i as follows:

$$\omega_i(X, R_i Y) = \omega_i(X, R_{i+1} R_{i+2} Y) = \omega_{i+2}(X, R_{i+2} Y) = \ldots = \omega_{i+1}(X, R_{i+1} Y) .$$

Note that g is non-degenerate because R_i is invertible and ω_i is non-degenerate. We prove invariance under the R_i using independence of i:

$$g(R_i X, R_i Y) = \omega_{i+1}(R_i X, R_{i+1} R_i Y) = -\omega_{i+1}(R_i X, R_{i+2} Y) = \omega_{i+2}(X, R_{i+2} Y) = g(X, Y) .$$

Finally, we prove symmetry using the invariance under R_i:

$$g(Y, X) = \omega_i(Y, R_i X) = -\omega_i(R_i X, Y) = \omega_i(R_i X, R_i^2 Y) = g(R_i X, R_i Y) = g(X, Y).$$

$\square$

The proposition shows that g is a pseudo-Riemannian metric compatible with the symplectic forms ω_i. As it is symmetric and non-degenerate, there must be tangent vectors X with $g(X, X) \neq 0$. Take such a vector X and consider also $R_1 X$, $R_2 X$ and $R_3 X$. By invariance of g we have $g(R_i X, R_i X) = g(X, X)$, and by the definition of g and the skew-symmetry of ω_i, the $R_i X$ are g-orthogonal to each other and to X. Replacing g by its negative if necessary, we find the following.

Corollary 3.1 *Every hyper-complex symplectic structure in complex dimension two is hyper-Kähler.*

Proof Indeed, the pseudo-Riemannian metric g is a definite Kähler metric compatible with the underlying hyper-complex structure, whose Kähler forms with respect to R_i are the ω_i (up to sign). $\square$

In higher dimensions, hyper-Kähler structures provide examples of hyper-complex symplectic structures for which the natural pseudo-Riemannian metric g is definite. However, there are many other examples, even on manifolds that do not support any Kähler structure, so that Corollary 3.1 does not generalise to higher dimensions.

3.3.2.2 Hypersymplectic Structures

Next we consider a triple of symplectic forms such that two recursion operators have square the identity, and one has square minus the identity. After renumbering we may assume $R_1^2 = -Id$ and $R_2^2 = R_3^2 = Id$. Then the cyclic relations $R_{i+2} \circ R_{i+1} \circ R_i = Id$ show that the R_i anti-commute and $R_2 R_1 = R_3$. It follows that $R_i \neq \pm Id$, so the trivial cases are excluded automatically.

We have the following result analogous to Proposition 3.8.

Proposition 3.9 *Let M be a manifold with three symplectic forms whose recursion operators satisfy* $R_1^2 = -Id$ *and* $R_2^2 = R_3^2 = Id$. *Then*

$$\omega_1(X, R_1Y) = -\omega_2(X, R_2Y) = -\omega_3(X, R_3Y),$$

and these expressions define a bilinear form $g(X, Y)$ *on* TM. *It is non-degenerate and symmetric, invariant under* R_1, *and satisfies* $g(R_iX, R_iY) = -g(X, Y)$ *for* $i = 2, 3$.

We omit the proof as it is literally the same as for Proposition 3.8.

Now in this case if we take a vector X with $g(X, X) \neq 0$, then $g(R_1X, R_1X) = g(X, X)$, and $g(R_2X, R_2X) = g(R_3X, R_3X) = -g(X, X)$, and the R_iX are g-orthogonal to each other and to X. Thus, we have a 4-dimensional subspace on which g is non-degenerate and has signature $(2, 2)$. Looking at the orthogonal complement of this subspace and proceeding inductively, we see that the metric g has neutral signature.

We can compare this data with the following definition due to Hitchin [32]; see also [18, 24].

Definition 3.29 A hypersymplectic structure on a manifold is a pseudo-Riemannian metric g of neutral signature, together with three endomorphisms I, S and T of the tangent bundle satisfying

$$I^2 = -\,\mathrm{Id}\,, \quad S^2 = T^2 = Id\,, \quad IS = -SI = T\,,$$

$$g(IX, IY) = g(X, Y)\,, \quad g(SX, SY) = -g(X, Y)\,, \quad g(TX, TY) = -g(X, Y)\,,$$

and such that the following three two-forms are closed:

$$\omega_I(X, Y) = g(IX, Y)\,, \quad \omega_S(X, Y) = g(SX, Y)\,, \quad \omega_T(X, Y) = g(TX, Y)\,.$$

Given a hypersymplectic structure in this sense, the recursion operators intertwining the three symplectic forms are, up to sign, precisely the endomorphisms I, S and T. Conversely, given three symplectic forms for which one of the pairwise recursion operators has square $-Id$ and the other two have square the identity, Proposition 3.9 shows that we can recover a uniquely defined hypersymplectic structure. Thus, we have proved the following corollary.

Corollary 3.2 *A hypersymplectic structure is equivalent to a unique triple of symplectic forms for which two of the recursion operators have square the identity, and one has square minus the identity.*

A hypersymplectic structure also defines a symplectic pair, and therefore a four-manifold with such a structure is symplectic for both choices of orientation. High-dimensional examples of hypersymplectic structures on closed manifolds have recently appeared in [1, 24].

3.3.3 Notes and Further Reading

We will see in Chap. 5 that the appearance of hypersymplectic structures plays an important role in applications to Monge–Ampère equations, operators and structures. Another interesting application of such a triple of endomorphisms was discussed in the framework of superstrings [13] where the authors provide a study of self-dual 4D space-time with (2,2) neutral metric and propose a real version of twistors relevant to this geometry. It would be interesting to compare this geometry within the general approach of Monge–Ampère structures.

3.4 Lecture Three: Symplectic Monge–Ampère Operators and Equations

This chapter contains a (hopefully) friendly introduction in the geometry of Monge–Ampère equations and operators. This geometric approach (in a wide contact geometric context) was originally proposed by V.V. Lychagin but philosophically and ideologically it has the origins in works of E. Cartan and his school.

Motivating by applications, we shall restrict our attention to a *symplectic Monge–Ampère equations and operators* (this notion and restriction will be explained below).

3.4.1 Monge–Ampère Equations

Loosely speaking, a **Monge–Ampère equation** is a second-order partial differential equation with determinant-like non-linearities. In particular, let M be an m-dimensional smooth manifold M with local coordinates $(q^1, \ldots, q^m)$. For $\phi \in C^\infty(M)$, we let

$$\text{hess}\,\phi := \begin{pmatrix} \phi_{q^1 q^1} & \cdots & \phi_{q^1 q^m} \\ \vdots & \ddots & \vdots \\ \phi_{q^m q^1} & \cdots & \phi_{q^m q^m} \end{pmatrix} \tag{3.17}$$

be the Hessian matrix. The subscripts indicate partial derivatives, for instance, $\phi_{q^1} := \frac{\partial \phi}{\partial q^1}$, etc. A **symplectic Monge–Ampère equation** is a linear combination of all the minors of the Hessian matrix with the coefficients being elements of $C^\infty(T^*M)$. In two dimensions, the general form thus is

$$A\phi_{q^1 q^1} + 2B\phi_{q^1 q^2} + C\phi_{q^2 q^2} + D(\phi_{q^1 q^1}\phi_{q^2 q^2} - \phi_{q^1 q^2}\phi_{q^2 q^1}) + E = 0, \tag{3.18}$$

where A, B, C, D, and E are smooth functions of $(q^1, q^2, \phi_{q^1}, \phi_{q^2}) \in T^*M$.

3.4.1.1 Monge–Ampère Operators

Lychagin [14] proposed a geometric approach to study these equations by making use of certain differential forms on the cotangent space (i.e. the phase space).

Suppose $\alpha \in \Omega^m(T^*M)$, and for a smooth function $\phi \in C^\infty(M)$ consider its differential $d\phi \in \Gamma(T^*M)$ given by $(q^1, \ldots, q^m) \mapsto (q^1, \ldots, q^m, \phi_{q^1}, \ldots, \phi_{q^m})$.

Definition 3.30 (*Lychagin* [14]) Let M be an m-dimensional manifold, $\alpha \in \Omega^m$ (T^*M), and $\varpi \in \Gamma(\det(TM))$. The differential operator $\Delta_{\alpha,\varpi} : C^\infty(M) \to C^\infty(M)$ defined by the pull-back

$$\Delta_{\alpha,\varpi}\phi := \iota_\varpi[(d\phi)^*\alpha] \qquad (3.19)$$

is called the **Monge–Ampère operator** associated with (α, ϖ).

Consequently, we may always associate with a Monge–Ampère operator the Monge–Ampère equation

$$\Delta_{\alpha,\varpi}\phi = 0. \qquad (3.20)$$

This equation does not depend on the choice of volume form $\varpi \in \Gamma(\det(TM))$, and, since we shall solely be dealing with this equation, we shall follow the literature and suppress the explicit appearance of ϖ and simply write Δ_α in the following.

To study this correspondence between differential forms and Monge–Ampère equations, we shall need the following definition.

Definition 3.31 (*Lychagin* [14]) Let (M, ω) be a symplectic manifold. A differential p-form α is said to be ω-**effective** if and only if $\iota_{\omega^{-1}}\alpha = 0$.

We shall use also a terminology ω-**primitive** or simply primitive form, to stress the parallels with Hodge theory on kähler varieties.

It is easy to see that for a differential m-form α on a $2m$-dimensional symplectic manifold (M, ω), the notion of ω-effectiveness is equivalent to saying that $\alpha \wedge \omega = 0$. Then, we have the following theorem (see also the text book [44] for a comprehensive treatment).

Theorem 3.6 (Hodge–Lepage–Lychagin [14]) *Let (M, ω) be a symplectic manifold. Then, any differential p-form $\alpha \in \Omega^p(M)$ has a unique decomposition*

$$\alpha = \alpha_0 + \alpha_1 \wedge \omega + \alpha_2 \wedge \omega \wedge \omega + \cdots \qquad (3.21)$$

into ω-effective differential $(p - 2k)$-forms $\alpha_k \in \Omega^{p-2k}(M)$. Furthermore, if two ω-effective p-forms vanish on the same p-dimensional Lagrangian submanifold, they must be proportional.

For any $\phi \in C^\infty(M)$, its graph

$$L_\phi := \{(q^1, \ldots, q^n, \phi_{q^1}, \ldots, \phi_{q^n})\} \subseteq T^*M \qquad (3.22)$$

is Lagrangian with respect to the standard symplectic form ω on T^*M, that is, we have $\omega|_{L_\phi} = 0$. Consequently, we have the following corollary of the above theorem.

Corollary 3.3 *Let M be a manifold. For the standard symplectic structure ω on T^*M, the correspondence between Monge–Ampère equations on M and conformal classes of ω-effective forms on T^*M is one-to-one.*

Let us pause for a moment and discuss two examples.

Example

Let M be 2-dimensional with local coordinates (q^1, q^2). Furthermore, let (q^1, q^2, p_1, p_2) be local coordinates on T^*M and consider the standard symplectic form $\omega = dq^1 \wedge dp_1 + dq^2 \wedge dp_2$ on T^*M. The Monge–Ampère equation (3.18) is associated with the ω-effective form

$$A dp_1 \wedge dq^2 + B(dq^1 \wedge dp_1 - dq^2 \wedge dp_2) + C dq^1 \wedge dp_2 + D dp_1 \wedge dp_2 + E dq^1 \wedge dq^2. \tag{3.23}$$

Example

Let $M = \mathbb{R}^3$ with coordinates (q^1, q^2, q^3) and let $(q^1, q^2, q^3, p_1, p_2, p_3)$ be coordinates on $T^*\mathbb{R}^3 \cong \mathbb{R}^6$ and consider the standard symplectic form $\omega = dq^1 \wedge dp_1 + dq^2 \wedge dp_2 + dq^3 \wedge dp_3$ on $T^*\mathbb{R}^3$. The Monge–Ampère equation

$$\det(\mathsf{Hess}(\phi)) = 1 \tag{3.24a}$$

is associated with the ω-effective form

$$dp_1 \wedge dp_2 \wedge dp_3 - dq^1 \wedge dq^2 \wedge dq^3, \tag{3.24b}$$

while the Monge–Ampère equation

$$\Delta\phi - \det(\mathsf{Hess}(\phi)) = 0, \tag{3.25a}$$

where Δ is the Laplacian, is associated with the ω-effective form

$$dp_1 \wedge dq^2 \wedge dq^3 + dq^1 \wedge dp_2 \wedge dq^3 + dq^1 \wedge dq^2 \wedge dp_3 - dp_1 \wedge dp_2 \wedge dp_3. \tag{3.25b}$$

3.4.1.2 Generalised Solutions

Recall that for any function ϕ on M, its graph $L_\phi = \{(q^1, \ldots, q^m, \phi_{q^1}, \ldots, \phi_{q^m})\}$ is Lagrangian with respect to the standard symplectic form ω on T^*M. Conversely, for any Lagrangian submanifold $L \subseteq T^*M$ there exists a function ϕ on M such that $L = L_\phi$. Moreover, it is clear that ϕ is a regular solution to the Monge–Ampère equation (3.20) if and only if also $\alpha|_{L_\phi} = 0$. We thus give the following definition.

Definition 3.32 (*Lychagin* [14]) Let ω be a symplectic structure on T^*M. A **generalised solution** to the Monge–Ampère equation (3.20) is a Lagrangian submanifold $L \subseteq T^*M$ with respect to ω (i.e. $\omega|_L = 0$) such that also $\alpha|_L = 0$.

3.4.1.3 Local Equivalence

A classical problem in the geometric study of differential equations is the problem of local equivalence up to a local change of dependent and independent coordinates. Put differently, when are two given differential equations equivalent up to the action of a local diffeomorphism on the phase space?

Let M be a manifold and Φ a diffeomorphism on T^*M. The natural action $\rhd$ of Φ on a Monge–Ampère operator Δ_α is given by

$$\Phi \rhd \Delta_\alpha := \Delta_{\Phi^*\alpha} . \tag{3.26}$$

Definition 3.33 (*Lychagin* [14]) Let M be an m-dimensional manifold and ω a symplectic form on T^*M. Furthermore, let α_1 and α_2 be two differential m-forms on T^*M. The two Monge–Ampère equations given by the operators Δ_{α_1} and Δ_{α_2} are called **symplectically equivalent** if and only if there is a local diffeomorphism $\Phi \in C^\infty(T^*M)$ such that $\Phi^*\omega = \omega$ and $\Phi^*\alpha_2 = \alpha_1$.

Note that diffeomorphisms $\Phi : (M, \omega) \to (M, \omega)$ on a symplectic manifold (M, ω) with $\Phi^*\omega = \omega$ are called **symplectomorphisms**.

Example

Consider $M = \mathbb{R}^2$ with coordinates (q^1, q^2). The Monge–Ampère equations

$$\det(\mathsf{Hess}(\psi)) = 1 \quad \text{and} \quad \Delta\phi = 0 \tag{3.27a}$$

correspond to the ω-effective forms

$$dp_1 \wedge dp_2 - dq^1 \wedge dq^2 \quad \text{and} \quad dp_1 \wedge dq^2 + dq^1 \wedge dp_2 , \tag{3.27b}$$

respectively, with $\omega = dq^1 \wedge dp_1 + dq^2 \wedge dp_2$. The partial Legendre transformation

$$\Phi(q^1, q^2, p_1, p_2) := (q^1, p_2, p_1, -q^2) , \tag{3.28}$$

which leaves ω invariant, yields

$$\Phi^*(\mathrm{d}p_1 \wedge \mathrm{d}p_2 - \mathrm{d}q^1 \wedge \mathrm{d}q^2) = \mathrm{d}p_1 \wedge \mathrm{d}q^2 + \mathrm{d}q^1 \wedge \mathrm{d}p_2 . \qquad (3.29)$$

Consequently, the two Monge–Ampère equations (3.27a) are symplectically equivalent. Note that this ceases to be true in higher dimensions.

Importantly, symplectic equivalence can be used to construct explicit solutions. The key observation is that symplectomorphisms preserve generalised solutions (see Definition 3.32) but not regular solutions: if Φ is a symplectomorphism and L a generalised solution to the Monge–Ampère equation given by $\Delta_{\Phi^*\alpha}$, then $\Phi(L)$ is a generalised solution to the Monge–Ampère equation given by Δ_α. To make this more transparent, it is instructive to discuss an explicit example.

Example

Consider the Laplace equation $\Delta\phi = 0$ on $\mathbb{R}^2$. Its generalised solutions are the complex curves of $T^*\mathbb{R}^2 \cong \mathbb{C}^2$. Upon applying (3.28) to the harmonic function

$$\phi(q^1, q^2) = \mathrm{e}^{q^1} \cos(q^2) , \qquad (3.30)$$

we obtain the generalised solution $\Phi(L_\phi)$. The identification of $\Phi(L_\phi)$ with the graph L_ψ of a function ψ on $\mathbb{R}^2$ then yields the regular solution

$$\psi(q^1, q^2) = q^2 \arcsin\left(\frac{q^2}{\mathrm{e}^{q^1}}\right) + \sqrt{\mathrm{e}^{2q^1} - (q^2)^2} \qquad (3.31)$$

to the Monge–Ampère equation $\det(\mathsf{Hess}(\psi)) = 1$.

3.4.2 Geometry of Differential Forms

Hence, as we have seen, the classical problem of local equivalence for Monge–Ampère equations can be understood as a problem of the Geometric Invariant Theory: the idea is to construct invariant structures which will characterise each equivalent class.

The first step of this approach is pointwise: we study the action of the symplectic group $Sp(n, \mathbb{R})$ on the space of primitive forms $\Lambda_0^n(\mathbb{R}^n)$. We will see next how one can 'integrate' such study.

3.4.2.1 The Bracket Structure

Let V^{2n} be a $2n$-dimensional real vector space. We fix a symplectic form Ω on V and the volume form

$$\text{vol} = \frac{\Omega^n}{n!}.$$

We denote by $\Lambda^n(V^*)$ the space of n-forms on V and by $\Lambda_0^n(V^*)$ the space of primitive n-forms, that is

$$\Lambda_0^n(V^*) = \{\omega \in \Lambda^n(V^*), \ \Omega \wedge \omega = 0\}.$$

We denote by $SL(2n)$ the group of automorphisms preserving the volume form vol and by $Sp(n, \mathbb{R})$ the group of automorphisms preserving the symplectic form Ω. Their Lie algebras are denoted by $sl(2n)$ and $sp(n, \mathbb{R})$.

Using the exterior product, we define an isomorphism $\mathcal{A} : \Lambda^{2n-1}(V^*) \to V$ by

$$< \alpha, \mathcal{A}(\theta) >= \frac{\alpha \wedge \theta}{\text{vol}}, \quad \text{for } \alpha \in \Lambda^1(V^*) \text{ and } \theta \in \Lambda^{2n-1}(V^*).$$

Definition 3.34 The bracket $\Phi : \Lambda^n(V^*) \times \Lambda^n(V^*) \to sl(V)$ is defined by

$$\Phi(\omega_1, \omega_2)(X) = \mathcal{A}\Big((\iota_X \omega_1) \wedge \omega_2 - (-1)^n \omega_1 \wedge (\iota_X \omega_2)\Big).$$

It is straightforward to check the two following lemmas.

Lemma 3.5 *This bracket is invariant under the action of $SL(2n)$, that is*

$$\Phi(F^*\omega_1, F^*\omega_2) = F^{-1} \circ \Phi(\omega_1, \omega_2) \circ F$$

for any $F \in SL(2n)$.

Lemma 3.6 *Let $\tilde{\Phi}$ be the bracket defined for the $(2n + 2)$-dimensional vector space $\tilde{V} = V \times \mathbb{R}_{t_1} \times \mathbb{R}_{t_2}$ endowed with the volume form*

$$\tilde{\text{vol}} = \text{vol} \wedge dt_1 \wedge dt_2.$$

Then the following relations hold:

1. $\tilde{\Phi}(\omega_1 \wedge dt_1, \omega_2 \wedge dt_2)(\partial_{t_1}) = -\tilde{\Phi}(\omega_1 \wedge dt_1, \omega_2 \wedge dt_2)(\partial_{t_2})$
2. $\tilde{\Phi}(\omega_1 \wedge dt_1, \omega_2 \wedge dt_2)(X) = \Phi(\omega_1, \omega_2)(X), \quad \forall X \in V.$

Note that this second lemma shows that Φ takes its values in $sl(2n)$.

In the case $n = 3$, the tensor $K_\omega = \frac{1}{2}\Phi(\omega, \omega)$ is the invariant constructed by Hitchin in [34], which can be easily extended to any odd n.

Proposition 3.10 (Hitchin) *When n is odd, the map $K : \Lambda^n(V^*) \to sl(2n)$, $\omega \mapsto \frac{1}{2}\Phi(\omega, \omega)$ is a moment map for the Hamiltonian action of $SL(2n)$ on $\Lambda^n(V^*)$ endowed with the symplectic form*

$$\Theta(\omega_1, \omega_2) = \frac{\omega_1 \wedge \omega_2}{vol}.$$

When n is even, the bracket Φ is antisymmetric and the situation is completely different. The analogue of Proposition 3.10 is the following, which is proved in [9] (unpublished).

Proposition 3.11 *We define on $\Lambda^n(V^*) \times sl(2n)$ the following bracket:*

1. $[A_1, A_2] = A_1 A_2 - A_2 A_1$,
2. $[A, \omega] = L_A(\omega)$ *and*
3. $[\omega_1, \omega_2] = \Phi(\omega_1, \omega_2)$,

for A, A_1 and A_2 in $sl(2n)$ and ω, ω_1 and ω_2 in $\Lambda^n(V^*)$.
Then $[,]$ is a Lie bracket.

The $Sp(n, \mathbb{R})$-version of these results is summed up in the following.

Proposition 3.12 1. *If ω_1 and ω_2 are primitive then $\Phi(\omega_1, \omega_2) \in sp(n, \mathbb{R})$.*
2. *If n is odd, then $K : \Lambda_0^n(V^*) \to sp(n, \mathbb{R})$, $\omega \mapsto \frac{1}{2}\Phi(\omega, \omega)$ is a moment map for the Hamiltonian action of $SP(n, \mathbb{R})$ on the symplectic subspace $\Lambda_0^n(V^*)$ of $\Lambda^n(V^*)$.*
3. *If n is even, the space $\Lambda_0^n(V^*) \oplus sp(n, \mathbb{R})$ is a Lie subalgebra of $\Lambda^n(V^*) \oplus sl(2n)$.*

Remark 3.12 When n is odd, the tensor K_ω defines a family of scalar invariants

$$a_k = \mathrm{Tr}(K_\omega^{2k}), \quad k \in \mathbb{N}$$

and a quadratic form (which was called for $n = 3$ the *Lychagin–Roubtsov quadratic form* in [6]):

$$q_\omega(X) = \Omega(K_\omega X, X).$$

When n is even, the adjoint operator $ad_\omega = [\omega, \cdot]$ defines an endomorphism

$$ad_\omega^2 : sp(n, \mathbb{R}) \to sp(n, \mathbb{R}),$$

which gives also a family of scalar invariants

$$a_k = \mathrm{Tr}(ad_\omega^{2k}), \quad k \in \mathbb{N}$$

and a symmetric polynomial of degree 4 defined by

$$q_\omega(X) = \mathrm{Tr}([ad_\omega^2(X \otimes \iota_X(\Omega))]^2).$$

3.4.2.2 Low-Dimensional Examples

The case $n = 2$.

The identity $\omega = \Omega(A_\omega \cdot, \cdot)$ gives an isomorphism between the space of 2-forms $\Lambda^2(\mathbb{R}^4)$ and the Jordan algebra $\mathrm{Jor}(\Omega)$ defined by

$$\mathrm{Jor}(\Omega) = \{A \in gl(4), \ \Omega(A\cdot, \cdot) = \Omega(\cdot, A\cdot)\}.$$

Our bracket Φ becomes then the usual bracket:

$$\Phi(\omega_1, \omega_2) = A_{\omega_1} A_{\omega_2} - A_{\omega_2} A_{\omega_1}.$$

We easily see then the isomorphism of Lie algebras

$$\Lambda_0^2(\mathbb{R}^4) \oplus sp(2, \mathbb{R}) = sl(4, \mathbb{R}).$$

Moreover, for $\omega \in \Lambda_0^2(\mathbb{R}^4)$, the endomorphism $ad_\omega^2 = ad_{A_\omega}^2 : sp(2, \mathbb{R}) \to sp(2, \mathbb{R})$ satisfies
$$\mathrm{Tr}(ad_\omega^2) = 16 \, \mathrm{pf} \, \omega,$$

where the pfaffian of ω is the classical invariant

$$\mathrm{pf} \, \omega = \frac{\omega \wedge \omega}{\Omega \wedge \Omega}.$$

The polynomial q_ω is the null polynomial.

The case $n = 3$.

It is proved in [34] that the action of $GL(6, \mathbb{R})$ on $\Lambda^3(\mathbb{R}^3)$ has two opened orbits separated by the hypersurface $\mathrm{HPf}(\cdot) = 0$ where

$$\mathrm{HPf}(\omega) = \frac{1}{6} \mathrm{Tr}(K_\omega^2).$$

Note that, for any 3-form the following holds:

$$K_\omega^2 = \mathrm{HPf}(\omega) \cdot \mathrm{Id}.$$

By analogy with the 2-dimensional case, we call this invariant the Hitchin pfaffian. A 3-form with a non-vanishing Hitchin pfaffian is said to be non-degenerate.

For a primitive form ω, we get a triple $(g_\omega, K_\omega, \Omega)$ with $g_\omega = \Omega(K_\omega \cdot, \cdot)$ the Monge–Ampère metric (see [12, 43]). This triple defines a ϵ-Kähler structure in the sense of [51], that is, the tensor K_ω satisfies, up a renormalization,

$$K_\omega^2 = \epsilon \, \mathrm{Id}, \ \text{with} \ \epsilon = 0, 1, -1.$$

Note that the Monge–Ampère metric has signature.

Moreover, in the non-degenerate case the form ω admits an unique dual form $\hat{\omega}$, such that $\omega + \sqrt{\epsilon}\hat{\omega}$ and $\omega - \sqrt{\epsilon}\hat{\omega}$ are the volume forms of the two eigenspaces of the Hitchin tensor K_ω. Saying differently, to each non-degenerate primitive forms corresponds a 'ϵ-Calabi–Yau' structure.

The case $n = 4$.

The Lie algebras $\Lambda^4(\mathbb{R}^8) \oplus sl(8, \mathbb{R})$ and $\Lambda_0^4(\mathbb{R}^8) \oplus sp(4, \mathbb{R})$ are known to be isomorphic to the exceptional Lie algebras E_7 and E_6 (see [55]). Moreover, it is proved in [38] that the family $\{a_k = \mathrm{Tr}(ad_\omega^{2k})\}_{k \in \mathbb{N}}$ forms a complete family of invariants.

Nevertheless, computations in these dimensions are extremely complicated. The author planes to implement an algorithm which could give in a reasonable time these invariants a_k.

It is worth mentioning that, on many examples, the symmetric polynomial q_ω of degree 4 is the square of a quadratic form. Is it always true ? A positive answer would be extremely useful to understand the geometry of PDE's of Monge–Ampère type in 4 variables.

3.4.2.3 Classifications Results

Monge–Ampère equations in two and three variables.

The action of the symplectic linear group on $2D$ and $3D$ symplectic Monge–Ampère equations with constant coefficients has a finite numbers of orbits and we know all of them as it is shown in Tables 3.1 and 3.2 (see [6, 43]).

Remark 3.13   1. In two variables, any SMAE with constant coefficients is linearisable, which is equivalent to a linear PDE. Moreover, the pfaffian distinguishes the different orbits.
2. In three variables, there exist non-linearisable SMAE with constant coefficients and they correspond to non-degenerate primitive 3-forms. Moreover, the Hitchin pfaffian does not distinguish the different orbits but so does the signature of the Monge–Ampère metric.

In $D = 4$, the action of the symplectic group is no more discrete and there is no hope to obtain an exhaustive classification list as in two or three variables. Instead of that, there are functional moduli in the classification problem. There are some partial results in [44] and in the PhD thesis of B.Banos. He has discovered that on many interesting examples, the associated geometry is completely degenerated. In other words, in the case of $D = 4$, it appears to be that there exist the notion of non-linear

Table 3.1 Classification of SMAE in two variables.

$\Delta_\omega = 0$	ω	$pf\omega$
$\Delta f = 0$	$dq_1 \wedge dp_2 - dq_2 \wedge dp_1$	1
$\Box f = 0$	$dq_1 \wedge dp_2 + dq_2 \wedge dp_1$	-1
$\frac{\partial^2 f}{\partial q_1^2} = 0$	$dq_1 \wedge dp_2$	0

Table 3.2 Classification of SMAE in three variables

	$\Delta_\omega = 0$	signature (q_ω)	$\lambda(\omega)$
1	hess $f = 1$	(3, 3)	1
2	$\Delta f - \text{hess } f = 0$	(0, 6)	-1
3	$\Box f + \text{hess } f = 0$	(4, 2)	-1
4	$\Delta f = 0$	(0, 3)	0
5	$\Box f = 0$	(2, 1)	0
6	$\Delta_{q_2, q_3} f = 0$	(0, 1)	0
7	$\Box_{q_2, q_3} f = 0$	(1, 0)	0
8	$\frac{\partial^2 f}{\partial q_1^2} = 0$	(0, 0)	0

Table 3.3 Examples of SMAE in four variables.

$\Delta_\omega = 0$	q_ω
Usual Monge–Ampère	$(dq_1 dp_1 + dq_2 dp_2 + dq_3 dp_3 + dq_4 dp_4)^2$
SLAG	$(dq_1^2 + dq_2^2 + dq_3^2 + dq_4^2 + dp_1^2 + dp_2^2 + dp_3^2 + dp_4^2)^2$
Plebanski I	0
Plebanski II	dq_1^4
Grant	0

but degenerated Monge–Ampère equation. In Table 3.3, we have the polynomial invariant q_ω computed by B.Banos for the following examples:

$$\text{hess } u = 1 \quad \text{(Usual Monge–Ampère equation)}$$

$$\text{hess } u - \left(\sum_{i<j} u_{q_i q_i} u_{q_j q_j} - u_{q_i q_j}^2\right) + 1 = 0 \quad \text{(4D Special Lagrangian equation)}$$

$$u_{q_1 q_2} u_{q_3 q_4} - u_{q_1 q_4} u_{q_2 q_3} = 1 \quad \text{(Plebanski I equation)}$$

$$u_{q_1 q_1} u_{q_3 q_3} - u_{q_1 q_3}^2 + u_{q_1 q_2} - u_{q_3 q_4} = 0 \quad \text{(Plebanski II equation)}$$

$$u_{q_1 q_1} + u_{q_1 q_4} u_{q_2 q_3} - u_{q_1 q_3} u_{q_2 q_4} = 0 \quad \text{(Grant equation)}$$

3.4.3 Notes and Further Reading

One can find some interesting partial classification results for $D = 4$ in [44]. We have remarked that for $n \geq 4$ the stabilisers of generic $SP(n, \mathbb{R})$−actions on $\Lambda_0(V^*)$ are trivial and we obtained a versal families of forms $\omega_\mu = \omega_0 + \sum_k \mu_k \omega_k$ with ω_0 as a representative of the generic orbit and the forms ω_k are base in the CokerB$_0$ with the operator $B_0 : S^2(V^*) \to \Lambda_0(V^*)$ given by ω_0 and the Spenser δ−differential (see [44] for details).

Another classification results in the dimension 4 we get (for primitive forms from $\Lambda_0^4(V^*)$ admitting so-called *symplectic transvections*) the list of normal forms which is a reduction of $3D$−classification [44].

We will see in Chap. 5 that the appearance of Plebansky 'heavenly' equations sometimes relates to a Monge–Ampère coupled structures. These equations play an important role in applications of Monge–Ampère equations in a construction of Hyper-Kähler metrics on holomorphic symplectic surfaces of K3 type (see [19, 26]).

3.5 Lecture Four: Monge–Ampère Structures

In this lecture, we begin by looking at a notion of *Monge–Ampère structure*. This notion quite naturally appears in the general framework of the classification problems which we have discussed in previous lectures. The first formal definition probably appeared in one of author's conference and colloquium talks, then in papers of his former PhD student B. Banos but ideologically it can be deduced from following [44]. The presentation in this chapter is based mainly on the review in progress [45].

3.5.1 General Properties

We shall introduce the notion of a Monge–Ampère structure in its maximally general form. Then we shall precise the specific properties in low dimensions repeating some definitions giving in the previous lectures.

Definition 3.35 Let M be a $2m$-dimensional manifold. A **Monge–Ampère structure** on M is a pair of differential forms $(\omega, \alpha) \in \Omega^2(M) \oplus \Omega^m(M)$ such that ω is symplectic and α is ω-effective.

3.5.1.1 4-Dimensional Monge–Ampère Geometry

In case of a 4-dimensional phase space, in the non-degenerate case, the geometry induced by a Monge–Ampère structure can be either complex or real and this distinction coincides with the usual distinction between elliptic and hyperbolic for differential equations in two variables.

To see this, let M be a 4-dimensional manifold. We define the Pfaffian pf $\alpha \in \mathbb{C}^\infty(M)$ and the endomorphism $R_\alpha : TM \to TM$ for a Monge–Ampère structure $(\omega, \alpha) \in \Omega^2(M) \oplus \Omega^2(M)$ by means of the equations [44]

$$\alpha \wedge \alpha = \text{pf } \alpha \, \omega \wedge \omega \quad \text{and} \quad \alpha(X, Y) = \omega(R_\alpha X, Y) \tag{3.32}$$

for all $X, Y \in \Gamma(TM)$. See also Chap. 3, Sect. 12.2.

Lemma 3.7 (Lychagin–Roubtsov–Chekalov [44]) *Let (ω, α) be a Monge–Ampère structure on a 4-dimensional manifold M. Then,*

$$\operatorname{tr}(R_\alpha) = 0 \quad \text{and} \quad R_\alpha^2 = -\operatorname{pf}\alpha \operatorname{Id} . \tag{3.33}$$

Proof [2] First, note that

$$\operatorname{tr}(R_\alpha) = \iota_{\omega^{-1}}\alpha . \tag{3.34}$$

Since α is ω-effective, this must vanish.

Next, let $\chi_\alpha(t) := \operatorname{pf}\alpha - t\omega$ for $t \in \mathbb{R}$ be the Pfaffian characteristic polynomial of α. Then, since α is ω-effective, we find

$$\operatorname{pf}\alpha - t\omega \, \omega \wedge \omega = (\alpha - t\omega) \wedge (\alpha - t\omega)$$
$$= \alpha \wedge \alpha + t^2 \omega \wedge \omega = (\operatorname{pf}\alpha + t^2)\,\omega \wedge \omega , \tag{3.35}$$

and from the non-degeneracy of ω, we conclude that

$$\chi_\alpha(t) = t^2 + \operatorname{pf}\alpha . \tag{3.36}$$

Furthermore, there is Pfaffian version of the Cayley–Hamilton theorem (see e.g. [39, Chap. 3.9]) so that $\chi_\alpha(R_\alpha) = 0$. This, in turn, implies (3.33). $\qquad\square$

This lemma allows us to define the endomorphism $J_\alpha : TM \to TM$ by

$$\frac{1}{\sqrt{|\operatorname{pf}\alpha|}}\alpha(X, Y) = \omega(J_\alpha X, Y), \quad \forall X, Y \in \Gamma(TM) \tag{3.37}$$

whenever $\operatorname{pf}\alpha \neq 0$. This is an almost-complex structure for $\operatorname{pf}\alpha > 0$ since then $J_\alpha^2 = -\operatorname{Id}$ and it is an almost product structure for $\operatorname{pf}\alpha < 0$ since then $J_\alpha^2 = \operatorname{Id}$, respectively. Since $\operatorname{tr}(J_\alpha) = 0$, the ± 1 eigenspaces of J_α are of the same dimension, and, consequently, the endomorphism J_α is, in fact, a para-complex structure. Furthermore, the above makes (J_α, ω) compatible in the sense of Definition 3.21. Consequently, we shall make use of Proposition 3.7 and write

$$\frac{\alpha}{\sqrt{|\operatorname{pf}\alpha|}} = \iota_{J_\alpha}\omega . \tag{3.38}$$

Notice that by virtue of Proposition 3.7, if the differential two-form $\alpha/\sqrt{|\operatorname{pf}\alpha|}$ is closed then J_α is integrable.

Returning to our discussion about Monge–Ampère structures on the cotangent bundle T^*M for a 2-dimensional manifold M, it is easy to see that the Monge–Ampère equation associated with Monge–Ampère operator Δ_α is elliptic

---

[2] I am grateful to M. Wolf for discussions which had improved this demonstration.

when pf $\alpha > 0$ and hyperbolic when pf $\alpha < 0$, respectively. When pf $\alpha = 0$ (i.e. in the degenerate case with $\alpha \wedge \alpha = 0$), we cannot define J_α, and in this situation, the Monge–Ampère equation associated with Δ_α is parabolic. Since in two dimensions (see Example 3.4.1.1) any elliptic (respectively, hyperbolic) Monge–Ampère equation with constant coefficients is symplectically equivalent (see Definition 3.33) to the Laplace (respectively, Klein–Gordon) equation, we have the following theorem.

Theorem 3.7 (Lychagin–Roubtsov–Chekalov [44]) *Let (ω, α) be a non-degenerate Monge–Ampère structure on T^*M for a 2-dimensional manifold M and let J_α be the almost-complex (respectively, para-complex) structure defined by*

$$\frac{\alpha}{\sqrt{\pm \mathrm{pf}\,\alpha}} = \iota_{J_\alpha} \omega \quad \text{with} \quad \alpha \wedge \alpha = \mathrm{pf}\,\alpha\, \omega \wedge \omega . \tag{3.39}$$

Then,

1. *in the elliptic case when pf $\alpha > 0$, the Monge–Ampère equation associated with the Monge–Ampère operator Δ_α is symplectically equivalent to*

$$\Delta \phi = 0 \tag{3.40a}$$

   if and only if
$$\mathrm{d}\Omega = 0 \quad \text{with} \quad \Omega = \omega - \mathrm{i}\,\iota_{J_\alpha} \omega ; \tag{3.40b}$$

2. *in the hyperbolic case when pf $\alpha < 0$, the Monge–Ampère equation associated with the Monge–Ampère operator Δ_α is symplectically equivalent to*

$$\Box \phi = 0 \tag{3.41a}$$

   if and only if
$$\mathrm{d}\Omega = 0 = \mathrm{d}\hat{\Omega} \quad \text{with} \quad \Omega = \omega + \iota_{J_\alpha} \omega \quad \text{and} \quad \hat{\Omega} = \omega - \iota_{J_\alpha} \omega . \tag{3.41b}$$

Proof ($\Rightarrow$) First, let ω be the standard symplectic structure on T^*M. If the Monge–Ampère equation associated with the Monge–Ampère operator Δ_α is symplectically equivalent to $\Delta \phi = 0$ (respectively, $\Box \phi = 0$), then the differential forms Ω and $\hat{\Omega}$ are closed.

($\Leftarrow$) To verify the converse, if Ω (and also $\hat{\Omega}$ when pf $\alpha < 0$) is closed then J_α is integrable by virtue of the corollary from Proposition 3.7.

Consequently, when pf $\alpha > 0$ there exists local coordinates $z^1 := q^1 - \mathrm{i} p_2$ and $z^2 := p_1 - \mathrm{i} q^2$ such that $\Omega = \mathrm{d}z^1 \wedge \mathrm{d}z^2$ locally and so

$$\begin{aligned} \omega &= \mathfrak{Re}\,\Omega = \mathrm{d}q^1 \wedge \mathrm{d}p_1 + \mathrm{d}q^2 \wedge \mathrm{d}p_2 , \\ \frac{\alpha}{\sqrt{\mathrm{pf}\,\alpha}} &= -\mathfrak{Im}\,\Omega = \mathrm{d}p^1 \wedge \mathrm{d}p^2 - \mathrm{d}q^1 \wedge \mathrm{d}q^2 . \end{aligned} \tag{3.42}$$

This yields the Laplace equation (see Example 3.4.1.3).

When pf $\alpha < 0$, there are local coordinates (q^1, q^2, p_1, p_2) such that $\Omega = 2\mathrm{d}q^1 \wedge \mathrm{d}p_1$ and $\hat{\Omega} = 2\mathrm{d}q^2 \wedge \mathrm{d}p_2$ locally and so

$$
\begin{aligned}
\omega &= \tfrac{1}{2}(\Omega + \hat{\Omega}) = \mathrm{d}q^1 \wedge \mathrm{d}p_1 + \mathrm{d}q^2 \wedge \mathrm{d}p_2, \\
\frac{\alpha}{\sqrt{-\,\mathrm{pf}\,\alpha}} &= \tfrac{1}{2}(\Omega - \hat{\Omega}) = \mathrm{d}q^1 \wedge \mathrm{d}p_1 - \mathrm{d}q^2 \wedge \mathrm{d}p_2.
\end{aligned}
\tag{3.43}
$$

This yields the Klein–Gordon equation in light-cone coordinates. □

Let us now show that a non-degenerate Monge–Ampère structure (α, ω) in four dimensions not just comes with one almost complex (respectively, para-complex) structure but, in fact, a whole two-sphere (respectively, hyperboloid) worth of such structures.

Lemma 3.8 *Let M be a 4-dimensional manifold equipped with a non-degenerate Monge–Ampère structure (ω, α). Furthermore, let J_α be the almost-complex (respectively, para-complex) structure (3.38). Then, there is a differential $(1,1)$-form Θ on M with respect to J_α such that $\Theta \wedge \Theta \neq 0$, $\Theta \wedge \omega = 0$, and $\Theta \wedge \iota_{J_\alpha} \omega = 0$.*

Proof The exterior product yields a non-degenerate metric g on $\bigwedge^2 TM$, so we know that there must be a differential two-form Θ such that $\Theta \wedge \Theta \neq 0$, $\Theta \wedge \omega = 0$, and $\Theta \wedge (\iota_{J_\alpha} \omega) = 0$. To be more explicit, note that ω and $\iota_{J_\alpha} \omega$ are linearly independent. Next, let $\rho \in \Omega^2(M)$ such that $\{\omega, \iota_{J_\alpha} \omega, \rho\}$ is linearly independent. By the Hodge–Lepage–Lychagin theorem (see Theorem 3.6), we have a unique decomposition $\rho = \rho_0 + \lambda_0 \omega$ with $\rho_0 \wedge \omega = 0$ and $\lambda_0 \in C^\infty(M)$. Since $(\iota_{J_\alpha} \omega) \wedge \iota_{J_\alpha} \omega) \neq 0$, we may again apply the Hodge–Lepage–Lychagin theorem to obtain the unique decomposition $\rho_0 = \rho_1 + \lambda_1 (\iota_{J_\alpha} \omega)$ with $\lambda_1 \in C^\infty(M)$ such that $\rho_1 \wedge (\iota_{J_\alpha} \omega) = 0$. Since $(\iota_{J_\alpha} \omega) \wedge \omega = 0$, we also have $\rho_1 \wedge \omega = 0$. Hence, $\{\omega, \iota_{J_\alpha} \omega, \rho_1\}$ is linearly independent, and we must also have that $\rho_1 \wedge \rho_1 \neq 0$ because of the non-degeneracy of g. In summary, we have thus obtained a $\Theta := \rho_1$ such that $\Theta \wedge \Theta \neq 0$, $\Theta \wedge \omega = 0$, and $\Theta \wedge \iota_{J_\alpha} \omega) = 0$.

Finally, since ω and $\iota_{J_\alpha} \Omega$ combine to give the differential $(2,0)$-forms Ω and differential $(0,2)$-forms $\hat{\Omega}$ defined by

$$
\Omega = \omega + \iota_{J_\alpha} \Omega; \quad \hat{\Omega} = \omega - \iota_{J_\alpha} \Omega,
\tag{3.44}
$$

and since $\Theta \wedge \omega = 0$ and $\Theta \wedge (\iota_{J_\alpha} \omega) = 0$, we conclude that $\Theta \wedge \Omega = 0$ and $\Theta \wedge \bar{\Omega} = 0$ (respectively, $\Theta \wedge \hat{\Omega} = 0$). Since $\Omega \wedge \bar{\Omega} \neq 0$ (respectively, $\Omega \wedge \hat{\Omega} \neq 0$), the differential two-form Θ must be of type $(1,1)$ with respect to J_α. This concludes the proof. □

Let V be a real vector space of dimension $2n$, with (almost) complex structure I (which defines a natural orientation on V) and compatible scalar product g. Remind then that the *Hodge star operator*

$$
\star : \Lambda^k(V^*) \to \Lambda^{2n-k}(V^*)
$$

has the following properties:

(a) If $e_1, \ldots, e_{2n}$ denotes an oriented g−orthonormal basis of V, then

$$\star(e_{i_1} \wedge \ldots \wedge e_{i_k}) = \epsilon e_{j_1} \wedge \ldots \wedge e_{j_{2n-k}},$$

where

$$\{i_1 \ldots, i_k, j_1 \ldots, j_{2n-k}\} = \{1, \ldots, 2n\}, \quad \epsilon = \operatorname{sgn}(i_1, \ldots, i_k, j_1, \ldots, j_{2n-k}).$$

(b)
$$\star \circ \star = (-1)^k;$$

(c)
$$g(\star\alpha, \star\beta) = g(\alpha; \beta);$$

(d)
$$g(\star\alpha, \beta) = (-1)^{\deg\alpha} g(\alpha, \star\beta).$$

This operation easily redefines in the exterior differential form algebra on an oriented (pseudo-)Riemannian manifold.

On a 4-dimensional manifold with a Riemannian (respectively, Kleinian) metric, the Hodge operator $\star_4$ on differential two-forms has the property $\star_4^2 = 1$. Therefore, the module of differential two-forms decomposes into a direct sum of the rank-3 module of self-dual differential two-forms (+ eigenspace with respect to $\star_4$) and the rank-3 module of anti-self-dual differential two-forms (− eigenspace with respect to $\star_4$). Furthermore, the eigenvalue equation $\star_4\rho = \pm\rho$ depends only on the conformal class of the metric. Making use of this, we arrive at the following result.

Proposition 3.13 *Let M be a 4-dimensional manifold equipped with a non-degenerate Monge–Ampère structure (ω, α). Furthermore, let J_α be the almost-complex (respectively, para-complex) structure (3.38). Then, there is a differential $(1, 1)$-form Θ on M with respect to J_α and a metric h_α on M both unique for pf $\alpha > 0$ and both unique up to a sign for pf $\alpha < 0$ such that*

1. *pf $\Theta = \operatorname{sgn}(\operatorname{pf}\alpha)$, $\Theta \wedge \omega = 0$, and $\Theta \wedge (\iota_{J_\alpha} \omega) = 0$,*
2. *Θ is the almost Kähler form for (h_α, J_α), that is, $\Theta(X, Y) = h_\alpha(J_\alpha X, Y)$ for all $X, Y \in \Gamma(TM)$ and*
3. *$(\omega, \iota_{J_\alpha} \omega, \Theta)$ are anti-self-dual with respect to h_α.*

Proof By virtue of Lemma 3.8, we already know that there is a differential $(1, 1)$-form K that satisfies the conditions $\Theta \wedge \Theta \neq 0$, $\Theta \wedge \omega = 0$, and $\Theta \wedge \iota_{J_\alpha} \omega) = 0$. Furthermore, on a 4-dimensional almost-complex (respectively, para-complex) manifold, it is always possible to pick a Riemannian (respectively, Kleinian) metric such that the differential $(2, 0)$-form, the differential $(0, 2)$-form, and one of the differential $(1, 1)$-forms are anti-self-dual, and since the differential $(1, 1)$-form is then proportional to the almost Kähler form for that metric, the scale of the metric is fixed by declaring the differential $(1, 1)$-form to be almost Kähler form.   □

In particular, for $(\omega, \iota_{J_\alpha} \omega, \Theta)$ on M, we may always pick a co-frame $\{e^i\}_{i=1,\dots,4}$ such that the differential $(2, 0)$-form and differential $(0, 2)$-form (3.44) are given by $\omega = e^1 \wedge e^3 + e^2 \wedge e^4$ and $J_\alpha \iota \omega = e^3 \wedge e^4 - \epsilon e^1 \wedge e^2$ with $\epsilon := \mathrm{sgn}(\mathrm{pf}\,\alpha)$. Note that an overall scale can always be absorbed in the definition of the basis $\{e^i\}_{i=1,\dots,4}$. Hence,

$$J_\alpha = \begin{pmatrix} 0 & 0 & 0 & \epsilon \\ 0 & 0 & -\epsilon & 0 \\ 0 & 1 & 0 & 0 \\ -1 & 0 & 0 & 0 \end{pmatrix} \tag{3.45}$$

in this basis. Furthermore, it is clear that ω and $J_\alpha \iota \omega$ are anti-self-dual for any metric in the conformal class of $\epsilon e^1 \otimes e^1 + e^2 \otimes e^2 + \epsilon e^3 \otimes e^3 + e^4 \otimes e^4$. The third anti-self-dual differential two-form is proportional to $\epsilon e^1 \wedge e^4 - e^2 \wedge e^3$ which, in turn, is proportional to the almost Kähler form. Hence, upon setting $\Theta = \epsilon e^1 \wedge e^4 - e^2 \wedge e^3$, we have fixed the scale of Θ to be pf $\Theta = \epsilon = \mathrm{sgn}(\mathrm{pf}\,\alpha)$, and, by declaring Θ to be the almost Kähler form, we have $\Theta(X, Y) = h_\alpha(J_\alpha X, Y)$ for all $X, Y \in \Gamma(TM)$ with $h_\alpha = \epsilon e^1 \otimes e^1 + e^2 \otimes e^2 + \epsilon e^3 \otimes e^3 + e^4 \otimes e^4$. Then, $\{e^i\}_{i=1,\dots,4}$ is an orthonormal co-frame. When $\epsilon = -1$, the above is unique only up to a sign since in this case $\Theta \mapsto -\Theta$ and $h_\alpha \mapsto -h_\alpha$ can also be used.

Remark 3.14 Notice that since α and $J_\alpha \iota \omega$ are proportional, also α is anti-self-dual with respect to the metric h_α.

Remark 3.15 Let $\varpi \in \Gamma(\bigwedge^4 TM)$ be a volume form normalised as

$$\tfrac{1}{2!} \iota_\varpi (\omega \wedge \omega) = -\mathrm{sgn}(\mathrm{pf}\,\alpha).$$

Then, in a coordinate-free notation, the metric h_α reads as

$$h_\alpha(X, Y) = \tfrac{1}{2} \iota_\varpi [(\iota_X \omega) \wedge (\iota_Y (\iota_{J_\alpha} \omega)) \wedge \Theta + (\iota_Y \omega) \wedge (\iota_X (\iota_{J_\alpha} \omega)) \wedge \Theta] \tag{3.46}$$

for all $X, Y \in \Gamma(TM)$. Furthermore, it is then easy to see that indeed $h_\alpha(J_\alpha X, Y) = \Theta(X, Y)$ and $h_\alpha(J_\alpha X, J_\alpha Y) = \epsilon h_\alpha(X, Y)$. Hence, h_α coincides with the metric (3.10).

Remark 3.16 Recall the definition of a generalised solution as given in Definition 3.32. Let L be a generalised solution for a non-degenerate Monge–Ampère structure (ω, α) on the cotangent bundle T^*M of a 2-dimensional manifold M. Since ω and $\iota_{J_\alpha} \omega$ vanish on L, we have $TL = \langle X, J_\alpha X \rangle$ for any $X \in \Gamma(TL)$. If we let $i : L \hookrightarrow T^*M$ be the embedding of L into T^*M, by virtue of Remark 3.15, it then immediately follows that

$$i^* h_\alpha = h_\alpha(i_* X, i_* X) \begin{pmatrix} 1 & 0 \\ 0 & \mathrm{sgn}(\mathrm{pf}\,\alpha) \end{pmatrix} \tag{3.47}$$

in the basis $\{X, J_\alpha X\}$. Hence, for pf $\alpha > 0$ the $i^* h_\alpha$ is Riemannian while for pf $\alpha < 0$ Kleinian.

Since h_α plays a special role the geometry of Monge–Ampère equations, we shall give it a name.

Definition 3.36 Let M be a 4-dimensional manifold equipped with a non-degenerate Monge–Ampère structure (ω, α). The metric h_α associated with (ω, α) by means of Proposition 3.13 is called the **Monge–Ampère metric**.

Upon recalling Propositions 3.5, 3.13 has the following immediate consequence.

Corollary 3.4 *Let M be a 4-dimensional manifold equipped with a non-degenerate Monge–Ampère structure (ω, α). Furthermore, let $(\omega, \iota_{J_\alpha} \omega)$, Θ, $h_\alpha)$ be as in Proposition 3.13. Then, M is almost hyper-complex for pf $\alpha > 0$ while almost hyper-paracomplex for pf $\alpha < 0$, respectively.*

Remark 3.17 Recalling the proof of Proposition 3.13, we let $\{e^i\}_{i=1,\dots,4}$ be an orthonormal co-frame such that

$$h_\alpha = \epsilon e^1 \otimes e^1 + e^2 \otimes e^2 + \epsilon e^3 \otimes e^3 + e^4 \otimes e^4 \tag{3.48}$$

and

$$\begin{aligned}
\omega := \omega_1 &= e^1 \wedge e^3 + e^2 \wedge e^4 , \\
\iota_{J_\alpha} \omega := \omega_2 &= e^3 \wedge e^4 - \epsilon e^1 \wedge e^2 , \\
\Theta := \omega_3 &= \epsilon e^1 \wedge e^4 - e^2 \wedge e^3 .
\end{aligned} \tag{3.49}$$

Next, we define the endomorphisms $\omega_i(X, Y) = h_\alpha(J_i X, Y)$ for $X, Y \in \Gamma(TM)$ as in (3.8). It is then a straightforward exercise to show that in this basis we have

$$J_1 = \begin{pmatrix} 0 & 0 & \epsilon & 0 \\ 0 & 0 & 0 & 1 \\ -\epsilon & 0 & 0 & 0 \\ 0 & -1 & 0 & 0 \end{pmatrix}, \quad J_2 = \begin{pmatrix} 0 & -\epsilon & 0 & 0 \\ 1 & 0 & 0 & 0 \\ 0 & 0 & 0 & 1 \\ 0 & 0 & -\epsilon & 0 \end{pmatrix},$$

$$J_3 = \begin{pmatrix} 0 & 0 & 0 & \epsilon \\ 0 & 0 & -\epsilon & 0 \\ 0 & 1 & 0 & 0 \\ -1 & 0 & 0 & 0 \end{pmatrix} \tag{3.50}$$

so that $J_1^2 = - \mathrm{Id}_4$, $J_2^2 = -\epsilon \, \mathrm{Id}_4 = J_3^2$, and $J_1 J_2 J_3 = -\epsilon \, \mathrm{Id}_4$.

Remark 3.18 Note that, in fact, any 4-dimensional almost-complex (respectively, para-complex) symplectic manifold admits an almost hyper-complex (respectively, hyper-para-complex) structure. Indeed, let (M, J, ω) be such a manifold. Then, we may set $\omega_1 := \omega$ and $\omega_2 := \iota_J \omega$. Since J is traceless, $\omega_1 \wedge \omega_2 = 0$ and since $J^2 = -\epsilon \, \mathrm{Id}$ with $\epsilon = \pm 1$ and ω non-degenerate, we also have $\omega_2 \wedge \omega_2 = \epsilon \omega_1 \wedge$

$\omega_1 \neq 0$. Hence, (ω_1, ω_2) constitutes a non-degenerate Monge–Ampère structure on M. Furthermore, by virtue of Proposition 3.13, there is a unique Hermitian (respectively, para-Hermitian) metric g and a $(1, 1)$-form ω_3 with respect to J such that $g(JX, Y) = \omega_3(X, Y)$ for $X, Y \in \Gamma(TM)$, the differential forms $(\omega_1, \omega_2, \omega_3)$ are anti-self-dual with respect to g, and $\omega_3 \wedge \omega_1 = 0 = \omega_3 \wedge \omega_2$ and $\omega_3 \wedge \omega_3 = \epsilon \omega_1 \wedge \omega_1$. Explicitly, g is given by (3.10). Consequently, the tuple $(\omega_1, \omega_2, \omega_3)$ constitutes an almost hyper-complex (respectively, hyper-para-complex) structure on M.

3.5.2 (4m + 2)-Dimensional MA Geometry

Let us now generalise the results of the previous section to higher dimensional Monge–Ampère geometry. To begin with, let M be a $(4m + 2)$-dimensional manifold. Motivated by Hitchin's discussion [6, 34] when M is 6-dimensional, we give the following two definitions for all $m \geq 1$.

Definition 3.37 Let M be a $(4m + 2)$-dimensional manifold and fix a volume form $\varpi \in \Gamma(\det(TM))$. For every $\alpha \in \Omega^{2m+1}(M)$, we define the endomorphism

$$A_{\alpha, \varpi} : TM \rightarrow TM,$$
$$X \mapsto \iota_\varpi[(\iota_X \alpha) \wedge \alpha)].$$
(3.51)

The quantity

$$\mathrm{HPf}(\alpha, \varpi) := -\frac{1}{4m + 2} \mathrm{Tr}(A_{\alpha, \varpi}^2) \in C^\infty(M)$$
(3.52)

is called the **Hitchin Pfaffian** of α.

Definition 3.38 Let M be a $(4m + 2)$-dimensional manifold. A differential $(2m + 1)$-form is said to be **Hitchin degenerate** if and only if its Hitchin Pfaffian vanishes. It is said to be **Hitchin non-degenerate** if and only if its Hitchin Pfaffian is non-zero.

Obviously, $\mathrm{Tr}(A_{\alpha, \varpi}) = 0$ and also $\mathrm{HPf}(\alpha, f\varpi) = f^2 \mathrm{HPf}(\alpha, \varpi)$ for all non-vanishing $f \in C^\infty(M)$. Hence, both the signs of the Hitchin Pfaffian and Hitchin (non-)degeneracy do not depend on the choice of volume form.

We should compare this notion with another one which we shall call a *weak degeneracy*.

Definition 3.39 A differential $k-$form $\alpha \in \Omega^k(M)$ is called a **weak non-degenerate** if and only if the map $\Gamma(TM) \mapsto \Omega^{k-1}(M)$ given by $X \rightarrow \iota_X \omega$ is injective.

We observe that Hitchin non-degeneracy is a stronger condition than the one given in Definition 3.39. For instance, when $m = 1$, Banos [12] and Bryant [15] proved that for any differential three-form on a 6-dimensional manifold M which is non-degenerate in the sense of Definition 3.39 there exist a basis $\{e^i\}_{i=1,\dots,6}$ of T^*M such that it is equal to either

$$e^{123} + e^{456}, \quad e^{156} \pm e^{264} \pm e^{345}, \quad \text{or} \quad e^{123} - e^{156} \pm e^{246} \mp e^{345}, \tag{3.53}$$

where $e^{123} := e^1 \wedge e^2 \wedge e^3$, etc. It is then rather straightforward to see that the Hitchin Pfaffian is zero for the second differential form in (3.53) while non-zero for the other two.

Definition 3.40 A Monge–Ampère structure (ω, α) on a $(4m + 2)$-dimensional manifold M is said to be

1. **Hitchin non-degenerate** if and only if α is Hitchin non-degenerate.
2. **Hitchin decomposable** if and only if α is of either form:
   a. $\alpha = \frac{1}{2}(\Omega - \hat{\Omega})$ for some $\Omega, \hat{\Omega} \in \Omega^{2m+1}(M)$ decomposable over $\mathbb{R}$ such that $\Omega \wedge \hat{\Omega} \neq 0$;
   b. $\alpha = \mathfrak{Im}\, \Omega$ for some $\Omega \in \Omega^{2m+1}(M) \otimes \mathbb{C}$ decomposable over $\mathbb{C}$ such that $\mathfrak{Re}\, \Omega \wedge \mathfrak{Im}\, \Omega \neq 0$.

Proposition 3.14 *Every Hitchin decomposable Monge–Ampère structure (ω, α) on a $(4m + 2)$-dimensional manifold M is Hitchin non-degenerate.*

Proof Let us first work in a complexified setting. In particular, let $\{E_i\}_{i=1,\dots,4m+2}$ be a basis of $T_{\mathbb{C}}M := TM \otimes \mathbb{C}$ and $\{e^i\}_{i=1,\dots,4m+2}$ a basis of $T_{\mathbb{C}}^*M := T^*M \otimes \mathbb{C}$ with $E_i \,\iota\, e^j = \delta_i{}^j$. Furthermore, let $\varpi_0 \,\iota\, (e^1 \wedge \dots \wedge e^{4m+2}) = 1$ and define

$$\alpha_0 := e^1 \wedge \dots \wedge e^{2m+1} + e^{2m+2} \wedge \dots \wedge e^{4m+2}. \tag{3.54}$$

Then,

$$A_{\alpha_0, f\varpi_0} E_i = \begin{cases} f E_i & \text{for } i = 1, \dots, 2m + 1 \\ -f E_i & \text{for } i = 2m + 1, \dots, 4m + 2 \end{cases} \tag{3.55}$$

for all non-vanishing $f \in C^\infty(M) \otimes \mathbb{C}$. Consequently,

$$A_{\alpha_0, f\varpi_0}^* \alpha_0 = f^{2m+1}(e^1 \wedge \dots \wedge e^{2m+1} - e^{2m+2} \wedge \dots \wedge e^{4m+2}) \tag{3.56}$$

and

$$\begin{aligned} \tfrac{1}{2}(\alpha_0 + f^{-2m-1} A_{\alpha_0, f\varpi_0}^* \alpha_0) &= e^1 \wedge \dots \wedge e^{2m+1}, \\ \tfrac{1}{2}(\alpha_0 - f^{-2m-1} A_{\alpha_0, f\varpi_0}^* \alpha_0) &= e^{2m+2} \wedge \dots \wedge e^{4m+2}, \\ \tfrac{1}{2} f^{-2m-1} A_{\alpha_0, f\varpi_0}^* \alpha_0 \wedge \alpha_0 &= e^1 \wedge \dots \wedge e^{4m+2}. \end{aligned} \tag{3.57}$$

Note also that α_0 is Hitchin non-degenerate with $\mathrm{HPf}(\alpha_0, f\varpi_0) = -f^2$.

Now, if $\alpha = \alpha_+ - \alpha_-$ with $\alpha_\pm \in \Omega^{2m+1}(M)$ so that $\alpha_\pm$ are decomposable over $\mathbb{C}$, then there exist $\{\xi^i \in \Omega_{\mathbb{C}}^1(M)\}_{i=1,\dots,4m+2}$ such that $\alpha_+ = \xi^1 \wedge \dots \wedge \xi^{2m+1}$ and $\alpha_- = \xi^{2m+2} \wedge \dots \wedge \xi^{4m+2}$, respectively. Moreover, the condition $\alpha_+ \wedge \alpha_- \neq 0$ implies that $\{\xi^i\}_{i=1,\dots,4m+2}$ forms a basis of $T_{\mathbb{C}}^*M$. Hence, there is a GL-transformation g such that $g \triangleright (\alpha, \varpi) = (\alpha_0, f\varpi_0)$, where $\varpi := g^{-1} \triangleright f\varpi_0$. Thus, α must be

Hitchin non-degenerate with Hitchin Pfaffian $\mathrm{HPf}(\alpha, \varpi)$. Furthermore, the algebraic conditions (3.57) hold on the GL-orbit so that

$$\alpha_{\pm} = \tfrac{1}{2}\left\{\alpha \pm [-\mathrm{HPf}(\alpha, \varpi)]^{-m-\frac{1}{2}} A^*_{\alpha, \varpi} \alpha\right\} \in \Omega^{2m+1}_{\mathbb{C}}(M) . \tag{3.58}$$

The above shows that for a Hitchin decomposable Monge–Ampère structure (ω, α), we must have either $\mathrm{HPf}(\alpha, \varpi) < 0$ or $\mathrm{HPf}(\alpha, \varpi) > 0$. This proves the assertion. $\qquad\square$

Remark 3.19 Note that when $\mathrm{HPf}(\alpha, \varpi) < 0$, we may set

$$\Omega := \alpha + |\mathrm{HPf}(\alpha, \varpi)|^{-m-\frac{1}{2}} A^*_{\alpha, \varpi} \alpha \quad \text{and} \quad \hat{\Omega} := -\alpha + |\mathrm{HPf}(\alpha, \varpi)|^{-m-\frac{1}{2}} A^*_{\alpha, \varpi} \alpha \tag{3.59a}$$

so that $\alpha = \tfrac{1}{2}(\Omega - \hat{\Omega})$ and $\Omega \wedge \hat{\Omega} \neq 0$ while for $\mathrm{HPf}(\alpha, \varpi) > 0$,

$$\Omega := |\mathrm{HPf}(\alpha, \varpi)|^{-m-\frac{1}{2}} A^*_{\alpha, \varpi} \alpha + i\alpha \tag{3.59b}$$

so that $\alpha = \mathfrak{Im}\,\Omega$ and $\mathfrak{Re}\,\Omega \wedge \mathfrak{Im}\,\Omega \neq 0$.

Remark 3.20 As proved by Hitchin and Banos [6, 34], for $m = 1$ the converse also holds, that is, any Hitchin non-degenerate Monge–Ampère structure is also Hitchin decomposable. The reason for that is that both the GL-orbit of (3.54) and the space of differential three-forms are 20-dimensional. In higher dimensions, this is no longer true as the dimension of GL-orbit is smaller than the dimension of the space of differential $(2m + 1)$-forms.

Corollary 3.5 *Let (ω, α) be a Hitchin decomposable Monge–Ampère structure on a $(4m + 2)$-dimensional manifold M. Then,*

$$A^2_{\alpha, \varpi} = -\mathrm{HPf}(\alpha, \varpi)\,\mathrm{Id} . \tag{3.60}$$

Proof Since any Hitchin decomposable Monge–Ampère structure lies in the GL-orbit of (3.54), we can deduce the general properties from (3.54) and so, the corollary then follows directly from (3.55). $\qquad\square$

Next, fix the volume form $\varpi \in \Gamma(\det(TM))$ such that $\iota_{\varpi} \frac{1}{(2m+1)!}\omega^{\wedge(2m+1)} = 1$, where $\frac{1}{(2m+1)!}\omega^{\wedge(2m+1)}$ is the Liouville volume form for a Hitchin decomposable Monge–Ampère structure (ω, α). We also introduce the symmetric bilinear form

$$g_{\alpha, \varpi}(X, Y) := \iota_{\varpi}[(\iota_X \alpha) \wedge (\iota_Y \alpha) \wedge \omega] \quad \text{for all} \quad X, Y \in \Gamma(TM) \tag{3.61}$$

which first appeared in [44] for $m = 1$. Since α is ω-effective, we may rewrite this as

$$g_{\alpha, \varpi}(X, Y) = \iota_{\varpi}[(\iota_X \alpha) \wedge \alpha \wedge (\iota_Y \omega)] \tag{3.62}$$

so that

$$g_{\alpha,\varpi}(X, Y) \; = \; A_{\alpha,\varpi} \, \iota_X \, \iota_Y \, \omega \; \equiv \; -\omega(A_{\alpha,\varpi} X, Y) \tag{3.63}$$

by virtue of (3.51). Thus, this bilinear form is non-degenerate due to Proposition 3.14 and Corollary 3.5. Generalising Definition 3.36, we give the following definition.

Definition 3.41 Let (ω, α) be a Hitchin decomposable Monge–Ampère structure on a $(4m + 2)$-dimensional manifold M and fix a volume form $\varpi \in \Gamma(\det(TM))$. The metric

$$h_{\alpha,\varpi}(X, Y) \; := \; \frac{\iota_\varpi[(\iota_X \alpha) \wedge (\iota_Y \alpha) \wedge \omega]}{\sqrt{|\mathrm{HPf}(\alpha, \varpi)|}} \quad \text{for all} \quad X, Y \in \Gamma(TM) \tag{3.64}$$

is called the **Monge–Ampère metric**.

Hence, we may set

$$J_{\alpha,\varpi} \; := \; \frac{\mathrm{sgn}(\mathrm{HPf}(\alpha, \varpi))}{\sqrt{|\mathrm{HPf}(\alpha, \varpi)|}} A_{\alpha,\varpi} \tag{3.65}$$

so that $J_{\alpha,\varpi}^2 = -\mathrm{sgn}(\mathrm{HPf}(\alpha, \varpi))\,\mathrm{Id}$ and $\mathrm{tr}(J_{\alpha,\varpi}) = 0$. Thus, $J_{\alpha,\varpi}$ is an almost-complex structure for $\mathrm{HPf}(\alpha, \varpi) > 0$ and an almost para-complex structure for $\mathrm{HPf}(\alpha, \varpi) < 0$, respectively.

Corollary 3.6 *Let (ω, α) be a Hitchin decomposable Monge–Ampère structure on a $(4m + 2)$-dimensional manifold M and let $J_{\alpha,\varpi}$ be as in (3.65). The pair $(J_{\alpha,\varpi}, \alpha)$ is compatible in the sense of Definition 3.21.*

Proof Similarly to Corollary 3.5, this can be easily deduced from (3.54) and (3.55), respectively.   □

By virtue of (3.65),

$$h_{\alpha,\varpi}(X, Y) \; = \; -\mathrm{sgn}(\mathrm{HPf}(\alpha, \varpi))\,\omega(J_{\alpha,\varpi} X, Y) \quad \Longleftrightarrow \quad \omega(X, Y) = h_{\alpha,\varpi}(J_{\alpha,\varpi} X, Y) \tag{3.66}$$

so that ω is the almost Kähler form for $(h_{\alpha,\varpi}, J_{\alpha,\varpi})$. As it is closed by assumption, we have the following result.

Proposition 3.15 *Let (ω, α) be a Hitchin decomposable Monge–Ampère structure on a $(4m + 2)$-dimensional manifold M. Moreover, let $h_{\alpha,\varpi}$ and $J_{\alpha,\varpi}$ be defined as in (3.64) and (3.65). For $\mathrm{HPf}(\alpha, \varpi) > 0$ (respectively, $\mathrm{HPf}(\alpha, \varpi) < 0$), the tuple $(M, h_{\alpha,\varpi}, J_{\alpha,\varpi}, \omega)$ is an almost Kähler (respectively, para-Kähler) manifold.*

Note that since ω is a differential $(1, 1)$-form with respect to $J_{\alpha,\varpi}$ and since α is ω-effective, the differential forms Ω and $\hat{\Omega}$ appearing in[3] Definition 3.40 are, in fact, differential $(2m + 1, 0)$- and $(0, 2m + 1)$-forms with respect to $J_{\alpha,\varpi}$, respectively. Since $(J_{\alpha,\varpi}, \alpha)$ is compatible in the sense of Definition 3.21 by virtue of Corollary

---

[3] See also (3.59).

3.6, Ω and $\hat{\Omega}$ are given in terms of α and $J_{\alpha,\varpi} \iota \alpha$ as in (3.44). This is completely analogous to the 4-dimensional case as discussed in Sect. 3.5.1.1.

Example

Consider $T^*\mathbb{R}^3$ with coordinates $(q^1, q^2, q^3, p_1, p_2, p_3)$ and the standard symplectic structure $\omega = dq^1 \wedge dp_1 + dq^2 \wedge dp_2 + dq^3 \wedge dp_3$.
    The Monge–Ampère equation

$$\epsilon_1 \phi_{q^1 q^1} + \epsilon_2 \phi_{q^2 q^2} + \epsilon_3 \phi_{q^3 q^3} - \epsilon_1 \epsilon_2 \epsilon_3 \det(\mathsf{Hess}(\phi)) = 0 \tag{3.67a}$$

with $\epsilon_i = \pm 1$ is associated with $\alpha = \mathfrak{Im}\,\Omega$, where

$$\Omega := (dq^1 + i\epsilon_1 dp_1) \wedge (dq^2 + i\epsilon_2 dp_2) \wedge (dq^3 + i\epsilon_3 dp_3) . \tag{3.67b}$$

Clearly, $\mathfrak{Re}\,\Omega \wedge \mathfrak{Im}\,\Omega \neq 0$. Hence, $J_{\alpha,\varpi}$ is an almost-complex structure. Since Ω is a closed differential $(3,0)$-form with respect to $J_{\alpha,\varpi}$, $J_{\alpha,\varpi}$ is integrable by virtue of Theorem 3.7.
    The Monge–Ampère equation

$$\det(\mathsf{Hess}(\phi)) = 1 \tag{3.68a}$$

is associated with $\alpha = \frac{1}{2}(\Omega - \hat{\Omega})$, where

$$\Omega := 2dp_1 \wedge dp_2 \wedge dp_3 \quad \text{and} \quad \hat{\Omega} := 2dq^1 \wedge dq^2 \wedge dq^3 . \tag{3.68b}$$

Note that $\Omega \wedge \hat{\Omega} \neq 0$. Consequently, $J_{\alpha,\varpi}$ is an almost para-complex structure. Since Ω and $\hat{\Omega}$ are closed differential $(3,0)$- and $(0,3)$-forms with respect to $J_{\alpha,\varpi}$, $J_{\alpha,\varpi}$ is integrable by virtue of Theorem 3.7.

Example

Consider the Chynoweth–Sewell equation [16] on $\mathbb{R}^3$

$$\phi_{q^1 q^1} \phi_{q^2 q^2} - \phi_{q^1 q^2}^2 + \phi_{q^3 q^3} = \gamma \quad \text{with} \quad \gamma \in \mathbb{R} . \tag{3.69}$$

The associated Monge–Ampère structure is

$$\begin{aligned} \omega &= dq^1 \wedge dp_1 + dq^2 \wedge dp_2 + dq^3 \wedge dp_3 , \\ \alpha &= dp_1 \wedge dp_2 \wedge dq^3 + dq^1 \wedge dq^2 \wedge (dp_3 - \gamma dq^3) . \end{aligned} \tag{3.70}$$

and since α is the sum of two real decomposable differential three-forms whose exterior product is non-vanishing, $J_{\alpha,\varpi}$ is an almost para-complex structure.

Furthermore, there is a symplectomorphism $\Phi : T^*\mathbb{R}^3 \to T^*\mathbb{R}^3$ defined by

$$\Phi^*(q^1, q^2, q^3, p_1, p_2, p_3) := (q^1, q^2, p_3, p_1, p_2, \gamma p_3 - q^3) \tag{3.71}$$

which has the property

$$\Phi^*\alpha = dp_1 \wedge dp_2 \wedge dp_3 - dq^1 \wedge dq^2 \wedge dq^3. \tag{3.72}$$

Consequently, the Chynoweth–Sewell equation is symplectially equivalent to (3.68a).

We then have the following classification generalising results of [44] and Banos' work [12].

Theorem 3.8 *Let M be a $(2m + 1)$-dimensional manifold and let (ω, α) be a Hitchin decomposable Monge–Ampère structure on T^*M. Fix a volume form $\varpi \in \Gamma(\det(T(T^*M)))$ and consider the almost-complex (respectively, para-complex) structure*

$$J_{\alpha,\varpi} X = \pm \frac{\iota_\varpi[(\iota_X \alpha) \wedge \alpha]}{\sqrt{\pm HPf(\alpha, \varpi)}} \tag{3.73a}$$

on T^*M, where

$$HPf(\alpha, \varpi) = -\frac{1}{4m+2} Tr\left\{ \left(X \mapsto \iota_\varpi[(\iota_X \alpha) \wedge \alpha] \right)^2 \right\}, \tag{3.73b}$$

together with the almost Kähler metric

$$h_{\alpha,\varpi}(J_{\alpha,\varpi} X, Y) = \omega(X, Y) \tag{3.74}$$

for all $X, Y \in \Gamma(T(T^*M))$. Then,

1. for $HPf(\alpha, \varpi) > 0$, the Monge–Ampère equation associated with the Monge–Ampère operator Δ_α is symplectically equivalent to

$$\sum_{i=1}^{2m+1} \epsilon_i PM_{[m]\setminus\{i\}}(hess\,\phi) + \sum_{i=1}^{m} (-1)^i \sum_{\sigma \in S_{2i+1}} \epsilon_\sigma PM_{[m]\setminus\sigma}(hess\,\phi) = 0, \tag{3.75a}$$

where $\epsilon_i = \pm 1$, S_{2i+1} is the set of order-$(2i + 1)$ subsets of $[m] := \{1, \ldots, 2m + 1\}$, $\epsilon_\sigma := \epsilon_{\sigma_1} \cdots \epsilon_{\sigma_{2i+1}}$ for $\sigma = \{\sigma_1, \ldots, \sigma_{2i+1}\} \in S_{2i+1}$, and $PM_{[m]\setminus\sigma}$ stands for the corresponding principal minor, if and only if

$$d\Omega = 0 \quad with \quad \Omega = \iota_{J_{\alpha,\varpi}} \alpha + i\alpha \tag{3.75b}$$

and the metric $h_{\alpha,\varpi}$ is flat;

2. *for* $HPf(\alpha, \varpi) < 0$, *the Monge–Ampère equation associated with the Monge–Ampère operator* Δ_α *is symplectically equivalent to*

$$\det(\mathsf{Hess}(\phi)) = 1 \tag{3.76a}$$

if and only if

$$d\Omega = 0 = d\hat{\Omega} \quad with \quad \Omega = \iota_{J_{\alpha,\varpi}} \alpha + \alpha \quad and \quad \hat{\Omega} = \iota_{J_{\alpha,\varpi}} \alpha - \alpha \tag{3.76b}$$

and the metric $h_{\alpha,\varpi}$ *is flat.*

Proof ($\Rightarrow$) First, let ω be the standard symplectic structure on T^*M. Then, it is clear that if the Monge–Ampère equation associated with the Monge–Ampère operator Δ_α is symplectically equivalent to one of the above equations, then Ω and $\hat{\Omega}$ are closed and $h_{\alpha,\varpi}$ is flat.

($\Leftarrow$) To prove the converse, consider first the case when $HPf(\alpha, \varpi) > 0$. The closure of Ω implies that α is closed thus making α into a $(2m + 1)$-plectic structure. In addition, it also makes $J_{\alpha,\varpi} \iota \alpha$ into a $(2m + 1)$-plectic structure and hence, by Proposition 3.7 implies the integrability of $J_{\alpha,\varpi}$. Hence, there are local complex coordinates $z^i = q^i + i\epsilon_i p_i$ for $i = \{1, \ldots, 2m + 1\}$ with $\epsilon_i = \pm 1$ such $\Omega = dz^1 \wedge \ldots \wedge dz^{2m+1}$ locally and so,

$$\begin{aligned}
\alpha &= \mathfrak{Im}\,\Omega \\
&= [\epsilon_1 dp_1 \wedge dq^2 \wedge \ldots \wedge dq^{2m+1} + \cdots + \epsilon_{2m+1} dq^1 \wedge \ldots \wedge dq^{2m} \wedge dp_{2m+1}] - \\
&\quad - [\epsilon_1 \epsilon_2 \epsilon_3 dp_1 \wedge dp_1 \wedge dp_3 \wedge dq^4 \wedge \ldots \wedge dq^{2m+1} + \cdots + \\
&\qquad + \epsilon_{2m-1} \epsilon_{2m} \epsilon_{2m+1} dq^1 \wedge \ldots \wedge dq^{2m-2} \wedge dp_{2m-1} \wedge dp_{2m} \wedge dp_{2m+1}] + \\
&\quad + \cdots + (-1)^m dp_1 \wedge \ldots \wedge dp_{2m+1}\,.
\end{aligned} \tag{3.77}$$

Since $h_{\alpha,\varpi}$ is flat, we also have

$$\begin{aligned}
\omega &= \tfrac{i}{2}\left(\epsilon_1 dz^1 \wedge d\bar{z}^{\bar{1}} + \cdots + \epsilon_{2m+1} dz^{2m+1} \wedge d\bar{z}^{\bar{2m+1}}\right) \\
&= dq^1 \wedge dp_1 + \cdots + dq^{2m+1} \wedge dp_{2m+1}\,.
\end{aligned} \tag{3.78}$$

It is a straightforward exercise to check that the Monge–Ampère structure (ω, α) then yields (3.75a); see also Example 3.5.2 for $m = 1$.

For $HPf(\alpha, \varpi) < 0$, since both Ω and $\hat{\Omega}$ are closed, $J_{\alpha,\varpi}$ is integrable. Hence, there exist local coordinates $(q^1, \ldots, q^{2m+1}, p_1, \ldots, p_{2m+1})$ such that $\Omega = 2dp_1 \wedge \ldots \wedge dp_{2m+1}$ and $\hat{\Omega} = 2dq^1 \wedge \ldots \wedge dq^{2m+1}$ locally and so,

$$\alpha = \tfrac{1}{2}(\Omega - \hat{\Omega}) = dp_1 \wedge \ldots \wedge dp_{2m+1} - dq^1 \wedge \ldots \wedge dq^{2m+1}\,. \tag{3.79}$$

Again, since $h_{\alpha,\varpi}$ is flat we also have

$$\omega = dq^1 \wedge dp_1 + \cdots + dq^{2m+1} \wedge dp_{2m+1} \,. \tag{3.80}$$

Again, it is easy to see that the Monge–Ampère structure (ω, α) then yields (3.76a); see also Example 3.5.2 for $m = 1$.   □

In analogy with the 4-dimensional case and for later convenience, we would like to give the following definition.

Definition 3.42 A Hitchin non-degenerate Monge–Ampère structure (ω, α) on a $(4m + 2)$-dimensional manifold M with Hitchin Pfaffian $\mathrm{HPf}(\alpha, \varpi)$ is called

1. **elliptic** if and only if $\mathrm{HPf}(\alpha, \varpi) > 0$;
2. **hyperbolic** if and only if $\mathrm{HPf}(\alpha, \varpi) < 0$;
3. **parabolic** if and only if $\mathrm{HPf}(\alpha, \varpi) = 0$.

Next, we recall the following definition.

Definition 3.43 (*Xu* [56, 57]) Let (M, g) be a $2m$-dimensional almost Hermitian (respectively, para-Hermitian) manifold with almost complex (respectively, para-complex) structure J and almost Kähler form ω. For J

1. an almost-complex structure, M is said to be a **nearly Calabi–Yau manifold** if and only if there is a differential $(m, 0)$-form Ω such that

$$d\omega = 0 \quad \text{and} \quad d\,\Im m\,\Omega = 0\,; \tag{3.81}$$

2. an almost para-complex structure, M is said to be a **nearly para Calabi–Yau manifold** if and only if there are a differential $(m, 0)$-form Ω and a differential $(0, m)$-form $\hat{\Omega}$ such that

$$d\omega = 0 \quad \text{and} \quad d(\Omega - \hat{\Omega}) = 0\,. \tag{3.82}$$

Hence, we have the immediate result.

Proposition 3.16 *Let (ω, α) be a Hitchin decomposable Monge–Ampère structure on a $(4m + 2)$-dimensional manifold M. Moreover, let $h_{\alpha,\varpi}$ and $J_{\alpha,\varpi}$ be defined as in (3.64) and (3.65). Suppose that α is closed. Then, for $\mathrm{HPf}(\alpha, \varpi) > 0$ (respectively, $\mathrm{HPf}(\alpha, \varpi) < 0$) the tuple $(M, h_{\alpha,\varpi}, J_{\alpha,\varpi}, \omega)$ is a nearly Calabi–Yau (respectively, nearly para Calabi–Yau) manifold.*

Example

Consider $T^*\mathbb{R}^3$ with coordinates $(q^1, q^2, q^3, p_1, p_2, p_3)$ and the standard symplectic structure $\omega = dq^1 \wedge dp_1 + dq^2 \wedge dp_2 + dq^3 \wedge dp_3$ and standard volume form $\varpi \in \Gamma(\det(T(T^*M)))$. Furthermore, let $f = f(q^1, q^2, q^3)$ be a non-vanishing function on $\mathbb{R}^3$ and define

$$\alpha := \tfrac{1}{2} f dq^1 \wedge dq^2 \wedge dq^3 - dp_1 \wedge dq^2 \wedge dq^3 - dq^1 \wedge dp_2 \wedge dq^3 - dq^1 \wedge dq^2 \wedge dp_3$$
(3.83)

It is easy to check that α is Hitchin decomposable and that $J_{\alpha,\varpi}$ and $h_{\alpha,\varpi}$ defined in (3.65) and (3.64) are given by

$$J_{\alpha,\varpi} = \begin{pmatrix} 0 & \sqrt{\tfrac{|f|}{2}} \, \mathrm{Id}_3 \\ -\mathrm{sgn}(f)\sqrt{\tfrac{2}{|f|}} \, \mathrm{Id}_3 & 0 \end{pmatrix} \text{ and } h_{\alpha,\varpi} = \begin{pmatrix} \mathrm{sgn}(f)\sqrt{\tfrac{|f|}{2}} \, \mathrm{Id}_3 & 0 \\ 0 & \sqrt{\tfrac{2}{|f|}} \, \mathrm{Id}_3 \end{pmatrix}.$$
(3.84)

In addition, for $f > 0$ we have

$$\mathfrak{Re}\,\Omega = \sqrt{\tfrac{2}{f}} \, dp_1 \wedge dp_2 \wedge dp_3 - $$
$$- \sqrt{\tfrac{f}{2}} (dq^1 \wedge dp_2 \wedge dp_3 + dp_1 \wedge dq^2 \wedge dp_3 + dp_1 \wedge dp_2 \wedge dq^3),$$

$$\mathfrak{Im}\,\Omega = \alpha,$$
(3.85a)

while for $f < 0$

$$\tfrac{1}{2}(\Omega + \hat{\Omega}) = \sqrt{-\tfrac{2}{f}} \, dp_1 \wedge dp_2 \wedge dp_3 - $$
$$- \sqrt{-\tfrac{f}{2}} (dq^1 \wedge dp_2 \wedge dp_3 + dp_1 \wedge dq^2 \wedge dp_3 + dp_1 \wedge dp_2 \wedge dq^3),$$

$$\tfrac{1}{2}(\Omega - \hat{\Omega}) = \alpha.$$
(3.85b)

Since α is closed, $(T^*\mathbb{R}^3, h_{\alpha,\varpi}, J_{\alpha,\varpi}, \omega, \alpha)$ is nearly Calabi–Yau. Finally, note that $J_{\alpha,\varpi}$ is integrable if and only if the function f is locally constant.

3.5.3 Explicite Examples of Generalised Almost Calabi–Yau on $T^*\mathbb{R}^3$ (After B.Banos)

3.5.3.1 Generalised Calabi–Yau Structures

Definition 3.44 A **generalised almost Calabi–Yau structure** on a 6-dimensional manifold X is a 5-uple $(g, \Omega, K, \alpha, \beta)$ where

1. g is a (pseudo) metric on X,
2. Ω is a symplectic on X,
3. K is a smooth section $X \to TX \otimes T^*X$ such that $K^2 = \pm Id$ and such that

$$g(U, V) = \Omega(KU, V)$$

for all tangent vectors U, V,

4. α and β are (eventually complex) decomposable 3-forms whose associated distributions are the distributions of K eigenvectors and such that

$$\frac{\alpha \wedge \beta}{\Omega^3} \quad \text{is constant.}$$

Definition 3.45 A generalised Calabi–Yau structure $(g, \Omega, K, \alpha, \beta)$ is said to be **integrable** if α and β are closed.

Note that a generalised Calabi–Yau structure is a Calabi–Yau structure if and only if the metric is definite positive and K is a complex structure.

Remark 3.21 The condition $d\alpha = d\beta = 0$ implies the integrability (in the Frobenius sense) of the distributions defined by the almost-complex structure or almost-product structure K. Therefore, according to the Newlander–Nirenberg theorem, it implies its integrability. For instance, when K is an almost-complex structure and g is definite positive , the almost Calabi–Yau structure $(g, \Omega, K, \alpha, \overline{\alpha})$ is integrable if and only if K is a complex structure and α is holomorphic.

Example

Each non-degenerate Monge–Ampère structure (Ω, ω_0) defines the generalised almost Calabi–Yau structure $(q_\omega, \Omega, K_\omega, \alpha, \beta)$ with

$$\omega = \frac{\omega_0}{\sqrt[4]{|\mathrm{HPf}(\omega_0)|)}}.$$

For instance, on $\mathbb{R}^6$, the generalised Calabi–Yau structure associated with the equation

$$\Delta(f) - \mathrm{hess}\, f = 0$$

is the canonical Calabi–Yau structure of $\mathbb{C}^3$

$$\begin{cases} g = -\sum_{j=1}^{3} dq^j.dq^j + dp_j.dp_j \\ K = \sum_{j=1}^{3} \frac{\partial}{\partial p_j} \otimes dq^j - \frac{\partial}{\partial q^j} \otimes dp_j \\ \Omega = \sum_{j=1}^{3} dq^j \wedge dp_j \\ \alpha = dz_1 \wedge dz_2 \wedge dz_3 \\ \beta = \overline{\alpha} \end{cases}$$

The generalised Calabi–Yau associated with the equation

$$\Box(f) + \text{hess } f = 0$$

is the pseudo Calabi–Yau structure

$$
\begin{cases}
q = dq^1.dq^1 - dq^2.dq^2 + dq^3.dq^3 + dp_1.dp_1 - dp_2.dp_2 + dp_3.dp_3 \\
K = \frac{\partial}{\partial q^1} \otimes dp_1 - \frac{\partial}{\partial p_1} \otimes dq^1 + \frac{\partial}{\partial p_2} \otimes dq^2 - \frac{\partial}{\partial q^2} \otimes dp_2 - \frac{\partial}{\partial p_3} \otimes dq^3 + \frac{\partial}{\partial q^3} \otimes dp_3 \\
\Omega = \sum_{j=1}^{3} dq^j \wedge dp_j \\
\alpha = dz_1 \wedge dz_2 \wedge dz_3 \\
\beta = \overline{\alpha}
\end{cases}
$$

The generalised Calabi–Yau structure associated with the equation

$$\text{hess } f = 1$$

is the 'real' Calabi–Yau structure

$$
\begin{cases}
g = \sum_{j=1}^{3} dq^j.dp_j \\
K = \sum_{j=1}^{3} \frac{\partial}{\partial q^j} \otimes dq^j - \frac{\partial}{\partial p_j} \otimes dp_j \\
\Omega = \sum_{j=1}^{3} dq^j \wedge dp_j \\
\alpha = dq^1 \wedge dq^2 \wedge dq^3 \\
\beta = dp_1 \wedge dp_2 \wedge dp_3
\end{cases}
$$

A manifold endowed with a 'real' Calabi–Yau structure is the analogue of a 'Monge–Ampère manifold' in the Kontsevich and Soibelman sense [40]. A Monge–Ampere manifold is an affine Riemannian manifold (M, g) such that locally

$$g = \sum_{i,j} \frac{\partial^2 F}{\partial q^i \partial q^j} dq^i.dq^j,$$

F being a smooth function satisfying

$$\det \left(\frac{\partial^2 F}{\partial q^i \partial q^j} \right) = \text{ constant.}$$

In the 'real' Calabi–Yau case we have such a potential F:

$$g = \sum_{i,j} \frac{\partial^2 F}{\partial q^i \partial p_j} dq^i . dp_j,$$

and det $\left(\frac{\partial^2 F}{\partial q^i \partial p_j} \right) = f(q)g(p)$ (see [12] for more details).

Let (Ω, ω) be a Monge–Ampère structure with $\mathrm{HPf}(\omega) = \pm 1$. Since $d\omega = d\hat{\omega}$ $= 0$ if and only if $d\alpha = d\beta = 0$, we have the obvious proposition.

Proposition 3.17 *A generalised almost Calabi–Yau structure associated with a non-degenerate Monge–Ampere structure is integrable if and only this Monge–Ampère structure is closed.*

3.5.3.2 Nondegenerate Monge–Ampère Equations

Let us come back now to the differential equation associated with a non-degenerate Monge–Ampère structure (Ω, ω) on a 6-dimensional manifold X. It is natural to ask if this equation is locally symplectically equivalent to one of these:

$$\begin{cases} \mathrm{hess}\ f = 1 \\ \Delta(f) - \mathrm{hess}\ f = 0 \\ \Box(f) + \mathrm{hess}\ f = 0 \end{cases}$$

According to Table 3.2, it will be the case if and only if (Ω, ω) is locally constant. The following theorem gives a criterion using the generalised Calabi–Yau structure associated.

Theorem 3.9 *A Monge–Ampère equation associated with a non-degenerate Monge–Ampère structure can be reduced by a symplectic change of coordinates to one of the following equations:*

$$\begin{cases} \mathrm{hess}\ f = 1 \\ \Delta(f) - \mathrm{hess}\ f = 0 \\ \Box(f) + \mathrm{hess}\ f = 0 \end{cases}$$

if and only if the generalised Calabi–Yau structure associated is integrable and flat.

We refer to [12] for the proof. The idea is that the integrability condition implies the existence of a 'generalised' Kähler potential and the flat condition allows us to choose a Darboux coordinates system in which this potential has a nice expression.

Lychagin and Roubtsov have proved an equivalent theorem in [44] using technics of formal integrability. Theorem 3.9 is more restrictive since it only concerns non-degenerate Monge–Ampère equations but it is worth mentioning that its statement and its proof are much more simple and that it has a nice geometric meaning. We

Table 3.4 (pseudo) Calabi–Yau and elliptic Monge–Ampère structures

Almost (pseudo) CY	Elliptic MA
(pseudo) CY	Closed elliptic MA
Flat (pseudo) CY	Locally constant elliptic MA

sum up in Table 3.4 the correspondence between (pseudo) Calabi–Yau structures and elliptic Monge–Ampère structures.

3.5.4 Notes and Further Reading

The additional information about classification results for symplectic Monge–Ampère structures, operators and equations (as it was mentioned in the text of the lecture) one can find in our book [22], in the papers of B. Banos [6, 12] and R. Bryant-Bryant. Another approach to integrability of symplectic Monge–Ampère operators was proposed by E. Ferapontov and his collaborators [20, 21]. Their concept of integrability is based on the integrability of systems so-called hydrodynamical type. We should stress that the Monge–Ampère equations and operators integrable in the hydrodynamical systems sense of Ferapontov correspond to degenerate ('linear') orbits in Banos–Lychagin–Roubtsov classification table. In its turn, the Monge–Ampère operators which correspond to Calabi–Yau-like structures (from our Table 3.4) are not integrable in hydrodynamical systems sense. These two different integrability phenomena for Monge–Ampère operators and their relations should be a subject of further studies.

3.6 Lecture Five

In this lecture, we shall discuss few recently discovered examples of various geometric approaches to interesting and important applied model Monge–Ampère equation appeared in geophysical dynamics—Dritschel–Viudez Monge–Ampère equation. This equation appears to be an example of illustration of power and richness of geometric methods described in previous chapters. We follow here are papers with B. Banos and I. Roulstone [7, 8], and use our results of [47]. The end of the lecture is devoted to some examples of Special Kähler and Special Lagrangian Monge–Ampère structures and to B.Banos [11] of examples generalise complex geometry applications to Jacobi first-order systems which are reduced in some cases to Monge–Ampère equations.

In contrary to our notation principle, we do not use the standard 'phase space notations' (q, p) for T^*M local coordinates in this chapter and keeps the notations which was originally used in the initial sources of examples.

3.6.1 Bi-Lagrangian, Special Lagrangian, Special Kähler and Monge–Ampère Equations

In [36], Theorem 2.4, Hitchin characterises special (pseudo) Kähler manifolds as manifolds that can be (locally) identified as **bi-Lagrangian** submanifolds of $V \times V$, where V is a symplectic vector space.

This is motivated by the example of special Kähler subvarieties—an analogue of special Lagrangian subvarieties in string theory.

We want to develop here an approach to splitting structures using a notion of bi-symplectic structure on cotangent bundle T^*M which is also can be enabled with an almost-complex structure.

Proposition 3.18 *Consider a pair of real MAO* Δ_{ω_i}, $i = 1, 2$. *They are equivalent (3.33) if they have the same bi-Lagrangian generalised solutions.*

We remind that the Hitchin's result (3.2.2.3) can be reinterpreted in terms of MAO. We shall identify $(V; \Omega)$ with $(\mathbb{C}^n, \Omega)$ with Ω is the canonical Kähler form and the real cotangent bundle $T^*\mathbb{C}^n$ with $\mathbb{C}^n \oplus \mathbb{C}^n$ correspondingly by the isomorphism $(x; y; p; q) \to (x + iy; p - iq)$.

3.6.1.1 Special Lagrangian Manifolds and Symplectic MAO

$n = 2$. Hyperkähler case

Let S be a K3 surface with a fixed complex structure I and Kähler metric g. Then it is a hyperkähler manifold with possible choice of two other complex structures J and K for which g is again a Kähler metric among the S^1-family $J \cos \theta + K \sin \theta$. The period form of S is a holomorphic (in I) 2-form $\Omega = \omega_2 + i\omega_3$, where $\omega_2(., .) = g(J., .)$ and $\omega_3(., .) = g(K., .)$. Recall that an oriented 2-dimensional submanifold $\Sigma \subset S$ is special Lagrangian with respect to the fixed symplectic two-form $\omega_1(., .) = g(I., .)$ if it is ω_1-Lagrangian ($\omega_1|_\Sigma = 0$) and $\omega_3|_\Sigma = 0$.

It follows from the seminal paper [30] that this property of Σ is equivalent to be a *holomorphic curve* with respect to the complex structure K.

As an example will consider a 'toy model' for the surface S its 'noncompact' analogue T^*M.

Example

Let Σ be the $U(1)$-invariant special Lagrangian manifold of $\mathbb{C}^2$ (see [30]) given by the common level of
$$z\bar{z} - w\bar{w} = C_1; \text{Re}(zw) = C_2.$$

Then in the real coordinates we have
$$\Sigma = \left\{ x^2 + p^2 - (y^2 + q^2) = C_1, xy - pq = C_2 \right\}, z = x + ip, w = y + iq.$$

We will parametrize Σ like a 2-dimensional surface

$$x = u, \, p = v, \, y^2 + q^2 = C_1 - (u^2 + v^2), \, uy - vq = C_2$$

such that the differentials are

$$dx = du, \, dp = dv, \, dy = -\frac{uv + qy}{vy + qy} du + \frac{q^2 - v^2}{yv + qu} dv, \quad dq = -\frac{u^2 - y2}{uv + qu} du - \frac{vu + yq}{uv + qu} dv.$$

Now we can easily check that

$$\omega_K|_\Sigma = \omega_I|_\Sigma = 0,$$

i.e. the surface Σ is a special Lagrangian submanifold for Ω and K. Performing a partial Legendre transformation in the plane x, p obtain a *dual* surface Σ^* such that

$$\Sigma^* = \left\{ z\bar{z} - w\bar{w} = C_1; \, \mathrm{Im}(zw) = C_2 \right\}$$

and which is also $\Omega = \omega_I$-Lagrangian and gives a solution to MA operator given by ω_J:

$$\omega_J|_{\Sigma^*} = \omega_I|_{\Sigma^*} = 0.$$

Remark that both surfaces Σ and Σ^* are invariant with respect to full Legendre transformations.

The other interesting feature of this solution surfaces is that they are served as momentum-level surfaces for hyperkähler momentum map. Namely, the group of unit quaternions (which we are thinking of as $SU(2)$) acts by right and by left multiplications. The first action commutes with I, J, K and the second rotates the basic forms. Then, it is easy to write the hyperkähler moment map for the quaternionic action in our coordinates:

$$\mu_\mathbb{H} = 1/2 l i \bar{l} = i\mu_1 + j\mu_2 + k\mu_3 = i\left\{z\bar{z} - w\bar{w}\right\} + j \,\mathrm{Im}(zw) - k \,\mathrm{Re}(zw).$$

We can recognise easily in the common momentum-level surfaces $(\mu_1 = C_1) \cap (\mu_3 = C_3)$ and $(\mu_1 = C_1) \cap (\mu_2 = C_3)$ our MA generalised solutions Σ and Σ^*.

We should stress that there are also generalised solutions which are not special Lagrangian with respect to Ω. The interesting example of this sort is provided by the following *Seiberg–Witten curve* $\Xi \subset \mathbb{C}^2$ such that $\Xi = \left\{ w^2 - ch(z) + \zeta = 0 \right\}$.

It is a straightforward to check that $\omega_K|_\Xi = 0$ (it is a *minimal* surface) but $\Omega|_\Xi \neq 0$.

$n = 3$. Special Lagrangian flat case

Let

$$g = \sum_{i=1}^{n} dz^i \circ d\bar{z}^i, \quad \Theta = \frac{\sqrt{-1}}{2} \sum_{i=1}^{n} dz^i \wedge d\bar{z}^i, \quad \Upsilon = dz^1 \wedge \ldots \wedge dz^n$$

be the standard Kähler metric, the Kähler 2-form and the holomorphic volume form on $\mathbb{C}^n$. The special unitary group $SU(n)$ action leaves all three forms invariant. We shall consider the identification $\mathbb{C}^n = \mathbb{R}^{2n}$ via $z_k = x_k + \sqrt{-1}y_k$; $k = 1, \ldots n$.

The forms $\omega_1 = \operatorname{Re} \Upsilon$ and $\omega_2 = \operatorname{Im} \Upsilon$ are real n–forms on $\mathbb{R}^{2n}$.

Let $n = 3$. The form Θ transforms in the natural real symplectic form on $T^*\mathbb{R}^3 = \mathbb{R}^3 \oplus \mathbb{R}^3$ and $\Upsilon = \omega_1 + i\omega_2$ is the holomorphic 3-form on $\mathbb{C}^3$. Let $f : \mathbb{R}^3 \to \mathbb{R}$ be a smooth function and $df : \mathbb{R}^3 \to T^*\mathbb{R}^3$ be the corresponding section. The graph of df is special Lagrangian if and only if f is a solution of the following pair of $Sp(3)$–equivalent equations with Monge–Ampère equations with operators $\Delta_{\omega_{1,2}}$:

$$\Delta_{\omega_1}(f) = (1 - \operatorname{Hess}_{yz}(f) - \operatorname{Hess}_{xz}(f) - \operatorname{Hess}_{xy}(f)) = 0$$

and

$$\Delta_{\omega_2}(f) = \Delta(f) - \operatorname{Hess}(f) = 0.$$

The $Sp(3)$-equivalence of the equations $\Delta_{\omega_{1,2}}(f) = 0$ is given by the Fourier-Legendre transformation $(x, y, z, p, q, r) \to (p, q, r, -x, -y, -z)$ of the corresponding 3-forms $\omega_{1,2}$:

$$\omega_1 = dx \wedge dy \wedge dz - dx \wedge dq \wedge dr + dy \wedge dp \wedge dr - dp \wedge dq \wedge dz$$

and

$$\omega_2 = dp \wedge dq \wedge dr - dp \wedge dy \wedge dz + dx \wedge dq \wedge dz - dx \wedge dy \wedge dr.$$

3.6.2 2d and 3d Rotating Stratified Flows—Dritschel–Viudez Diagnostic MAEs

Recently, a new approach to modelling stably stratified geophysical flows was proposed in [22, 23]. This approach is based on the explicit conservation of potential vorticity and uses a change of variables from the usual primitive variables of velocity and density to the components of ageostrophic horizontal vorticity. This change results in a Monge–Ampère-like non-linear equation with non-constant coefficients. The equation gives the conditions for static and inertial stability and changes the type from elliptic to hyperbolic.

3.6.2.1 2d Diagnostic Dritchel–Viduez MAE

It is written as

$$E\left(\phi_{xx}\phi_{zz} - \phi_{xz}^2\right) + A\phi_{xx} + 2B\phi_{xz} + C\phi_{zz} + D = 0, \tag{3.86}$$

with

$$E = 1, \quad A = 1 + \theta xz, \quad B = \frac{1}{2}(\theta_{zz} - \theta_{xx})$$

$$C = 1 - \theta_{xz}, \quad D = \theta_{xx}\theta_{zz} - \theta_{xz}^2 - \varpi,$$

where θ is a given potential and the dimensionless PV anomaly ϖ may be also considered as a given quantity (see the details in [22]).

The corresponding Monge–Ampère structure is

$$\begin{cases} \Omega = dx \wedge dp + dz \wedge dr \\ \omega = E dp \wedge dr + A dp \wedge dz + B(dx \wedge dp - dz \wedge dr) + C dx \wedge dr + D dx \wedge dz. \end{cases}$$

The pfaffian is pf $\omega = R$ with R the Rellich's parameter:

$$R = AC - ED - B^2 = 1 + \varpi - \left(\frac{\Delta\theta}{2}\right)^2.$$

Moreover, a direct computation gives

$$d\omega = d\left(\frac{\Delta\theta}{2}\right) \wedge \Omega. \tag{3.87}$$

3.6.2.2 3d Rotating Stratified Flows—Dritschel–Viudez MAE

Let us consider the 3d- Dritschel–Viudez equation [22]

$$E\left(\Phi_{zz}\left(\Phi_{xx} + \Phi_{yy}\right) - \Phi_{xz}^2 - \Phi_{yz}^2\right) + A\left(\Phi_{xx} + \Phi_{yy}\right)$$
$$+ 2B_1\Phi_{xz} + 2B_2\Phi_{yz} + C\Phi_{zz} + D = 0$$

with

$$A = 1 + \Theta_z \qquad B_1 = \tfrac{1}{2}\Delta\varphi - \Theta_x \qquad B_2 = \tfrac{1}{2}\Delta\psi - \Theta_y$$

$$C = 1 - \Theta_z \quad D = \Theta_x\Delta\varphi + \Theta_y\Delta\psi - |\nabla\Theta|^2 - \varpi \qquad E = 1$$

where φ and ψ are potentials, $\Theta = \varphi_x + \psi_y$ and ϖ is the dimensionless PV anomaly (see details in [22]).

The associated primitive form is

$$\omega = E(dx \wedge dq \wedge dr - dy \wedge dp \wedge dr) + A(dy \wedge dz \wedge dp - dx \wedge dz \wedge dq)$$
$$+ B_1(dx \wedge dy \wedge dp + dy \wedge dz \wedge dr) + B_2(dx \wedge dy \wedge dq - dx \wedge dz \wedge dr)$$
$$+ Cdx \wedge dy \wedge dr + Ddx \wedge dy \wedge dz$$

We can check directly that, independently of the coefficients,

$$\lambda(\omega) = 0.$$

In other words, the Dritschel–Viudez equation is 'degenerate' (the underlying Monge–Ampère geometry is degenerated) and it is, at each point, equivalent to a linear equation.

Consider the *generalising Rellich's invariant* $R = AC - ED - B_1^2 - B_2^2$. As in the $2d$ case, this parameter separates elliptic and hyperbolic cases.

Indeed, the non-zero eigenvalues of g_ω are

$$\lambda_1 = 2R$$
$$\lambda_2 = S + \sqrt{S^2 - 4(1 + A^2)R}$$
$$\lambda_3 = S - \sqrt{S^2 - 4(1 + A^2)R}$$

with $S = R + 1 + A^2 + B_1^2 + B_2^2$. Therefore,

- if $R > 0$, then g_ω has signature $(3, 0)$ and $3d$-Dritschel–Viudez equation is elliptic equivalent in each point to $\Delta u = 0$;
- if $R < 0$, then g_ω has signature $(1, 2)$ and $3d$-Dritschel–Viudez equation is hyperbolic, equivalent in each point to $\Box u = 0$.

There is an interesting and open question: what is an invariant meaning of the $3d$ generalising Rellich's invariant $R = AC - ED - B_1^2 - B_2^2$. Remind that $2d$-Rellich's invariant $R = AC - ED - B^2$ of the Monge–Ampère equation $\Delta_{omega} = 0$ coincides with $Pf(\omega)$.

Those readers who are interested in additional geometric investigation of $2d$ and $3d$ diagnostic Dritchel–Viduez MAEs to address our paper [8] in this volume.

3.6.3 Generalised Complex Geometry and Monge–Ampère Structures

Proposition 3.19 *To any 2-dimensional symplectic Monge–Ampère equation of divergent type $\Delta_\omega = 0$ corresponds a Hitchin pair (ω, Ω) and therefore a 4-dimensional generalised complex structure.*

Remark 3.22 Let $L^2 \subset M^4$ be a 2-dimensional submanifold. Let $T_L \subset T$ be its tangent bundle and $T_L^0 \subset T^*$ its annihilator. L is a generalised complex submanifold (according to the terminology of [28]) or a generalised Lagrangian submanifold (according to the terminology of [14]) if $T_L \oplus T_L^0$ is closed under $\mathbb{J}$. When $\mathbb{J}$ is defined by (3.13), this is equivalent to saying that L is a Lagrangian with respect to Ω and closed under R, that is, L is a generalised solution of $\Delta_\omega = 0$.

3.6.3.1 2d—Dritschel–Viudez Equation and Underlying Generalised Complex Geometry

Relation (3.87) implies that $2d$- Dritschel–Viudez equation is of divergent type since

$$d(\omega + \lambda\Omega) = 0 \quad \text{with} \quad \lambda = -\frac{\Delta\varpi}{2}$$

and to any $2d$ Monge–Ampère equation $\Delta_\omega = 0$ of divergent type corresponds an integrable generalised complex structure $\mathbb{J}_\omega : M^4 \to \text{End}(TM \oplus T^*M)$ (see [12]). It is defined by

$$\mathbb{J}_\omega = \left(\begin{array}{c|c} A_\omega - \lambda & \Omega^{-1} \\ \hline 2\lambda\omega - (1 - R + \lambda^2)\Omega & \lambda - A_\omega^* \end{array} \right)$$

This geometry, introduced by N. Hitchin and M. Gualtieri, generalises both complex and symplectic geometries. It does not provide particular coordinates system but, int the context of Monge–Ampère equations, it gives informations on conservation law, which are a natural generalisation of first integral for differential equations.

A conservation law of $\Delta_\omega = 0$ is a 1-form $\alpha \in \Omega^1(M^4)$ such that $d\alpha|_L = 0$ on any generalised solution L.

The Hodge-Lepage-Lychagin theorem implies that α is a conservation law if and only if $d\alpha = f\omega + g\Omega$. The function f is called a generating function with conjugate g and

$$L = (f + ig)^{-1}(c) \text{ is a generalised solution of } \Delta_\omega = 0$$

It is proved in [12] that a function f is a generating function if and only it is pluriharmonic on $(M, \mathbb{J}_\omega)$, that is

$$\partial_\omega \bar{\partial}_\omega f = 0$$

Example

$f(x, z, p, r) = x$ is a generating function for $2d$-Dritschel–Viudez equation with conjugate function

$$g(x, z, p, r) = \varpi_x + z + r.$$

3.6.3.2 Generalised Complex Geometry and Jacobi Systems

Let us consider now an Hitchin pair of bivectors (π, Π) in dimension 4. Since Π is non-degenerate, it defines two 2-forms ω and Ω, which are not necessarily closed, and related by the tensor A. A generalised Lagrangian surface is a surface closed under A, or equivalently, bi-Lagrangian: $\omega|_L = \Omega|_L = 0$. Locally, L is defined by two functions u and v satisfying a first-order system

$$\begin{cases} a + b\frac{\partial u}{\partial x} + c\frac{\partial u}{\partial y} + d\frac{\partial v}{\partial x} + e\frac{\partial v}{\partial y} + f \det J_{u,v} = 0 \\ A + B\frac{\partial u}{\partial x} + C\frac{\partial u}{\partial y} + D\frac{\partial v}{\partial x} + E\frac{\partial v}{\partial y} + E \det J_{u,v=0} \end{cases}$$

with

$$J_{u,v} = \begin{pmatrix} \frac{\partial u}{\partial x} & \frac{\partial u}{\partial y} \\ \frac{\partial v}{\partial x} & \frac{\partial v}{\partial y} \end{pmatrix}$$

Such a system generalises both Monge–Ampère equations and Cauchy–Riemann systems and is called Jacobi system (see [22]).

With the help of Hitchin's formalism, we understand now the integrability condition (3.14) as a 'divergent type' condition for Jacobi equations.

3.6.4 Notes and Further Reading

The reference for $2d$ and $3d$ Dritchel–Viduez equation and the geophysical origins of the diagnostic problems I refer [22, 23]. The details of Generalised Complex geometry applications investigated and studied by B. Banos in [11].

Some other geometric methods in incompressible fluid dynamics can be found in [7, 46, 48, 49].

One of the most interesting questions surrounding the subject is to uncover the 'Higher' geometry that lies behind the Monge–Ampère structures. A proposal for a framework unifying the symplectic, multisymplectic and deformational methods in the Monge–Ampère structures as well as a huge variety of geophysical applications of these methods will be put forward in [45].

In a different direction, the relation between Lie and Courant algebroid structures arising within the Monge–Ampère was constructed in [41] and in the PhD thesis of P. Antunes (Ecole Polytechnique, 2010) [2].

References

1. A. Andrada, I.G. Dotti: Double products and hypersymplectic structures on $\mathbb{R}^{4n}$, Commun. Math. Phys. **262** (2006), 1–16.
2. P. Antunes, J. Nunes da Costa: Nijenhuis and compatible tensors on Lie and Courant algebroids, Journal of Geometry and Physics, 65 (2013) 66–79.
3. M. Audin: Lagrangian submanifolds, in "Symplectic geometry of integrable Hamiltonian systems", (Barcelona, 2001), pp. 1–83, Adv. Courses Math. CRM Barcelona, Birkhäuser, Basel, 2003.
4. G. Bande, P. Ghiggini, D. Kotschick: Stability theorems for symplectic and contact pairs, Int. Math. Res. Not. **2004:68** (2004), 3673–3688.
5. G. Bande, D. Kotschick: The geometry of symplectic pairs, Trans. Amer. Math. Soc. **358** (2006), 1643–1655.
6. B. Banos, Non-degenerate Monge-Ampère Structures in dimension 3, Letters in Mathematical Physics, 62 (2002) 1–15.
7. B. Banos, V. Roubtsov, I. Roulstone, Monge-Ampèrte structures and the geometry of incompressible flows, J. Phys. A, 49 (2016) 244003.
8. B. Banos, V. Roubtsov, I. Roulstone, On the geometry arising in some meteorological models in 2 and 3 dimension, This volume.
9. B. Banos, V. Rubtsov Géométries intégrables et équations aux dérivées partielles non linéaires, in preparation.
10. B. Banos, A. Swann: Potentials for hyper-Kähler metrics with torsion, Class. Quantum Grav. 21, 2004, 3127–3135.
11. B. Banos: Monge-Ampère equations and generalised complex geometry, the 2-dimensional case, Journal of Geometry and Physics.
12. B. Banos: On symplectic classification of effective 3-forms and Monge-Ampère equations, Diff. Geometry and its Applications, 19, (2003) 147–166.
13. J. Barrett, G.W. Gibbons, M.J. Perry, C.N. Pope, P. Ruback: Kleinian geometry and the $N = 2$ superstring, Int. J. Mod. Phys. **A4** (1994), 1457–1494.
14. O. Ben-Bassat, M. Boyarchenko, Submanifolds of generalised complex manifolds, J. Symplectic Geom. 2, 2004, No 3, 309–355.
15. R. Bryant: On the geometry almost-complex 6-manifolds, Asian J. Math. vol. 10, 2006, 561.
16. S. Chynoweth, M.J. Sewell: Dual variables in semigeostrophic theory, Proc. Roy. Soc. A, 424, (1989), 155.
17. M. Crainic, Generalized complex structures and Lie brackets, 2004, math.DG/0412097.
18. A.S. Dancer, H.R. Jørgensen, A. F. Swann, Metric geometries over the split quaternions, Rend. Sem. Mat. Univ. Pol. Torino **63** (2005), 119–139.
19. S.K. Donaldson: Two-forms on four-manifolds and elliptic equations, Preprint arXiv:math.DG/0607083 v1 4Jul2006.
20. B. Doubrov, B.E. Ferapontov, B. Kruglikov, V. Novikov: On a class of integrable systems of Monge-Ampère type, J. Math. Phys. 58 (2017), no. 6, 063508, 12 pp.
21. B. Doubrov, E. Ferapontov: On the integrability of symplectic Monge-Ampère equations J. Geom. Phys. 60 (2010), no. 10, 1604–1616.
22. D.G. Dritschel, A. Viudez: A balanced approach to modelling rotating stably-stratified geophysical flows, Journ. Fluid Mech. (2003).
23. D.G. Dritschel, A. Viudez: An explicit potential vorticity conserving approach to modelling nonlinear internal gravity waves, Journ. Fluid Mech. (2000).
24. A. Fino, H. Pedersen, Y.-S. Poon, M.W. Sørensen, Neutral Calabi–Yau structures on Kodaira manifolds, Commun. Math. Phys. **248** (2004), 255–268.
25. D. Freed: Special Kähler Manifolds, Comm. Math. Phys., vol. 203:1 (1999), p. 31–52.
26. H. Geiges, Symplectic couples on 4-manifolds, Duke Math. J **85** (1996), 701–711.
27. G. Grantcharov, Y.S. Poon: Geometry of hyper-Kähler connections with torsion Comm. Math. Phys. 213 , 2000, No 1, 19–37.

28. M. Gualtieri, Generalized complex geometry, 2004, math.DG/0401221.
29. M. Gualtieri: Generalized complex geometry, 2004, math.DG/0401221.
30. R. Harvey and H.B. Lawson, Calibrated geometries, Acta. Math.148, p. 47–157, (1982).
31. N.J. Hitchin, The self-duality equations on a Riemann surface, Proc. London Math. Soc. **55** (1987), 59–126.
32. N.J. Hitchin: Hypersymplectic quotients, Atti della Accademia delle Scienze di Torino, Classe di Scienze Fisiche, Matematiche e Naturali, 124 Supp. (1990), 169–180.
33. N.J. Hitchin: Metrics on moduli spaces, in "The Lefschetz centennial conference, Part I (Mexico City, 1984)", Contemp. Math., vol. 58, Amer. Math. Soc., Providence, R.I., (1986), 157–178.
34. N. Hitchin: The geometry of three-forms in six and seven dimensions, Journal of Differential Geometry, 56 (2001).
35. N.J. Hitchin: The Moduli Space of Complex Lagrangian Submanifolds, Asian J. Math. vol. 3:1, 1999, 77–92.
36. B. Hoskins: The geostrophic Momentum Approximation and the Semi-geostrophic Equations, Journal of the atm. sciences, vol 32, n° 2 (1975).
37. D.D. Joyce, Compact manifolds with special holonomy, Oxford Math. Monographs, Oxford University Press 2000.
38. A.A. Katanova, Explicit form of certain multivector invariants, Advances in Soviet mathematics, vol 8, 1992.
39. M.A. Knus: Quadratic and Hermitian forms over rings, Springer Verlag, Berlin, 1980.
40. M. Kontsevich, Y. Soibelman: Affine structures and non-Archimedean analytic spaces, in "The unity of mathematics", pp. 321–385, Progr. Math., 244, Birkhäuser Boston, Boston, MA, 2006.
41. Y. Kosmann-Schwarzbach, V. Rubtsov: Compatible structures on Lie algebroids and Monge-Amp'ere operators, Acta Appl. Math., 109 (2010), no. 1, 101–135.
42. D. Kotschick, Orientations and geometrisations of compact complex surfaces, Bull. London Math. Soc. **29** (1997), 145–149.
43. Lychagin V.V., Rubtsov V.N. and Chekalov I.V.: A classification of Monge-Ampère equations, Ann. scient. Ec. Norm. Sup., 4 ème série, t.26, 1993, 281–308.
44. V.V. Lychagin, V.N. Roubtsov and I.V. Chekalov, A classification of Monge-Ampère equations, Ann. scient. Ec. Norm. Sup., 4 ème série, t.26, 1993, 281–308.
45. J. McYorist, V. Roubtsov, I. Roulstone, M. Wolf, On the Monge-Amère Geometry of Incompressible Fluid Flows, in progress, 2018–2019.
46. V. Roubtsov, I. Roulstone: Examples of quaternionic and Kähler structures in Hamiltonian models of nearly geostrophic flow., J. Phys. A, 30 (1997), no. 4, L63–L68.
47. V. Roubtsov, I. Roulstone: Holomorphic structures in hydrodynamical models of nearly geostrophic flow, Proc. R. Soc. Lond. A, 457 (2001).
48. I. Roulstone, B. Banos, J.D. Gibbon, V. Roubtsov: A geometric interpretation of coherent structures in Navier-Stokes flows, Proc. R. Soc. A, 465 , (2009,) No 1, 2015.
49. I. Roulstone, B. Banos, J.D. Gibbon, V. Roubtsov: Kähler Geometry and Burgers? Vortices, 2009, Proceedings of Ukrainian National Academy Mathematics, 16 (2). pp. 303–321. (in honour of the 60th birthday of I. S. Krasil?schik.)
50. S. Salamon, Riemannian geometry and holonomy groups, Pitman Res. Notes, Longman 1989.
51. L. Schäfer, Géométrie tt^* et applications pluriharmoniques", DPhil Thesis, Nancy.
52. M.Stenzel: Ricci-flat metrics on the complexification of a compact rank one symmetric space., Manuscripta Math. 80 (1993), no. 2, 151–163.
53. F.J. Turiel, Classification locale simultanée de deux formes symplectiques compatibles, Manunscripta Math. **82** (1994), 349–362.
54. M. Verbitsky, A minicourse on Kähler geometry(Tel-Aviv University, December, 2010) Slides are here http://verbit.ru.
55. E.B. Winberg, V.V. Gorbatsevich and A.L. Onishchik, Lie groups and Lie algebras III, structure of Lie groups and Lie algebras Encyclopedia of mathematical sciences, vol 41, 1994.
56. Xu, Feng: Geometry of SU(3) manifolds. Thesis Ph.D., Duke University., 2008. 118 pp.
57. Xu, Feng: On instantons on nearly Kähler 6-manifolds, Asian J. Math. 13 (2009), no. 4, 535–567.

Chapter 4
Introduction to Symbolic Computations in Differential Geometry with Maple

Sergey N. Tychkov

4.1 Introduction

We discuss here computations in differential geometry with Maple. Our primary tool for that will be the package Differential Geometry (DG for short), which contains a lot of facilities to perform computations with vector fields, differential forms, tensors, Lie algebras etc.

We use a built-in version of the package, which is a bit out of date, though its more optimized and faster than the latest version. The latter can be found on the web page [1].

All Maple codes presented in these lectures is typed in the classic worksheet mode.

A comprehensive introduction to programming in Maple will not be given in these lectures. We discuss only the necessary functions briefly when needed.

However, at the end of each section, there are few exercises, which require knowledge of the Maple programming language. Nevertheless, we recommend to do them.

All commands begin after the standard Maple prompt >.

4.2 Basic Setup

The first command in all our programs will be the following:

```
> restart;
```

S. N. Tychkov (✉)
V. A. Trapeznikov Institute of Control Sciences of Russian Academy of Sciences,
65 Profsoyuznaya street, Moscow 117997, Russia
e-mail: sergey.lab06@gmail.com; sergey.lab06@yandex.ru

© Springer Nature Switzerland AG 2019
R. A. Kycia et al. (eds.), *Nonlinear PDEs, Their Geometry, and Applications*,
Tutorials, Schools, and Workshops in the Mathematical Sciences,
https://doi.org/10.1007/978-3-030-17031-8_4

Generally, it is a good idea to put command `restart` in the beginning of a worksheet, because it guarantees to achieve reproducible results on each rerun of a worksheet.

```
> with(DifferentialGeometry);
```

The command `with` imports functions and submodules from a given package (DifferentialGeometry in our case). After execution of this command, Maple shows the list of all imported objects. These objects are now accessible for immediate usage. If we do not want to import objects into the scope of our program, we use the following syntax `package[object]` or `package:-object` to access `object`.

Sometimes, it is convenient to hide the output of a command, i.e., the list of imported functions and modules, which can be rather long. In such case, we put the colon instead of the semicolon at the end of line.

In order to start computations with the Differential Geometry package, we should initialize DG by calling function `DGsetup`. It has many forms, but now we need only the simplest one.

The following command defines a manifold M with local coordinates x and y.

```
> DGsetup([x, y], M, verbose);
```
With the given parameters, it generates the output:

$$\textit{The following coordinates have been protected}:$$
$$[x, y]$$
$$\textit{The following vector fields have been defined and protected}:$$
$$[D\_x, D\_y]$$
$$\textit{The following differential } 1-\textit{forms have been defined and protected}:$$
$$[dx, dy]$$
$$\textit{frame name}: M$$

The first argument is the list of local coordinates. The second argument is a name of the manifold M (actually it is a name of a frame). The optional `verbose` argument asks DG to show additional info. The `quiet` option is also possible (it is set by default). The names x, y are now protected and cannot be assigned. After this command, the prompt changes to > M, where M indicates the current frame.

Since `DGsetup` protects names for coordinates, the following command fails.

```
M > x := 1;
```

Error, attempting to assign to 'x' which is protected. Try declaring 'local x'.

Unfortunately, the following commands do not raise an error. So, we should be careful and not to make such assignments.

```
M > dx := 0;
M > D_x := 0;
M > M := 0;
```

The symbols D_x, D_y denote the basis vectors ∂_x and ∂_y, respectively. And the symbols dx, dy denote the basis of the cotangent space T^*M. Please note that, we must use the function evalDG() to create a Differential Geometry object.

For example,

```
M > X := evalDG(D_x - D_y);
    R := evalDG(-y * D_x + x * D_y);
    S := evalDG(x * D_x + y * D_y);
```

$$X := D\_x - D\_y$$
$$R := -yD\_x + xD\_y$$
$$S := xD\_x + yD\_y$$

We can look into the internal representation of objects with the procedure lprint (linear print).

```
M > lprint(X);
```

$$\_DG([["vector", M, []], [[[1], 1], [[2], -1]]])$$

As we see, DG objects are implemented as nested lists with the prefix _DG. Let us try to subtract ∂_x from the vector X.

```
M > Z0 := X - D_x;
```

$$Z0 := D\_x - D\_y - D\_x$$

The result is not simplified and moreover, it is not calculated correctly.

```
M > lprint(Z0);
```

$$\_DG([["vector", M, []], [[[1], 1], [[2], -1]]]) - \_DG([["vector", M, []], [[[1], 1]]])$$

As lprint shows us that we did not get a proper DG object. To make DG perform operations correctly, the command evalDG() must be used.

```
M > Z1 := evalDG(X - D_x);
```

$$Z1 := -D\_y$$

On the other hand, the command

```
M > Y0 := X - X;
```

$$Y0 := 0$$

evaluates correctly (even without evalDG). But the rule of thumb is to use evalDG always.

Now, we define a covector field

```
M > omega1 := dx;
```

$$\omega 1 := dx$$

lprint shows that we created an object with type *form*.
```
M > lprint(omega1);
```

$$\_DG([["form", M, 1], [[[1], 1]]])$$

Of course, we can reuse names x and y for coordinates in the other frame.
```
M > DGsetup([x, y, z], E, quiet);
```

$$frame\ name : E$$

Suppose we try to use an assigned name as coordinate in DGsetup.
```
E > w := 1;
E > DGsetup([x, y, w], C, quiet);
```

$$Error,\ (in\ DifferentialGeometry : -DGsetup)\ invalid\ arguments$$

Error, because we have assigned a value to the variable w.

4.2.1 *Subpackage* Tools

Here, we discuss the subpackage Tools, which contains commands useful for the development our own functions or packages on top of the DG package.

The first command we discuss is DGinfo, which gives information about all kinds of DG objects. The functions have two parameters. The first one is a DG object and the second is a keyword string.

First, let us look at the ways of obtaining information about frames. Consider the following command.

```
E > frames := Tools:-DGinfo("FrameNames");
```

$$frames := [E, M]$$

This returns a list of frame names we have defined before. Note that in this case, only one parameter for the keyword is passed.

Get the current frame we work with.

```
E > currframe := DGinfo("CurrentFrame");
```

$$currframe := E$$

Get dimension of the base

```
E > DGinfo(M, "FrameBaseDimension");
```

$$2$$

Of course, commands may be combined

```
E > DGinfo(DGinfo("CurrentFrame"), "FrameBaseDimension");
```

$$3$$

Switching between frames is possible with the command ChangeFrame. It returns the name of the previous frame.

```
E > ChangeFrame(M);
```

$$E$$

We can remove the frame with the command RemoveFrame, which returns as a result in the number of frames left after removal.

```
M > RemoveFrame(M);
```

$$1$$

Many forms of DGinfo work without providing the frame. In this case, it retrieves information about the current frame. For example, keyword FrameBaseVectors obtains the list of basis vectors.

```
E > DGinfo("FrameBaseVectors");
```

$$[D\_x, D\_y, D\_z]$$

Command evalDG is not the only way to define a DG object. The command DGzip() is a convenient tool to create DG objects, especially, large ones. The first argument is a list of coefficients, the second one is a list of DG objects such as vectors, and the third is a name of the operation. The possible operation names are "plus", "wedge", and "tensor".

```
E > DGzip( [1, 2, 3], [dx, dy, dz], "plus");
    Omega := DGzip([dz, dy, dx], "wedge");
    T := DGzip([dx + dy, dz], "tensor");
```

$$dx + 2dy + 3dz$$
$$\Omega := -dx \wedge dy \wedge dz$$
$$T := dxdz + dydz$$

If we omit the last parameter DGzip will use "plus" by default.

Instead of enumerating a covector basis, we can use a command such as `DGinfo("FrameBaseForms")`.

```
E > alpha := DGzip( [f, g, h], DGinfo("FrameBaseForms"));
```

$$\alpha := f\,dx + g\,dy + h\,dz$$

There is also a function for creating DG objects with zero coordinates. To create a zero vector use `DGzero("vector")`, the keywords `"form"` and `"tensor"` are possible too.

`DGinfo` can obtain data from a particular DG object. We can get the *list* of coefficients of a given form, vector, or tensor using the keyword `Coefficient-List`.

For example, the command

```
E > DGinfo(alpha, "CoefficientList", [dx, dz]);
```

$$[f, h]$$

returns the list containing coefficients of the form α at the basis forms dx and dz.

If we need all coefficients we use the keyword `"all"` instead of a list.

```
E > DGinfo(V, "CoefficientList", "all");
```

$$[f, g, h]$$

In the conclusion of this section, we consider the function `GetComponents`, which calculates the coefficients of a DG object with respect to a list of DG objects of the same type.

Let X be a list of vector fields defined as follows.

```
E > X := evalDG([D_x+D_z, D_x-x**2*D_y, y**2*D_y]);
```

$$X := [D\_x + D\_z, \ D\_x - x^2 D\_y, \ y^2 D\_y]$$

Then, the components of a vector Y with respect to the vectors X[i] can be calculated with the command `GetComponents(Y, X)`.

For example,

```
E > GetComponents(evalDG((x**2-y)*D_x + y*D_z), X);
```

$$\left[y, \ x^2 - 2y, \ \frac{(x^2 - 2y)x^2}{y^2} \right]$$

Also, we can decompose a list of DG objects at once.

Consider a list of differential 2-forms f.

```
E > f := evalDG([ z * dx &w dy + x * dz &w dx,
x * dy &w dz - y * z * dx &w dz,
x * y * dz &w dx + z * dx &w dy   ]);
```

$$f := \left[dx \wedge dy, \; -y^2 dx \wedge dz + xdy \wedge dz, \; 7dx \wedge dy - z^3 dx \wedge dz\right]$$

The command

```
E > GetComponents(evalDG([dz &w dx, dx &w dy, dy &w dz]), f);
```

computes coordinates in the given basis of a 2-forms list

$$[[\frac{y}{z\,(y-1)}, 0, -\frac{1}{z\,(y-1)}],$$

$$[\frac{yz}{x^2\,(y-1)}, x^{-1}, -\frac{yz}{x^2\,(y-1)}],$$

$$[-\frac{1}{x\,(y-1)}, 0, \frac{1}{x\,(y-1)}]].$$

? Exercises

1. Implement a function Sum, which calculates the sum of two DG objects. (Hint: use DGinfo with the keyword CoefficientList.)
2. Write a function, which for a list of vectors X and a symbol a generates the vector a1*X[1]+...+an*X[n]. (Hint: the commands seq and add may be useful.)
3. Using commands DGzip, DGinfo implement a function hook(vec, covec), which applies a differential form covec to a vector vec.
4. Implement a function extDiff, which calculates the differential of a given function.
5. Extend the function extDiff from the previous exercise so it can calculate the exterior differential of a differential 1-form.
6. Write a function commut, which calculates the commutator of two given vector fields.
7. Using results of the previous and third exercise implement calculations of the Lie derivative of functions and 1-forms.
8. There is a function DGvolume, which generates a volume form, i.e., a form of maximal degree with a given expression as a coefficient. Using functions evalDG and DGinfo implement such function by yourself.

9. Write your version of the function `GetComponents`, (Hint: the built-in command `solve` can be useful.)

4.3 Calculations with Vectors and Forms

In this section, we discuss tools for computations with vector fields and differential 1-forms.

Let us define a three-dimensional manifold M,

```
DGsetup([x, y, z], M):
```

$$frame\ name : M$$

two lists of vector fields:

```
M > X := evalDG([ D_x, D_y, D_z,
                  x * D_y - y * D_x,
                  y * D_z - z * D_y,
                  z * D_x - x * D_z,
                  x * D_x + y * D_y + z * D_z ]);
      Y := evalDG([ (y + z)**2 * D_x,
                    (x + y)**2 * D_z,
                    (z + x)**2 * D_y ]);
```

$$X := [\, D\_x,\, D\_y,\, D\_z,\, xD\_y - yD\_x,$$
$$yD\_z - zD\_y,\, zD\_x - xD\_z,\, xD\_x + yD\_y + zD\_z\,]$$
$$Y := \left[\, (y+z)^2 D\_x,\, (x+y)^2 D\_z,\, (z+x)^2 D\_y \,\right],$$

and a list of differential 1-forms:

```
M > w := evalDG([ dz - y * dx, dy + u(x, y) * dz,
                  DGzip([a,b,c](x,y,z), [dx, dy, dz]) ]):
```

$$w := [\, -ydx + dz,\, dy + u(x, y)dz,\, a(x, y, z)dx + b(x, y, z)dy + c(x, y, z)dz \,]$$

It is worth to note the syntax we used to define a list of arbitrary functions. Namely, `[a,b,c](x,y,z)` is equivalent to `[a(x,y,z), b(x,y,z), c(x,y,z)]`.

The Lie bracket of two vector fields can be computed with the command `LieBracket`.

```
M > LieBracket(X[1], X[3]);
M > LieBracket(Y[1], Y[2]);
```

$$0 \, D\_x$$
$$-2(x + y)^2 (y + z) D\_x + 2(y + z)^2 (x + y) D\_z$$

Note that, a null vector is denoted by $0 \, D\_x$, not by 0.

The function `ExteriorDerivative` which calculates the exterior derivative of a differential form and a function.

For example,

```
M >  df  := ExteriorDerivative(f(x,y,z,p));
```

$$df := \left(\frac{\partial}{\partial x} f(x, y, z, p) \right) dx + \left(\frac{\partial}{\partial y} f(x, y, z, p) \right) dy + \left(\frac{\partial}{\partial z} f(x, y, z, p) \right) dz$$

Recall that a built-in function `map` transforms a list `[a1, ..., an]` with a given function `f` into the list `[f(a1), ..., f(an)]`.

The following command calculates the exterior derivative of a list of forms `w`.

```
M >  h  := map(ExteriorDerivative, w);
```

$$h := \left[dx \wedge dy, \ \left(\frac{\partial}{\partial x} u(x, y) \right) dx \wedge dz + \left(\frac{\partial}{\partial y} u(x, y) \right) dy \wedge dz, \right.$$

$$\left(\frac{\partial}{\partial x} b(x, y, z) - \frac{\partial}{\partial y} a(x, y, z) \right) dx \wedge dy +$$

$$\left(\frac{\partial}{\partial x} c(x, y, z) - \frac{\partial}{\partial z} a(x, y, z) \right) dx \wedge dz +$$

$$\left. \left(\frac{\partial}{\partial y} c(x, y, z) - \frac{\partial}{\partial z} b(x, y, z) \right) dy \wedge dz \right]$$

We obtained a list of 2-forms, which we can differentiate in the same way as we did before and get zeroes.

```
M >  h2  := map(ExteriorDerivative, h);
```

$$h2 := [0 \, dx \wedge dy \wedge dz, \ 0 \, dx \wedge dy \wedge dz, \ 0 \, dx \wedge dy \wedge dz]$$

Now, we are going to study how to define our own functions on the top of the DG package. Let us define our own function, which maps a vector field X to the commutator $[X, a(x)\partial_x + b(y)\partial_y + c(z)\partial_z]$, where a, b, and c are arbitrary functions.

```
M >  customLieBracket  :=
X -> LieBracket(X, a * D_x + b * D_y + c * D_z):
```

We called our function `customLieBracket`. Let us check how it works.

```
M > customLieBracket(X[1]);
M > customLieBracket(X[2]);
M > customLieBracket(X[7]);
```

$$0\,D\_x$$
$$0\,D\_x$$
$$-a\,D\_x - b\,D\_y - c\,D\_z$$

Now instead of calling our function with each element of X, we can find all commutators with the list X at once using the command map.

```
M > X0 := map(customLieBracket, X);
```

$$X0 := [0D\_x,\ 0D\_x,\ 0D\_x,\ bD\_x - aD\_y,$$

$$cD\_y - bD\_z,\ -cD\_x + aD\_z,\ -aD\_x - bD\_y - cD\_z]$$

There are many null vectors in the list Y. So let us make a function that filters out zeroes from a list of vectors. We call this function allButZero and here is the definition.

```
M > allButZero := T -> remove(X ->
Tools:-DGequal( [Tools:-DGzero("vector")], [X] ), T):
```

Applying function allButZero to the list X0, we should obtain a list with null vectors.

```
M > allButZero(X0);
```

$$[bD\_x - aD\_y,\ cD\_y - bD\_z,\ -cD\_x + aD\_z,\ -aD\_x - bD\_y - cD\_z]$$

Now, let us develop a function that computes all pairwise Lie brackets of a given list of vector fields. For the obvious reasons, it is called derive0.

```
M > derive0 := V -> ([ seq(seq(LieBracket(V[k], V[j]),
j = 1..k-1), k = 1..nops(V)) ]):
```

Test it on different input data.

```
M > derive0([]);
M > derive0([D_x]);
M > derive0([D_x, D_y]);
M > derive0(X);
```

$$[]$$
$$[]$$
$$[0D\_x]$$
$$[0D\_x,\ 0D\_x,\ 0D\_x,\ -D\_y,\ D\_x,\ 0D\_x,\ 0D\_x,\ -D\_z,\ D\_y,\ zD\_x-$$
$$xD\_z,\ D\_z,\ 0D\_x,\ -D\_x,\ zD\_y-yD\_z,\ -yD\_x+$$
$$xD\_y,\ -D\_x,\ -D\_y,\ -D\_z,\ 0D\_x,\ 0D\_x,\ 0D\_x]$$

? Exercises

Explain, why the second call gets an empty list and not a one element list with zero in it.

Now using `allButZero`, we can implement a slightly better version of the function `derive`. The new version is called `derive1`.

```
M > derive1 := A -> allButZero(derive0(A)):
```

Let us see how it works.

```
M > derive1(evalDG([D_x, D_y, x * D_x + y * D_y]));
M > derive1(X);
```

$$[-D\_x,\ -D\_y]$$

$$[-D\_y,\ D\_x,\ -D\_z,\ D\_y,\ zD\_x-xD\_z,$$

$$D\_z,\ -D\_x,\ zD\_y-yD\_z,\ -yD\_x+xD\_y,\ -D\_x,\ -D\_y,\ -D\_z]$$

Clearly, the function `derive1` works better than `derive0`, but it has an obvious shortcoming. Its result is not necessarily a list of linearly independent vectors. To overcome this difficulty, we are going to use the function `DGbasis`, which for a list of vectors L calculates a sublist S such that $span(S) = span(L)$ and vectors `S[1],...,S[n]` are linearly independent. Also we add the option `method="real"`, which asks `DGbasis` to consider vectors over real numbers.

```
M > DGbasis(derive1(X), method = "real");
```

$$[-D\_y,\ D\_x,\ -D\_z,\ zD\_x-xD\_z,\ zD\_y-yD\_z,\ -yD\_x+xD\_y]$$

Please note that, `DGbasis` will produce a drastically different result without this option provided.

```
M > DGbasis(derive1(X));
```

$$[-D\_y,\ D\_x,\ -D\_z]$$

Now, we are ready to write the final version of the `derive` function.

```
M > derive := A -> DGbasis(derive1(A), method = "real"):
```

Once again test it.

```
M > derive(X);
```

$$[-D\_y, \ D\_x, \ -D\_z, \ zD\_x - xD\_z, \ zD\_y - yD\_z, \ -yD\_x + xD\_y]$$

Now, we can apply it twice to the list X to get the second derivative of the Lie algebra generated by the vector fields X.

```
M > derive(derive(X));
```

$$[D\_z, \ D\_x, \ D\_y, \ -yD\_x + xD\_y, \ -zD\_y + yD\_z, \ zD\_x - xD\_z]$$

Now, we should make the next step and implement a function that applies `derive` on a list T of vectors n times (n > 0).

The idea is to use the obvious definition:

```
deriveN(T, n)  =  derive(T), if n = 1
deriveN(T, n)  =  deriveN(derive(T), n - 1), otherwise.
```

Hence, implementation is straightforward

```
M > deriveN := (T, n) ->
'if'(n = 1, derive(T), deriveN(derive(T), n - 1));
```

Let us see how it works

```
M > deriveN(evalDG([D_x, D_y]), 1);
M > deriveN(X, 2);
M > deriveN(X, 20);
```
$$[]$$

$$[D\_z, \ D\_x, \ D\_y, -y \, D\_x + x \, D\_y, -z \, D\_y + y \, D\_z, z \, D\_x - x \, D\_z]$$

$$[D\_z, \ D\_x, \ D\_y, -y \, D\_x + x \, D\_y, -z \, D\_y + y \, D\_z, z \, D\_x - x \, D\_z]$$

Note that, the third and second outputs are identical.

Now, recall that the Lie derivative of a function f along the vector field X can be calculated as follows $X(f) = \sum_{i=1}^{n} X^i f_{x^i}$. At first, we consider a naive implementation called `LieDer`.

Also, we import the package `Tools` for convenience.

```
M > with(Tools):
M > LieDer:= (X, f) ->foldl('+', 0, op( zip('*',
DGinfo(X, "CoefficientList", DGinfo("FrameBaseVectors")),
```

```
map( v -> diff(f, v),DGinfo("FrameIndependentVariables")))));
```

To make this implementation more clear, we should recall how the functions `foldl`, `op` and `zip` work. `foldl` has an arbitrary number of arguments, but not less then three. Given arguments `f,x0,...,xn`, it evaluates `f(f...(f(x0, x1)...)`, `xn)`. The command `op` transforms any list, set, or expression into the sequence. For example, `op([a,b,c])` will give `a,b,c`. `zip` does "gluing" of two lists into one using a given binary function. For example, `zip(f, [a,b,c], [x,y,z])` will produce `[f(a,x), f(b,y), f(c,z)]`.

Note that, `LieDer` is written in a quite universal way. It uses the command `DGinfo` to obtain all the needed data, so it does not depend on an actual frame.

Let us test how it works.

```
M > LieDer(evalDG(D_x+D_y+D_z), z*y*x);
```

$$yz + zx + zy$$

The command `Hook` calculates the interior product of a vector (or a list of vectors) and a differential form.

For example, calculate $dx(\partial_x)$ and $dx(\partial_y)$

```
M > Hook( D_x, dx );
M > Hook( D_y, dx );
```

$$1$$
$$0$$

Let us look at a more general example. Calculate the value of a form $f_1 dx + f_2 dy + f_3 dz$ on a vector $a_1 \partial_x + a_2 \partial_y + a_3 \partial_z$

```
Hook( DGzip([a1, a2, a3], [D_x, D_y, D_z]),
DGzip([f1, f2, f3], [dx, dy, dz]) );
```

$$a1\, f1 + a2\, f2 + a3\, f3$$

The command `Hook` can calculate the internal product of a vector with a 2-form and a 3-form.

```
Hook(evalDG(a*D_x + b*D_y), evalDG(dx &w dy));
Hook(evalDG(a*D_x + b*D_y + c*D_z), evalDG(dx &w dy &w dz));
```

$$-b\,dx + a\,dy$$
$$c\,dx \wedge dy - b\,dx \wedge dz + a\,dy \wedge dz$$

It also possible to insert a list of vectors into a differential form.

```
Hook(evalDG([a*D_x, b*D_y, c*D_z]), DGvolume("form", 1, M));
```

The latter command is equivalent to the following:

```
Hook(evalDG(c*D_z), Hook(evalDG(b*D_y),
Hook(evalDG(a*D_x), DGvolume("form", 1, M))));
```

The result is the same abc in both cases, of course.

After acquainting with the command `Hook`, we are able to write a function for computing of the Lie derivative of a 1-form. We will use Cartan's formula $L_X(\omega) = \iota_X(d\omega) + d(\iota_X f)$. We implement this formula as function `LieDerForm`.

```
LieDerForm := (X, f) ->
evalDG(Hook(X, ExteriorDerivative(f)) +
ExteriorDerivative(Hook(X, f)));
```

Note that, we have to use `evalDG` because of the addition.

A simple test:

```
LieDerForm(evalDG(x*D_x+y*D_y),evalDG(f(x,y)*dy+g(y,z)*dz));
```

$$\left(\frac{\partial}{\partial y}f(x,y)\right)y + f(x,y) + \left(\frac{\partial}{\partial x}f(x,y)\right)x\,dy + \left(\frac{\partial}{\partial y}g(y,z)\right)y\,dz$$

The DG package has its own function `LieDerivative` for calculating Lie derivatives of all kinds of tensors.

```
LieDerivative(x * D_x + y * D_y, f(x,y) * dy + g(y,z) * dz);
```

This instruction produces the same result as our function `LieDerForm`.

As an example, let us apply all vectors from the list X to an arbitrary n-form ($n = 3$ in our case).

```
map(T->LieDerivative(T, DGvolume("form", f(x,y,z), M)), X);
```

$$[\frac{\partial}{\partial x}f\,dx \wedge dy \wedge dz, \frac{\partial}{\partial y}f\,dx \wedge dy \wedge dz,$$

$$\frac{\partial}{\partial z}f\,dx \wedge dy \wedge dz, (-y\frac{\partial}{\partial x}f + x\frac{\partial}{\partial y}f)\,dx \wedge dy \wedge dz,$$

$$(-z\frac{\partial}{\partial y}f + y\frac{\partial}{\partial z}f)\,dx \wedge dy \wedge dz,$$

$$(z\frac{\partial}{\partial x}f - x\frac{\partial}{\partial z}f)\,dx \wedge dy \wedge dz,$$

$$(3f + x\frac{\partial}{\partial x}f + y\frac{\partial}{\partial y}f + z\frac{\partial}{\partial z}f)\,dx \wedge dy \wedge dz]$$

Define the metric tensor

```
g0 := evalDG(dx &s dx + dy &s dy + dz &s dz):
```
Note that to define symmetric product, we use symbol &s.
Hook can be used to calculate the scalar product of two vectors.
```
Hook(evalDG([a*D_x+b*D_z, c*D_y + d*D_z]), g0);
```

$$d\, b$$

4.3.1 Computing Symmetries

Consider the following problem. Let M be a spherical layer $M = S^2 \times \mathbb{R}$ with the metric $g = dx^2 + sin^2x\, dy^2 + dz^2$. We want to find conformal symmetries of this metric, i.e., such vector fields T that $L_T g = \lambda\, g$ for some function λ.

The first thing we should do is to define the metric.

```
M> gs := evalDG(dx &s dx + sin(x)**2 * dy &s dy + dz &s dz):
```

And an unknown vector field and the Lie derivative of the metric along with it.

```
M> T  := DGzip([a, b, c](x,y,z), [D_x, D_y, D_z]):
M> LD := LieDerivative(T, gs):
```

Then, the condition we want to be satisfied is of the form

```
M> cond := evalDG(LD - lambda(x,y,z) * gs);
```

Now, we must extract coefficients from the expression cond. We should use the command DGinfo with the keyword "CoefficientSet" because we do not want identical coefficients.

```
M> condsys := DGinfo(cond, "CoefficientSet", "all");
```

Thus, we obtained the system of six equations

$$condsys := \left\{ \sin(x)\left(2\,\frac{\partial b}{\partial y}\sin x - \lambda \sin x + 2\,a \cos x\right), \right.$$

$$\frac{\partial}{\partial x}c + \frac{\partial}{\partial z}a,\ -\lambda + 2\,\frac{\partial}{\partial x}a,\ -\lambda + 2\,\frac{\partial}{\partial z}c,\ -\frac{\partial b}{\partial x}\cos^2 x + \frac{\partial}{\partial y}a + \frac{\partial}{\partial x}b,$$

$$\left. -\frac{\partial b}{\partial z}\cos^2 x + \frac{\partial}{\partial y}c + \frac{\partial}{\partial z}b \right\}$$

This is an overdetermined PDE system, and if we try to solve it with the command pdsolve:

```
M > pdsolve(condsys, [a,b,c](x,y,z));
```

we get the message that this system is inconsistent. Unfortunately, pdsolve is wrong, because in many cases, it cannot solve systems with a functional parameter. To avoid this obstacle, we have to directly eliminate λ from the system.

```
M > lambdarestr := solve(condsys[1], lambda(x,y,z));
condsys2 := eval(condsys, lambdarestr);
```

$$condsys2 := \left\{ 0, \frac{\left(\frac{\partial}{\partial y}b(x, y, z)\right)\sin(x) + a(x, y, z)\cos(x)}{\sin(x)} - \frac{\partial}{\partial x}a(x, y, z), \right.$$

$$-2\frac{\left(\frac{\partial}{\partial y}b(x, y, z)\right)\sin(x) + a(x, y, z)\cos(x)}{\sin(x)} + 2\frac{\partial}{\partial z}c(x, y, z),$$

$$\frac{\partial}{\partial x}c(x, y, z) + \frac{\partial}{\partial z}a(x, y, z), -\left(\frac{\partial}{\partial x}b(x, y, z)\right)(\cos(x))^2 +$$

$$\frac{\partial}{\partial y}a(x, y, z) + \frac{\partial}{\partial x}b(x, y, z), -\left(\frac{\partial}{\partial z}b(x, y, z)\right)(\cos(x))^2 +$$

$$\left. \frac{\partial}{\partial y}c(x, y, z) + \frac{\partial}{\partial z}b(x, y, z)\right\}$$

Now, the solution is easy to obtain.

```
M > solution := pdsolve(op(2..6, condsys2));
```

$$solution := \left\{ a(x, y, z) = (\_C6\sin x + (\_C7\sin y + \_C8\cos y)\cos x)\,e^{-z} + \right.$$

$$e^z(\_C4\sin y + \_C5\cos y)\cos x + e^z\sin x\_C3 + \_C2\cos y + \_C1\sin y,$$

$$b(x, y, z) = \frac{\_C4\,e^z\cos y - e^{-z}\sin y\_C8 + e^{-z}\cos y\_C7 - e^z\sin y\_C5}{\sin x} +$$

$$\frac{\_C1\cos y - \_C2\sin y}{\tan x} + \_C9, c(x, y, z) = ((\_C7\sin y + \_C8\cos y)\sin x$$

$$-\_C6\cos x)\,e^{-z} - e^z(\_C4\sin y + \_C5\cos y)\sin x + e^z\cos x\_C3 + \_C10\right\}$$

Here, $\_C1, \ldots, \_C10$ are arbitrary constants. Note that ,pdsolve uses prefix '$\_$' to distinguish them from user-defined names. Thus, we obtained a ten-dimensional Lie algebra of symmetries.

Now, we can compute the infinitesimal symmetry.

```
Tsol := simplify(eval(T, solution));
```

Let us check if we got the conformal symmetry we looked for.

```
M > test := evalDG(LieDerivative(Tsol, gs)-lambda(x,y,z)*gs);
M > testCoeffs := DGinfo(test, "CoefficientList", "all");
M > simplify(eval(testCoeffs, eval(lambdarestr, solution)));
```

$$[0, 0, 0]$$

We omitted the output of all but the last command. The obtained list of zeroes means that we have solved the PDE system correctly.

? Exercises

1. One can note that our implementation of deriveN will not work correctly if we pass 0 as the second argument. Fix the implementation of the function so it can process 0 correctly. Then, rewrite this function so it can deal with any possible expressions as parameters. (Hint: use the built-in commands error and type.)

2. If we try to run the command deriveN(T, 200), it will take a considerable amount of time to finish calculations. Propose methods to improve the performance of deriveN.

3. Given a list of vectors T = [T1, ..., Tn], which form a Lie algebra, write a function that tests if the Lie algebra is nilpotent. (Hint: the Lie algebra is not solvable if nops(deriveN(T, k)) = nops(deriveN(T, k + 1)) for some k > 0.

4. Write your own implementation (based on LieDer) of ExteriorDerivative, which works only with functions. Try to make it as short as possible.

5. Implement the function LieDer in a more natural way. (Hint: use Hook.)

6. Try to write a function, which can determine if a given list of vector fields forms a Lie algebra.

4.4 Transformations

In this section, we discuss such DG objects as transformations. We begin with a simple example of transformations defined by the vector fields.

Define a coordinate frame E2 on a two-dimensional plane.

```
> with(DifferentialGeometry):
> DGsetup([x, y], E2);
```

$$frame\ name : E2$$

Consider the following vector fields on the 2D plane.

```
E2 > S := evalDG( x * D_x + y * D_y );
     R := evalDG( x * D_y - y * D_x );
```

$$S := xD\_x + yD\_y$$
$$R := -yD\_x + xD\_y$$

With the command `Flow`, we can calculate the one-parameter group of diffeomorphisms of a vector field. The first argument of `Flow` is a vector field, the second is a symbol denoting parameter of the transformation.

```
E2 > Sc := Flow(S, t);
     Rt := Flow(R, t);
```

$$Sc := \left[x = x\,e^t,\ y = y\,e^t\right]$$
$$R := [x = -y\sin(t) + x\cos(t),\ y = y\cos(t) + x\sin(t)]$$

The command `Flow` creates a DG object of the transformation type. Though the result can look like a list of equalities, it has a complex internal structure. Let us look into this object with `lprint`.

```
E2 > lprint(Sc);
```

$\_DG([["transformation", [[E2, 0], [E2, 0]], [], [Matrix(2, 2,$
$(1, 1) = \exp(t), (2, 2) = \exp(t), datatype = anything, storage$
$= rectangular, order = Fortran\_order, shape = [])]],$
$[[x * \exp(t), x], [y * \exp(t), y]]])$

We see that `Sc` is a DG object of the type `"transformation"`, which present a mapping from the manifold E2 into itself (Fig. 4.1).

We can calculate the image of a point under the action of the transformation. with the command `ApplyTransformation` (Fig. 4.2).

Consider examples:

1. rotate the point $(1, 1)$ on the angle $\frac{\pi}{4}$,

```
E2 > eval(ApplyTransformation(Rt, [1, 1]), t=Pi/4);
```

Fig. 4.1 Trajectories of the vector field `Sc`

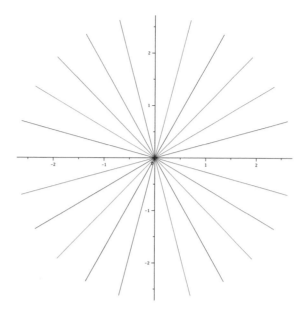

Fig. 4.2 Trajectories of the vector field `Rt`

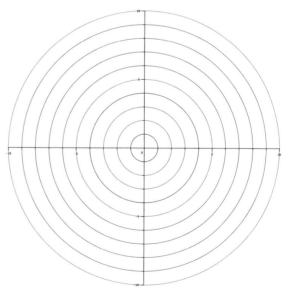

$$\left[0, \sqrt{2}\right]$$

2. scale the point $(0, 0)$

```
E2 > ApplyTransformation(Sc, [0, 0]);
```

$$[0, 0]$$

3. scale the point $(1, 1)$

```
E2 > ApplyTransformation(Sc, [1, 1]);
```

$$\left[e^t, e^t\right]$$

With `ApplyTransformation`, we are able to plot trajectories of vector fields. Recall that a call `map2(f, a, [a1,..., an])` evaluates to the list `[f(a,a1),..., f(a,an)]`

```
E2 > trajSc := map2(ApplyTransformation, Sc,
[seq([cos(p*2*Pi/24), sin(p*2*Pi/24)], p=0..23)]):
E2 > plot( map( tr->[op(tr),t=-infinity..1], trajSc) );

E2 > trajRt := map2(ApplyTransformation, Rt,
[seq([p, 0], p=0..10)]):
E2 > plot(map(tr->[op(tr), t=0..2*Pi], trajRt));
```

Define a vector field X as a linear combination of the vector fields S and R.

```
E2 > X := evalDG(S/3 + R);
```

$$X := \left(\frac{x}{3} - y\right) D\_x + \left(x + \frac{y}{3}\right) D\_y$$

Calculate a flow of X.

```
E2 > Xt := Flow(X, t);
```

$$Xt := \left[x = e^{t/3} \left(-y \sin(t) + x \cos(t)\right), y = -e^{t/3} \left(-y \cos(t) - x \sin(t)\right) \right]$$

```
E2 > trajXt := map2(ApplyTransformation, Xt,
[seq([0, p], p=0..10)]):
E2 > plot(map(tr->[op(tr), t=0..7], trajXt));
```

There is a command `InfinitesimalTransformation`, which is inverse to the function `Flow`. With it is possible to compute the Lie algebra of infinitesimal

Fig. 4.3 Trajectories of the linear combination of `Rt` and `Sc`

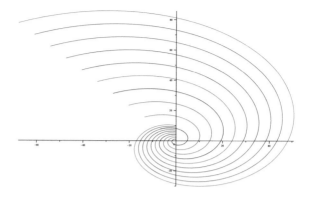

generators for an action of a Lie group on a manifold. The first argument is a transformation object, the second one is a list of symbols representing parameters of the transformation (Fig. 4.3).

```
E2 >  InfinitesimalTransformation(Xt, [t]);
```

$$X := \left(\frac{x}{3} - y\right) D\_x + \left(x + \frac{y}{3}\right) D\_y$$

```
E2 >  Phi  := Transformation(E2, E2, [ x=exp(t)*
(cos(theta)*x - sin(theta)*y) + a, y=exp(s)*(sin(theta)
*x + cos(theta)*y) + b ]);
```

$$\Phi = \left[x = e^{t} \left(\cos\left(\theta\right) x - \sin\left(\theta\right) y\right) + a, \; y = e^{s} \left(\sin\left(\theta\right) x + \cos\left(\theta\right) y\right) + b \right]$$

4.4.1 Operations on Transformations

The same result can be achieved with the function `ComposeTransformations`. This command creates the composition of an arbitrary number of transformations given as arguments. Suppose that $A : M_1 \rightarrow M_2$, $B : M_2 \rightarrow M_3$, then `Compose Transformations(B, A)` creates a DG object representing the transformation $B \circ A : M_1 \rightarrow M_3$.

Let us define five transformations `Phi1,...,Phi5`.

```
E2 > Phi1 := Transformation(E2, E2, [x = x + a, y = y]):
E2 > Phi2 := Transformation(E2, E2, [x = x, y = y + b]):
E2 > Phi3 := Transformation(E2, E2, [x = x * exp(t), y = y]):
E2 > Phi4 := Transformation(E2, E2, [x = x, y = y * exp(s)]):
E2 > Phi5 := Transformation(E2, E2, [x = cos(theta)
* x - sin(theta) * y, y = sin(theta) * x + cos(theta) * y]):
```

With the following instructions, we create the transformation Phi0 : Phi1 o Phi2 o Phi3 o Phi4 o Phi5

```
E2 > Phi0 := ComposeTransformations(Phi1, Phi2, Phi3,
Phi4, Phi5);
```

$$\Phi 0 = \left[x = e^t \left(\cos(\theta)x - \sin(\theta)y \right) + a, \ y = e^s \left(\sin(\theta)x + \cos(\theta)y \right) + b \right]$$

The equivalence of transformations can be verified with the command DGequal.

```
E2 > Tools:-DGequal(Phi, Phi0);
```

true

Again with InfinitesimalTransformation, we obtain the infinitesimal generators of the corresponding Lie algebra. But this time, we have to pass a list of all five parameters.

```
E2 > InfinitesimalTransformation( Phi, [a, b, t, s,
theta] );
```

$$[D\_x, D\_y, xD\_x, yD\_y, -yD\_x + xD\_y]$$

Now to explore other capabilities of DG regarding pushing and pulling all kinds of tensors, we need to define another 2D manifold S2 with coordinates u and v.

```
E2 > DGsetup([u, v], S2);
```

Also, we define the following transformation E2 → S2, which is called stereographic projection.

```
S2 > Sp := Transformation(E2, S2, [
u = 2 * x / (1 + x**2 + y**2),
v = 2 * y / (1 + x**2 + y**2) ]);
```

$$Sp := \left[u = \frac{2x}{x^2 + y^2 + 1}, \ v = \frac{2y}{x^2 + y^2 + 1} \right]$$

For the further discussion, we need to compute the inverse transformation of Sp, which can be done with the command InverseTransformation.

```
S2 > Ps := convert(InverseTransformation(Sp), radical);
```

$$Ps := \left[x = \frac{u \left(1 + \sqrt{-u^2 - v^2 + 1} \right)}{u^2 + v^2}, \ y = \frac{\left(1 + \sqrt{-u^2 - v^2 + 1} \right) v}{u^2 + v^2} \right]$$

Note that, we use the command `convert` with the option `radical` to suppress appearance of `RootOf` expressions.

Now check if the inverse transformation was computed correctly by composing it with the original transformation.

```
S2 > simplify( ComposeTransformations(Ps, Sp), symbolic );
```

$$[x = x, \ y = y]$$

```
S2 > simplify(ComposeTransformations(Sp, Ps));
```

$$[u = u, \ v = v]$$

Instead of using the commands `lprint` or `op`, the properties of transformations can be obtained with the command `DGinfo`.

For example, `DGinfo` with the keyword `"JacobianMatrix"` returns th

```
S2 > with(Tools):
S2 > SpJcbn := simplify(DGinfo(Sp, "JacobianMatrix"));
```

$$SpJcbn := \begin{bmatrix} \dfrac{-2x^2+2y^2+2}{\left(x^2+y^2+1\right)^2} & -\dfrac{4xy}{\left(x^2+y^2+1\right)^2} \\[4mm] -\dfrac{4xy}{\left(x^2+y^2+1\right)^2} & \dfrac{2x^2-2y^2+2}{\left(x^2+y^2+1\right)^2} \end{bmatrix}$$

```
S2 > PsJcbn := simplify(DGinfo(Ps, "JacobianMatrix"));
```

$$PsJcbn := \begin{bmatrix} -\dfrac{\left(1+\sqrt{-u^2-v^2+1}\right)\left(-\sqrt{-u^2-v^2+1}\,v^2+u^2\right)}{\left(u^2+v^2\right)^2\sqrt{-u^2-v^2+1}} & \dfrac{uv\left(1+\sqrt{-u^2-v^2+1}\right)^2}{\left(u^2+v^2\right)^2\sqrt{-u^2-v^2+1}} \\[6mm] -\dfrac{uv\left(1+\sqrt{-u^2-v^2+1}\right)^2}{\left(u^2+v^2\right)^2\sqrt{-u^2-v^2+1}} & \dfrac{\left(1+\sqrt{-u^2-v^2+1}\right)\left(\sqrt{-u^2-v^2+1}\,u^2-v^2\right)}{\left(u^2+v^2\right)^2\sqrt{-u^2-v^2+1}} \end{bmatrix}$$

Now, let us verify that the matrices PsJbcn and SpJcbn are mutually inverse. But at first, we must write both matrices using the same coordinates. This can be achieved with the command `Pullback`, which is used to pullback differential forms. Recall that matrix multiplication in Maple is denoted by ".".

```
S2 > simplify(Pullback(Sp, PsJcbn).SpJcbn, symbolic);
```

$$\begin{bmatrix} 1 & 0 \\ 0 & 1 \end{bmatrix}$$

Here, we used the `Pullback` command, which is used to pullback differential forms (including functions). Let `f0`, `f1`, and `f2` be the following forms on `E2`.

```
S2 > f0 := evalDG( x**2 * dx + y * dy );
S2 > f1 := evalDG( dx &w dy );
S2 > f2 := evalDG( dx &s dx + dy &s dy );
```

$$f0 := x^2 dx + ydy$$

$$f1 := dx \wedge dy$$

$$f2 := dxdx + dydy$$

Now with `Pullback`, we find the image of `f0`, `f1`, and `f2` on the manifold `S2`.

```
S2 > simplify( Pullback(Ps, [f0, f1, f2]), symbolic );
```

[This output is long so it was omitted.]

Consider the example of computation the area bounded by a curve. By virtue of Green's theorem, this problem is equivalent to computation of a integral of a 1-form.

```
S2 > DGsetup([t], R1);
```

$$frame\,name : R1$$

Define a 1-form, whose exterior differential equals to the volume form on the plain.

```
R1 > omega1 := evalDG( 1/2 * (x*dy-y*dx) );
```

$$\omega1 := -\frac{ydx}{2} + \frac{xdy}{2}$$

The curve is defined as a transformation.

```
R1 > C := Transformation(R1, E2, [x = cos(t)
* (5 - sin(20*t) / 10), y = sin(t) * (5 - sin(20*t)
/ 10)]);
```

$$C := [x = \cos(t)(5 - 1/10\sin(20t)), \, y = \sin(t)(5 - 1/10\sin(20t))]$$

```
R1 > plot([op(ApplyTransformation(C, [tau])), tau=0..2*Pi]);
```

Then, we pull back the form `omega1` (Fig. 4.4).

Fig. 4.4 Graph of the curve C

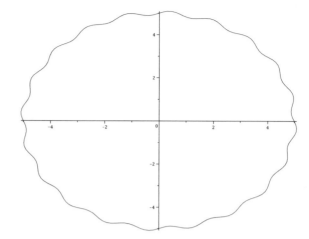

```
R1 >  alpha1 := Pullback(C, omega1):
```

The expression for alpha1 is rather long so it is omitted.

The DG package has the built-in command IntegrateForm for evaluating an integral of an *n*-form on a *n*-dimensional manifold.

```
R1 >  IntegrateForm(alpha1, t=0..2*Pi):
```

$$\frac{5001\pi}{200}$$

? Exercises

1. With the command Pushforward find the image of the infinitesimal generators

$$[D\_x, D\_y, x D\_x, y D\_y, -y D\_x + x D\_y]$$

   on the manifold S2.
2. Write a function, which draws trajectories of a given vector field passing through a given list of points.
3. Try to implement your own versions of Pullback and Pushforward working only with vector fields and 1-forms. (Hint: use DGinfo to obtain all necessary data from DG objects.)
4. Write your implementation of the functions Flow and Infinitesimal Transformation, which can work with a 1-parametric transformations. (Hint: use the functions dsolve, DGinfo and DGzip.)

Reference

1. Anderson, Ian M. and Torre, Charles G., *The Differential Geometry Package* (2016). Downloads. Paper 4. http://digitalcommons.usu.edu/dg_downloads/4.

Part II
Participants Contributions

This part contains original research articles of participants of the school. They serve as examples of how the theory presented in the previous part can be applied to various disciplines.

Chapter 5
On the Geometry Arising in Some Meteorological Models in Two and Three Dimensions

Bertrand Banos, Volodya Roubtsov and Ian Roulstone

5.1 Introduction

This paper can be considered as an additional note to lectures delivered by two of us (I.R. and V.R.) during the *Summer school Wisła -18 "Nonlinear PDEs, their geometry, and applications"* of Bałtycki Instytut Matematyki , in Wisła , Poland, 20–30 August 2018.

We start with two introductive sections. First, we briefly describe balanced meteorological models and their place in the problems of atmospheric dynamics. This section aims to explain to mathematicians the basic terminology and can be considered as a basic guide in the literature on this subject.

Then we describe (in down-to-the-earth form) the geometric approach to the Monge–Ampère equations (MAE) a non-linear second-order partial differential equations naturally appeared in mathematical studies of the balanced meteorological models. This approach was developed by V.V. Lychagin and his school and can

B. Banos
Faculté des sciences et sciences de l'ingénieur Bureau B125 - Sciences II,
Université de Bretagne Sud, Lorient, France
e-mail: bertrand.banos@univ-ubs.fr

V. Roubtsov (✉)
Maths Department, University of Angers, 2, Lavoisier Boulevard, CEDEX 01,
49045 Angers, France
e-mail: volodya@univ-angers.fr

Theory Division, Math. Physics Lab, ITEP, 25, Bol. Tcheremushkinskaya,
117218 Moscow, Russia

I. Roulstone
Department of Mathematics, University of Surrey,
GU2 7XH Guildford, Surrey, United Kingdom
e-mail: I.Roulstone@surrey.ac.uk

© Springer Nature Switzerland AG 2019   185
R. A. Kycia et al. (eds.), *Nonlinear PDEs, Their Geometry, and Applications*,
Tutorials, Schools, and Workshops in the Mathematical Sciences,
https://doi.org/10.1007/978-3-030-17031-8_5

serve a clear and transparent example of rich and fruitful interaction of geometry and theoretical meteorology designed in the papers of British geophysicists B. Hoskins, M. McIntyre et al.

The material of these sections is not new at all but we decided to add it to make easy the further reading both for terminological and from technical reasons. The next section is devoted to MA geometric description of a non-balanced geophysical model—Dritschel–Viudez $2d$ rotating stratified flows. This is the main novel result which is based on a wonderful appearance of the Hitchin hypersymplectic stucture which gives (under some additional assumptions) a reduction of the diagnostic MAE of Dritschel–Viudez to linear Laplace or Tricomi type equations.

The last sections are written in a form of illustrative examples (which are anyway new up to our best knowledge) and we recommend to potential readers to use these examples as illustrative part of V.R. lectures published in this volume.

5.2 A Brief Guide in Balanced Meteorological Models

The topic of meteorology may appear, in the first instance, as a rather arcane choice of application for the subject of Monge–Ampère equations and their associated geometric properties. For that reason, we shall be begin by explaining why the study of the fluid dynamics of the atmosphere provides a rich arena for displaying the theory and applications of Monge–Ampère geometry.

Considering the dynamics of the atmosphere as a problem in fluid mechanics, in which weather systems are governed by the Navier–Stokes equations coupled to moist thermodynamics, we learn from the classic textbooks of meteorology that a number of important physical constraints facilitate the introduction of systems of partial differential equations that are, in many ways, simpler and more tractable than Navier–Stokes-based models. The first such constraint relates to stratification in the atmosphere: the troposphere—the bottom layer of the atmosphere in which most weather occurs—is about 10 km deep, and this should be contrasted with the radius of the Earth, which is of the order of 6×10^3 km. Therefore, the troposphere can be considered as a very thin shell of fluid surrounding a body that rotates about a central axis. The rotation and stratification of the atmosphere lead to the notions of geostrophic and hydrostatic balance, respectively.

Geostrophic balance is the approximate equality between the horizontal pressure gradient force and the Coriolis acceleration. It is well known that on large scales (of the order of 10^2–10^3 km), air flows more-or-less parallel to isobars of constant pressure, and this is the geostrophic effect. In the vertical, the pressure gradient is balanced to a very good approximation by gravity, and this means the acceleration of air in the vertical is usually (especially on large scales) very much smaller than the acceleration due to gravity. This is hydrostatic balance.

The notions of geostrophic and hydrostatic balance can be incorporated, via rigorous asymptotic analysis, into the governing equations, and leading order terms are deemed to capture the salient dynamics of the atmosphere on sufficiently large scales.

The resulting models, which can be used to study phenomena such as jet stream meander, and the formation of weather fronts, bear the names quasi-geostrophic theory and semi-geostrophic theory. For a detailed introduction to this subject matter see for example, the article by A.A. White in [13].

When atmospheric flows are considered to be close to geostrophic, it is possible to show that the coupling between the momentum and temperature distributions, and conserved quantities such as potential vorticity (see A.A. White *op. cit.*), is governed by a non-linear partial differential equation of Monge–Ampère type. Imposing hydrostatic balance leads to models in either two or three spatial dimensions, and in turn, this leads to the study of Monge–Ampère equations in two or three independent variables. The effect of the Earth's rotation invariably manifests itself as a constraint that keeps such equations elliptic (in the 2d case) or keeps the pressure distribution convex (in the 3d case). The breakdown of convexity can be interpreted, for example as the formation of a weather front.

There is another interesting type of geophysical dynamical models—so-called rotating stratified flows.

This model is characterized by an explicit conservation of potential vorticity and one can use a change of variables from the usual primitive variables of velocity and density to the components of ageostrophic horizontal vorticity. It is quite amazing, that being quite different by geophysical properties from balanced models both 2*d*- and 3*d*-rotating stratified flow models are governed by a similar geometry which permits to apply the Monge–Ampère structure approach to their studies and to get some new special exact solutions using the Hitchin hypersymplectic structure which naturally appears in the framework of the MAE geometry.

5.3 Monge–Ampère Geometry

A (symplectic) Monge–Ampère equation (MAE) is a second-order PDE with a determinant-like non-linearity. For instance, a MAE in two variables can be written as follows:

$$A\phi_{xx} + 2B\phi_{xy} + C\phi_{yy} + D(\phi_{xx}\phi_{yy} - \phi_{xy}^2) + E = 0, \tag{5.1}$$

where A, B, C, and D are smooth functions of $(x, y, \phi_x, \phi_y) \in T^*\mathbb{R}^2$.

In dimension n, a Monge–Ampère equation is a linear combination of the minors of the hessian matrix $\left(\phi_{x_i x_j}\right)_{i,j=1...n}$.

! Comment

We will denote by hess(ϕ) the determinant of the hessian matrix of ϕ. For example, in two variables, hess$(\phi) = \phi_{xx}\phi_{yy} - \phi_{xy}^2$.

5.3.1 Monge–Ampère Operators

Lychagin has proposed a geometric approach of these equations, using differential forms on the cotangent space (i.e., the phase space). The idea is to associate with a form $\omega \in \Omega^n(T^*\mathbb{R}^n)$ the MAE equation $\Delta_\omega = 0$ where $\Delta_\omega : C^\infty(\mathbb{R}^n) \to \Omega^n(\mathbb{R}^n) \cong C^\infty(\mathbb{R}^n)$ is the differential operator defined by

$$\Delta_\omega(\phi) = (d\phi)^*\omega$$

with $d\phi : \mathbb{R}^n \to T^*\mathbb{R}^n$, $q \mapsto (q, \phi_q)$.

A form $\omega \in \Omega^n(T^*\mathbb{R}^n)$ is said to be primitive if $\omega \wedge \Omega = 0$ with Ω the canonical symplectic form on $T^*\mathbb{R}^n$)

The so-called Hodge–Lepage–Lychagin theorem asserts that this correspondence is a one-to-one correspondence between MAE and conformal classes of primitive forms

For instance, MAE (5.1) is associated with the primitive form

$$Adp \wedge dy + B(dx \wedge dp - dy \wedge dq) + Cdx \wedge dq + Ddp \wedge dq + Edx \wedge dy,$$

where (x, y, p, q) is the symplectic system of coordinates of $T^*\mathbb{R}^2$. In three dimensions, the "real MAE"

$$\mathrm{hess}(\phi) = 1$$

is associated with the effective form

$$dp \wedge dq \wedge dr - dx \wedge dy \wedge dz$$

and the "special lagrangian equation"

$$\Delta\phi - \mathrm{hess}(\phi) = 0$$

is associated with the form

$$dp \wedge dy \wedge dz + dx \wedge dq \wedge dz + dx \wedge dy \wedge dr - dp \wedge dq \wedge dr.$$

5.3.2 Generalized Solutions

For any function ϕ on $\mathbb{R}^n$, its graph

$$L_\phi = \{(q, \phi_q), q \in \mathbb{R}^n\} \subset T^*\mathbb{R}^n$$

is lagrangian ($\Omega|_L = 0$) and conversely for any lagrangian graph $L^n \subset T^*\mathbb{R}^\kappa$, it exists ϕ such that $L = L_\phi$. Moreover ϕ is a (regular) solution of $\Delta_\omega = 0$ if and only if $\omega|_{L_\phi} = 0$.

We define then a generalized solution of the MAE $\delta_\omega = 0$ as a Lagrangian submanifold $L^n \subset (M^{2n}, \Omega)$ on which vanishes ω:

$$\Omega|_L = 0 \quad \text{and} \quad \omega|_L = 0.$$

5.3.3   The Problem of Local Equivalence

A classical problem in the geometric study of differential equations is the problem of local equivalence: two given differential equations are they equivalent up to a local change of dependent and independent coordinates (that is, up to the action of a local diffemorphism of the phase space)? The diffeomorphisms which preserve the MAE family are the diffeomorphisms, which preserve the symplectic form (the symplectomorphisms) and their action, is in the Lychagin correspondence, the natural action on the differential forms

$$F \cdot \Delta_\omega = \Delta_{F^*\omega}.$$

We will say then that two Monge–Ampère equations $\Delta_{\omega_1} = 0$ and $\Delta_{\omega_2} = 0$ are symplectically equivalent if there exists a (local) symplectomorphism $F : (T^*\mathbb{R}^n, \Omega) \to (T^*\mathbb{R}^n, \Omega)$ such that

$$F^*\omega_2 = \omega_1.$$

For instance, the partial Legendre transformed $F : T^*\mathbb{R}^2 \to T^*\mathbb{R}^2$ defined by

$$F(x, y, p, q) = (x, q, p, -y) \tag{5.2}$$

exchanges the forms $dp \wedge dq - dx \wedge dy$ and $dp \wedge dy + dx \wedge dq$. Therefore, the MAE hess$(\phi) = 1$ and the Laplace equation $\Delta\phi = 0$ are symplectically equivalent in dimension 2. (This is not true in higher dimensions).

How can we use this symplectic equivalence to construct explicit solutions? The point is that a symplectomorphism will not preserve regular solutions but generalized solutions:

If L is a generalized solution of $\Delta_{F^*\omega}$ then $F(L)$ is a generalized solution of $\Delta_\omega = 0$.

Example

The generalized solution of the Laplace equation $\phi_{xx} + \phi_{yy} =$ are the complex curves of $T^*\mathbb{R}^2 = \mathbb{C}^2$. Applying (5.2) to the harmonic function

$$\phi(x + iy) = e^x \cos(y)$$

we obtain then a generalized solution $L = F(L_\phi)$ and identifying it as a graph $L = L_\psi$, a regular solution

$$\psi(x, y) = y \arcsin(\frac{y}{e^x}) + \sqrt{e^{2x} - y^2}$$

of the MAE hess$(\psi) = 1$.

5.3.4 Monge–Ampère Structures

We define then a Monge–Ampère structure on a $2n$-manifold M (locally $M = T^*\mathbb{R}^n$) as a pair of differential forms

$$(\Omega, \omega) \in \Omega^2(M) \times \Omega^n(M)$$

such that

1. Ω is symplectic that is nondegenerate and closed ($\Omega^2 \neq 0$ and $d\Omega = 0$),
2. ω is effective that is $\omega \wedge \Omega = 0$.

5.3.4.1   2d MAE and 4d Monge–Ampère Geometry

In four dimensions ($n = 2$), in the nondegenerate case, this geometry can be either complex or real and this distinction coincides actually with the usual distinction elliptic/hyperbolic for differential equations in two variables.

We define the pfaffian pf$(\omega) \in C^\infty(M)$ and the tensor $A_\omega : M \to TM \otimes T^*M$ by

$$\omega \wedge \omega = \text{pf}(\omega) \, \Omega \wedge \Omega \qquad \omega(\cdot, \cdot) = \Omega(A_\omega \cdot, \cdot)$$

and they satisfy

$$A_\omega^2 = -\,\text{pf}(\omega)\text{Id}$$

Hence, the tensor $I_\omega = \dfrac{A_\omega}{\sqrt{|\text{pf}(\omega)|}}$ is an almost complex structure if $\Delta_\omega = 0$ is elliptic (pf$(\omega) > 0$) and an almost product structure if $\Delta_\omega = 0$ is hyperbolic (pf$(\omega) < 0$).

Lychagin and Roubtsov have explained in [9, 10] the link there is between the problem of local equivalence of MAE in two variables and the integrability problem of this tensor I_ω. They have actually proved the following:

Proposition 5.1 *The three following assertions are equivalent:*

1. $\Delta_\omega = 0$ *is locally equivalent to one of the two equations*

$$1.\ \Delta\phi = 0$$
$$2.\ \Box\phi = 0$$

2. the almost complex or almost product structure I_ω is integrable

3. the form $\dfrac{\omega}{\sqrt{|\operatorname{pf}(\omega)|}}$ *is closed.*

5.3.4.2 3d MAE and 6d Monge–Ampère Geometry

In six dimensions ($n = 3$), there is again a correspondence between real/complex geometry and "nondegenerate" Monge–Ampère structures.

N. Hitchin ([7]) has associated with each 3form on M^6 a pfaffian $\lambda(\omega) \in C^\infty(M)$ and a tensor $A_\omega : M \to TM \otimes T^*M$ satisfying

$$A_\omega^2 = \lambda(\omega)\mathrm{Id}$$

Moreover he proved that any nondegenerate 3-form (that is $\lambda(\omega) \neq 0$) writes as the sum of two decomposable forms:

1. $\lambda(\omega) > 0$ if and only if

$$\omega = \alpha_1 \wedge \alpha_2 \wedge \alpha_3 + \beta_1 \wedge \beta_2 \wedge \beta_3$$

2. $\lambda(\omega) < 0$ if and only if

$$\omega = (\alpha_1 + i\beta_1) \wedge (\alpha_2 + i\beta_2) \wedge (\alpha_3 + i\beta_3) + (\alpha_1 - i\beta_1) \wedge (\alpha_2 - i\beta_2) \wedge (\alpha_3 - i\beta_3)$$

In the nondegenerate case, it exists then a dual form $\hat\omega$ such that $\omega + \hat\omega$ and $\omega - \hat\omega$ or $\omega + i\hat\omega$ and $\omega - i\hat\omega$ are decomposable.

Moreover, Lychagin and Roubtsov ([10]) have defined a symmetric tensor g_ω on M satisfying the following:

1. g_ω is non degenerate if and only $\lambda(\omega) \neq 0$
2. it has signature $(3, 3)$ if $\lambda(\omega) > 0$ and signature $(6, 0)$ or $(4, 2)$ if $\lambda(\omega) < 0$.
3. the triple $\left(g_\omega, \Omega, I_\omega = \dfrac{A_\omega}{\sqrt{|\lambda(\omega)|}} \right)$ is compatible that is

$$g_\omega(\cdot, \cdot) = \Omega(J_\omega \cdot, \cdot)$$

So, one can associate with a nondegenerate Monge–Ampère structure (Ω, ω) on a *six*-dimensional manifold an almost (pseudo para) Kähler structure with Kähler form Ω. Moreover this almost Kähler structure is "normalized" by two decomposable 3-forms: we use then the terminology of "generalized Calabi-Yau structure" (see [1]).

For example, the Monge–Ampère structure associated with the "real" MAE in three variables (x, y, z)

$$\mathrm{hess}(\phi) = 1$$

is the real Calabi–Yau structure on $T^*\mathbb{R}^3$

$$g_\omega = \begin{pmatrix} 0 & Id \\ Id & 0 \end{pmatrix} \quad K_\omega = \begin{pmatrix} Id & 0 \\ 0 & -Id \end{pmatrix}$$

$$\alpha = dx \wedge dy \wedge dz \qquad \beta = dp \wedge dq \wedge dr$$

The Monge–Ampère structure associated with the special lagrangian equation

$$\Delta\phi - \mathrm{hess}(\phi) = 0$$

is the canonical Calabi–Yau structure on $T^*\mathbb{R}^3 = \mathbb{C}^3$

$$g_\omega = \begin{pmatrix} Id & 0 \\ 0 & Id \end{pmatrix} \quad K_\omega = \begin{pmatrix} 0 & Id \\ Id & 0 \end{pmatrix}$$

$$\alpha = dz_1 \wedge dz_2 \wedge dz_3 \qquad \beta = d\overline{z_1} \wedge d\overline{z_2} \wedge d\overline{z_3}$$

As in the *four*-dimensional case, there is a strong link between the problem of local equivalence of MAE in three variables and the integrability problem of generalized Calabi–Yau structures on $\mathbb{R}^6$ (see [1]):

Proposition 5.2 *The two following assertions are equivalent:*

1. *the MAE $\Delta_\omega = 0$ is locally equivalent to one of the three equations*

    1. $\mathrm{hess}(\phi) = 1$
    2. $\Delta\phi - \mathrm{hess}(\phi) = 0$
    3. $\Box\phi + \mathrm{hess}(\phi) = 0$

2. *the forms* $\dfrac{\omega}{\sqrt[4]{|\lambda(\omega)|}}$ *and* $\dfrac{\hat{\omega}}{\sqrt[4]{|\lambda(\omega)|}}$ *are closed and g_ω is flat.*

! Comment

It is important to note that the geometry associated with a MAE $\Delta_\omega = 0$ of real type ($\lambda(\omega) > 0$) is essentially real but it is very similar to the classic Kähler geometry. In particular, when this geometry is integrable, there exists a potential Φ and a coordinate systems $(x_i, p_i)_{i=1,2,3}$ on $T^*\mathbb{R}^6$ such that

$$g_\omega = \sum_{i,j} \frac{\partial^2 \Phi}{\partial x_i \partial p_j} dx_i \cdot dp_j$$

and

$$\det \left(\frac{\partial^2 \Phi}{\partial x_j \partial p_j} \right) = f(x)g(p).$$

5.4 2d Rotating Stratified Flows—Dritschel–Viudez MAE

Recently, a new approach to modelling stably stratified geophysical flows was proposed in [5, 6]. This approach is based on the explicit conservation of potential vorticity and uses a change of variables from the usual primitive variables of velocity and density to the components of ageostrophic horizontal vorticity. This change results in a Monge–Ampère-like non-linear equation with non-constant coefficients. The equation gives the conditions for static and inertial stability and changes the type from elliptic to hyperbolic.

It is written as

$$E\left(\phi_{xx}\phi_{zz} - \phi_{xz}^2\right) + A\phi_{xx} + 2B\phi_{xz} + C\phi_{zz} + D = 0 \qquad (5.3)$$

with

$$E = 1, \quad A = 1 + \theta xz, \quad B = \frac{1}{2}(\theta_{zz} - \theta_{xx})$$

$$C = 1 - \theta_{xz}, \quad D = \theta_{xx}\theta_{zz} - \theta_{xz}^2 - \varpi$$

where θ is a given potential and the dimensionless PV anomaly ϖ may be also considered as a given quantity (see the details in [6]).

The corresponding Monge–Ampère structure is

$$\begin{cases} \Omega = dx \wedge dp + dz \wedge dr \\ \omega = E dp \wedge dr + A dp \wedge dz + B(dx \wedge dp - dz \wedge dr) + C dx \wedge dr + D dx \wedge dz \end{cases}$$

The pfaffian is $\mathrm{pf}(\omega) = R$ with R the Rellich's parameter:

$$R = AC - ED - B^2 = 1 + \varpi - \left(\frac{\Delta\theta}{2}\right)^2.$$

Moreover, a direct computation gives

$$d\omega = d\left(\frac{\Delta\theta}{2}\right) \wedge \Omega. \tag{5.4}$$

5.4.1 Integrability of the Complex/Product Structure

Lychagin–Roubtsov criteria says that $2d$ Dritschel–Viudez equation is locally equivalent to a Monge–Ampére equation with constant coefficients if and only if

$$\begin{cases} \Delta\theta = 2c_1 \\ R = c_2 \end{cases}$$

In that case, for example for $R > 0$, we see that

$$\omega + i\sqrt{R}\,\Omega = du \wedge dv$$

with

$$\begin{cases} u = x - (c_1 - ic_2)z - \theta_z + p \\ v = -(c_1 + ic_2)x + \theta_x + r. \end{cases}$$

In other words, if $\theta = 2c_1$ and $R = c_2 > 0$ then $2d$ Dritschel–Viudez equation is equivalent to Laplace equation

$$\theta_{xx} + \theta_{zz} = 0$$

modulo the Legendre transform

$$F(x, z, p, r) = \frac{1}{\sqrt{R}}(x - c_1z - \theta_z, c_2z, -c_2x, -c_1x + \theta_x + r).$$

5.4.2 Underlying Hypersymplectic Geometry

We will apply to this equation a recent observation: if our $four$-dimensional manifold M, endowed with the Monge–Ampère structure (Ω, ω) admits a Lagrangian fibration (main example: M is the cotangent bundle of a smooth $2d$-manifold), then it exists as a conformal split metric on M^4.

When the corresponding Monge–Ampère equation is given by (5.3), this metric is written as

$$g = C(dx)^2 - 2Bdxdz + A(dz)^2 + E/2(dpdx + drdz), \qquad (5.5)$$

Using this metric, we get an additional 2-form $\hat{\omega}$ defined by

$$\hat{\omega}(\cdot, \cdot) = g(A_\omega \cdot, \cdot) \quad \text{with} \quad \omega(\cdot, \cdot) = \Omega(A_\omega \cdot, \cdot)$$

In coordinates,

$$\hat{\omega} = \left(-2AC + 2B^2 + D\right) dx \wedge dz - Bdx \wedge dp - Cdx \wedge dr + Adz \wedge dp$$
$$+ Bdz \wedge dr - dp \wedge dr.$$

Introducing $\Theta = \dfrac{\Omega}{\sqrt{|R|}}$, we get an hypersymplectic triple $(\Theta, \omega, \hat{\omega})$ satisfying

$$\omega^2 = -\hat{\omega}^2 = \pm\Theta^2,$$
$$\omega \wedge \hat{\omega} = \omega \wedge \Theta = \hat{\omega} \wedge \Theta = 0.$$

Equivalently, we obtain three tensors I, S, and T satisfying

$$I^2 = -1, \; S^2 = 1, \; T^2 = 1$$
$$ST = -TS = -I,$$
$$TI = -IT = S,$$
$$IS = -SI = T.$$

Moreover, we have

$$d\hat{\omega} = -d\left(\frac{\Delta\theta}{2}\right) \wedge \Omega.$$

Hence, when $\Delta\theta = 0$, then ω and $\hat{\omega}$ are closed and satisfy $\omega^2 = -\hat{\omega}^2$: they define then an integrable product structure. Indeed, in the new coordinates:

$$X = \int R(x, z)dx, \;\; U = x - \theta_z + p,$$

$$Z = z, \qquad\quad V = z + \theta_x + r.$$

we see that

$$\begin{cases} \omega = dU \wedge dV - dX \wedge dZ \\[2ex] \hat{\omega} = -(dU \wedge dV + dX \wedge dZ) \\[2ex] \Omega = \dfrac{1}{R}(dX \wedge dU - SdZ \wedge dU + RdZ \wedge dV) \quad \text{with } S = \displaystyle\int R_z dx. \end{cases}$$

In other words, when θ is harmonic, a submanifold

$$L = \left\{ (\psi_Z, Z, U, \psi_U), \ (Z, U) \in \mathbb{R}^2 \right\}$$

is a generalized solution of 2d-Dritschel Viudez equation if and only if

$$\psi_{ZZ} + R\psi_{UU} = S.$$

5.4.3 2d-Diagnostic Equation of Dritschel–Viudez: Special Choice of Constant Coefficients

In the following partial case, this diagnostic equation which corresponds to the choice

$$A = 1, \ B_1 = B_2 = 0, \ C = E = \epsilon^2, \ D = -\varpi, \tag{5.6}$$

becomes two-dimensional:

$$\phi_{xx} + \phi_{zz} + \phi_{xx}\phi_{zz} - \phi_{xz}^2 = \varpi. \tag{5.7}$$

It describes a geostrophically balanced steady $2d$ flow which is closed to the QG models described above.

It is interesting to observe that the corresponding effective form

$$\omega = dp \wedge dx + dr \wedge dz + dp \wedge dr - \varpi \, dx \wedge dz$$

has the Pfaffian $Pf(\omega) = 1 + \varpi$. In the classical notations, the Pfaffian is nothing but the Rellich's parameter $R = AC - DE - B^2$.

We denote as usually by Π the dimensionless potential vorticity which relates to the PV anomaly ϖ as

$$\varpi \equiv \Pi - 1,$$

hence we had obtained the following meaning of the Pfaffian for $2d$ flow MAE

$$Pf(\omega) = \Pi,$$

or, in the case of the $2d$ diagnostic equation this metric depends on the potential θ and on the PV anomaly ϖ:

$$g = (1 - \theta_{xz})(dx)^2 - (\theta_{zz} - \theta_{xx})dxdz + (1 + \theta_{xz})(dz)^2 + 1/2(dpdx + drdz).$$

5.4.4   Reduction to Constant Coefficients

Now, we will discuss a reducibility of this equation to a normal form with constant coefficients.

Let us calculate the symplectic invariant of the (3.32) for the given θ and ϖ and check the criteria of ([9]). The direct computation gives that

$$dw = 1/2d(\Delta\theta) \wedge \Omega.$$

The criteria ([9]) shows that if the given potential function θ is a harmonic ($\Delta\theta = 0$), then the $2d$ diagnostic equation of Dritschel and Viudez is reducible by a local symplectomorphism to an MA equation with constant coefficients.

We can see that the Pfaffian (which is also is the Rellich invariant of the diagnostic equation) in the case of harmonic potential θ is equal to $1 + \varpi = \Pi$. This is exactly the same value as it was in the above-mentioned $2d$ model with constant coefficients.

Example

It is interesting to study a "Legendre-dual" condition on the potential : hess$(\theta) = 1$ provides us also with an example of reducible to constant coefficients MAE diagnostic equation. The coefficient D in this case simplifies: $D = 1 - \varpi$ and the Pfaffian in this case (like in the general $2d$ diagnostic equation) is equal to

$$Pf(\omega) = 1 + \varpi - \frac{(\Delta\theta)^2}{4} = \Pi - \frac{(\Delta\theta)^2}{4}$$

If we suppose that the potential function $\theta(x, z)$ satisfies to conditions of the famous Jorgens theorem ([4]), then we can assume that

$$\theta(x, z) = \alpha x^2 + 2\beta xz + \gamma z^2 + l,$$

where $l = l(x, z)$ denotes linear and constant terms with constant

$$\alpha, \beta, \gamma, \alpha\beta - \gamma^2 = 1/4.$$

The Laplacian value in this case is $2(\alpha + \gamma)$ and the Pfaffian is equal to $\Pi - (\alpha + \gamma)^2 = \Pi$.

5.4.5 Variation of the Potential and Hyper-Kähler Metrics in 4d

Another interesting class of exact solutions is given by variations of different partial choices of the potential $\theta(x, z)$ which correspond to some different "geometries" in $4d$.

We will start with some general facts.

Let us consider a $4d$ metric (5.5) and compare it with the following:

$$g = -2\Theta_{qq}(dx)^2 + 4\Theta_{pq}dxdz - 2\Theta_{pp}(dz)^2 + 4\Theta_{pq}(dpdx + dqdz), \quad (5.8)$$

where $\Theta = \Theta(x, z, p, q)$ is a (complex valued) function on the phase space T^*M of the initial $2d$ configuration base. We suppose in addition that Θ is a solution of the following $4d$ MAE (so-called Plebanski second "Heavenly equation".)

$$Hess_{pq}\Theta + \Theta_{xp} + \Theta_{qz} = \Theta_{pp}\Theta_{qq} - \Theta_{pq}^2 + \Theta_{xp} + \Theta_{qz} = 0. \quad (5.9)$$

The interest of this special choice of the metric is based on the fact that it is an example of Hyper-Kähler metric in $4d$.

There are many examples of such metrics and our idea is, starting from one of it, to find a solution for the diagnostic MAE. We consider the most simple example which can be reduced to a constant coefficient case (if the potential vorticity ϖ is constant). Then the corresponding MAE $2d$ diagnostic type equation reads as

$$\left(\phi_{zz}\phi_{xx} - \phi_{xz}^2\right) - 2\Theta_{pp}\phi_{xx} - 2\Theta_{pq}\phi_{xz} - 2\Theta_{qq}\phi_{zz} + D = 0, \quad (5.10)$$

where D denotes the "low-order" part which has no influence on the $4d$ (pseudo)-metric. Its Pfaffian equals to

$$Pf(\omega) = Hess_{pq}\Theta - D = -(\Theta_{xp} + \Theta_{zq}) - D.$$

Now, starting with any given solution of $4d$ MAE (Plebansky Second Heavenly) we can construct a $2d$ PDE with *non-constant* coefficients and sometimes we are able to solve it or to say something about its solutions.

It follows that for the diagnostic $2d$ MAE, we should have

$$\Theta_{pp} = -\frac{1}{2}(1 - \theta_{xz}),$$

$$\Theta_{qq} = -\frac{1}{2}(1 + \theta_{xz}),$$

$$\Theta_{pq} = -\frac{1}{4}(\theta_{zz} - \theta_{xx}).$$

First one can remark that in the case of $2d$ diagnostic MAE choice of the coefficients immediately means that $\Theta_{pq} = \frac{1}{8}$ which implies in its own turn that

$$\theta_{xx} - \theta_{zz} = -\frac{1}{2}.$$

The Eq. (5.9) has, in this case, the following form:

$$Hess_{pq}\Theta + \Theta_{xp} + \Theta_{qz} = \frac{1}{4}(1 - \theta_{xz}^2) - \frac{1}{64} + \Theta_{xp} + \Theta_{qz} = 0. \qquad (5.11)$$

We have also the following compatibility conditions on the function Θ :

$$\Theta_{xpp} + \Theta_{pqz} = 0, \ \Theta_{pqx} + \Theta_{qqz} = 0. \qquad (5.12)$$

But $\Theta_{xpp} = \Theta_{pqz} = \Theta_{pqx} = \Theta_{qqz} = 0$ because of $\Theta_{pq} = \frac{1}{8}$.
Then, the Heavenly equation for the *diagnostic* metric (5.5) reads

$$(1 - \theta_{xz}^2) - \frac{1}{16} = 0 \qquad (5.13)$$

and $\theta_{xz} = \pm\frac{\sqrt{15}}{4}$.

It immediately follows from the relations between the derivatives of Θ and θ that $\theta_{xxz} = \Theta_{ppx} = 0$ and $\theta_{xzz} = \Theta_{qqz} = 0$.

The solution $\Theta(p, q, x, z)$ of the Eq. (5.13) is a quadratic in p and q :

$$\Theta = \alpha p^2 + 2\beta pq + \gamma q^2 + \delta(x, z),$$

where

$$2\alpha = \Theta_{pp} = -\frac{1}{2}(1 - \theta_{xz}) = \frac{1}{2}(1 \pm \frac{\sqrt{15}}{4}),$$

$$2\gamma = \Theta_{qq} = -\frac{1}{2}(1 + \theta_{xz}) = \frac{1}{2}(1 \mp \frac{\sqrt{15}}{4}).$$

Hence

$$\Theta(x, z, p, q) = \frac{1}{2}(1 \pm \frac{\sqrt{15}}{4})p^2 + \Theta_{pq} = \frac{1}{4}pq + \frac{1}{2}(1 \mp \frac{\sqrt{15}}{4})q^2 + \delta(x, z).$$

It immediately follows that the potential $\theta(x, z)$ is quadratic linear

$$\theta(x, z) = Ax^2 \pm \frac{\sqrt{15}}{4}xz + (A + \frac{1}{2})z^2 + l_1 x + l_2 z + l_3$$

and its Hessian can be easily expressed as

$$\text{hess}_{xz}\,\theta = A\left(A + \frac{1}{2}\right) - \frac{15}{16}.$$

The Eq. (5.10) becomes

$$\left(\phi_{zz}\phi_{xx} - \phi_{xz}^2\right) + \left(1 \pm \frac{\sqrt{15}}{4}\right)\phi_{xx} - \frac{1}{4}\phi_{xz} + \left(1 \mp \frac{\sqrt{15}}{4}\right)\phi_{zz} + A\left(A + \frac{1}{2}\right) - \frac{15}{16} - \varpi = 0.$$

$$(5.14)$$

Pfaffian of the corresponding effective form is

$$Pf(\omega) = \text{hess}_{pq}\,\Theta - D = -(\Theta_{xp} + \Theta_{zq}) - D = -D = \varpi - A\left(A + \frac{1}{2}\right) - \frac{15}{16}.$$

The Eq. (5.14) is equivalent to a linear and is reduced to elliptic or hyperbolic depending on the sign of $\varpi - A\left(A + \frac{1}{2}\right) - \frac{15}{16}$.

5.5 Some Examples of 3d-Geostrophic Models

We study in this section some examples of Monge–Ampère equations in three variables which come from some geostrophic models.

5.5.1 The Birkett and Thorpe Equation ([3])

This equation can be written as

$$\frac{1}{f^3 N^2}\,\text{hess}(u) + \frac{1}{f}(u_{xx} + u_{yy} + \frac{f^2}{N^2}u_{zz}) + \frac{1}{fN^2}\left\{(u_{xx} + u_{yy})u_{zz} - u_{xz}^2 - u_{yz}^2\right\}$$
$$+ \frac{1}{f^3}(u_{xx}u_{yy} - u_{xy}^2) = s$$

$$(5.15)$$

We assume that f and N are constant.
The effective form associated with is

$$\omega = \frac{1}{f^3 N^2}dp \wedge dq \wedge dr + \frac{1}{f}(dp \wedge dy \wedge dz + dx \wedge dq \wedge dz + \frac{f^2}{N^2}dx \wedge dy \wedge dr)$$
$$+ \frac{1}{fN^2}dp \wedge dy \wedge dr + \frac{1}{fN^2}dx \wedge dq \wedge dr + \frac{1}{f^3}dp \wedge dq \wedge dz - sdx \wedge dy \wedge dz.$$

One can check that the Hitchin pfaffian is positive:

$$\lambda(\omega) = \frac{(f+s)^2}{f^6 N^4}.$$

So ω is the sum of two decomposable three forms when $s + f \neq 0$:

$$\omega = \frac{1}{NF}(f dx + \frac{1}{f}dp) \wedge (f dy + \frac{1}{f}dq) \wedge (Ndz + \frac{1}{N}dr) - (f+s)dx \wedge dy \wedge dz.$$

and therefore (5.15) is equivalent to

$$\text{hess}(u) = N^2 f^3 (f + s)$$

modulo the action of the Legendre transformed

$$F(x, y, z, p, q, r) = (x, y, z, p - f^2 x, q - f^2 y, r - N^2 z).$$

It is worth mentioning that this equivalence is independent from s but the underlying real Calabi–Yau is integrable if and only if s is constant.

5.5.2 The Hoskins Equation ([8])

This equation originates in the study of potential vorticity in semi-geostrophic theory and takes the form

$$\frac{1}{f^2}(u_{xx} + u_{yy}) + \frac{1}{N^2}u_{zz} - \frac{1}{f^4}(u_{xx} - u_{xy}^2) = 1. \tag{5.16}$$

The associated effective form is

$$\omega = \frac{1}{f^2}(dp \wedge dy \wedge dz + dx \wedge dq \wedge dz) + \frac{1}{N^2}dx \wedge dy \wedge dr - \frac{1}{f^4}dp \wedge dq \wedge dz$$
$$- dx \wedge dy \wedge dz.$$

One again, this form is the sum of two decomposable real forms:

$$\omega = \frac{1}{N^2}dx \wedge dy \wedge dr - (dx - \frac{1}{f^2}dp) \wedge (dy - \frac{1}{f^2}dq) \wedge dz.$$

and (5.16) is then equivalent to

$$\text{hess}(u) = f^4 N^2$$

up the symplectomorphism

$$F(x, y, z, p, q, r) = (p, q, z, -x + f^2 p, -y + f^2 q, r).$$

5.5.3 The McIntyre-Roulstone Equation ([12])

This equation originates from a study which seeks to derive more accurate approximations to the semi-geostrophic equations (i.e., models that resemble more closely the underlying Navier–Stokes-based model) that share some of the elegant mathematical properties of semi-geostrophic theory. The equation for the vorticity of a one-parameter family of models takes the form

$$1 + \Delta u + (1 - c^2) \, \text{hess}_{x,y}(u) = \frac{\zeta^C}{f}, \tag{5.17}$$

where assume that c and f are constant, and c is the parameter that distinguishes various asymptotic flow regimes. The corresponding effective form is

$$\omega = dp \wedge dy \wedge dz + dx \wedge dq \wedge dz + dx \wedge dy \wedge dr$$
$$+ (1 - c^2) dp \wedge dq \wedge dz + (1 - \frac{\zeta^C}{f}) dx \wedge dy \wedge dz$$

and is still of real type:
$$\lambda(\omega) = (1 - c^2)^2$$

Actually,

$$\omega = dx \wedge dy \wedge \left(dr + (\beta - \frac{1}{\alpha})dz\right) + \frac{1}{\alpha}(dx + \alpha dp) \wedge (dy + \alpha dq) \wedge dz$$

with

$$\begin{cases} \alpha = 1 - c^2 \\ \beta = 1 - \frac{\zeta^C}{f} \end{cases}$$

Note that the form $dr + (\beta - \frac{1}{\alpha})dz$ is closed if and only if $\beta = \beta(z)$. Under this assumption, (5.17) is equivalent to

$$\text{hess}(u) = -\frac{1}{(1 - c^2)^3}$$

up the symplectomorphism

$$F(x, y, z, p, q, r) = (-\alpha p, -\alpha q, z, p + \frac{1}{\alpha}q, q + \frac{1}{\alpha}y, r - \int_{z_0}^{z} (\beta(t) - \frac{1}{\alpha})dt).$$

5.5.4   The Snyder–Skamarock–Rotunno Equation ([16])

This equation is

$$\frac{1}{q_{HG}}\phi_{zz} + \phi_{xx} + \phi_{yy} - 4R_0 \, \text{hess}_{x,y} \, \phi - \frac{3R_0^2}{q_{HG}} \, \text{hess}(\phi) = 0. \qquad (5.18)$$

The associated effective form is

$$\omega = \frac{1}{q_{HG}}dx \wedge dy \wedge dr + dp \wedge dy \wedge dz + dx \wedge dq \wedge dz - 4R_0 dp \wedge dq \wedge dz - \frac{3R_0^2}{Q}dp \wedge dq \wedge dr.$$

Once again, the underlying geometry is real since the Hitchin paffian is positive:

$$\lambda(\omega) = 4\frac{R_0^2}{q_{HG}^2}.$$

Therefore, ω is the sum of two decomposable 3-forms

$$\omega = \frac{R_0^2}{2q_{HG}}(\frac{1}{R_0}dx - dp) \wedge (\frac{1}{R_0}dy - dq) \wedge (\frac{q_{HG}}{R_0}dz + 3dr)$$
$$+ \frac{16}{R_0}(\frac{3}{4}R_0 dp - \frac{1}{4}dx) \wedge (\frac{3}{4}R_0 dq - \frac{1}{4}dy) \wedge (-\frac{1}{2}dz - \frac{R_0}{2q_{HG}}dr).$$

Hence, (5.18) is equivalent to

$$\text{hess}(\phi) = -\frac{R_0^3}{32q_{HG}}$$

modulo the symplectomorphism

$$F(x, y, z, p, q, r)$$
$$= (3R_0 x + p, 3R_0 y + q, -\frac{R_0}{q_{HG}}z + r, \frac{1}{2}x + \frac{1}{2R_0}p, \frac{1}{2}y + \frac{1}{2R_0}q, -\frac{3}{2}z + \frac{q_{HG}}{2R_0}r).$$

5.6 3d Rotating Stratified Flows—Dritschel–Viudez MAE

Let us consider the $3d$- Dritschel–Viudez equation ([6])

$$
E\left(\Phi_{zz}\left(\Phi_{xx}+\Phi_{yy}\right)-\Phi_{xz}^2-\Phi_{yz}^2\right)+A\left(\Phi_{xx}+\Phi_{yy}\right)
$$
$$
+2B_1\Phi_{xz}+2B_2\Phi_{yz}+C\Phi_{zz}+D=0
$$

with

$$
A=1+\Theta_z \qquad B_1=\tfrac{1}{2}\Delta\varphi-\Theta_x \qquad B_2=\tfrac{1}{2}\Delta\psi-\Theta_y
$$

$$
C=1-\Theta_z \quad D=\Theta_x\Delta\varphi+\Theta_y\Delta\psi-|\nabla\Theta|^2-\varpi \qquad E=1
$$

where φ and ψ are potentials, $\Theta=\varphi_x+\psi_y$ and ϖ is the dimensionless PV anomaly (see details in [6]).

The associated primitive form is

$$
\begin{aligned}
\omega={}&E(dx\wedge dq\wedge dr-dy\wedge dp\wedge dr)+A(dy\wedge dz\wedge dp-dx\wedge dz\wedge dq)\\
&+B_1(dx\wedge dy\wedge dp+dy\wedge dz\wedge dr)+B_2(dx\wedge dy\wedge dq-dx\wedge dz\wedge dr)\\
&+Cdx\wedge dy\wedge dr+Ddx\wedge dy\wedge dz
\end{aligned}
$$

We can check directly that, independently of the coefficients,

$$
\lambda(\omega)=0.
$$

In other words, the Dritschel–Viudez equation is "degenerate" (the underlying Monge–Ampère geometry is degenerated) and it is, at each point, equivalent to a linear equation.

Consider the Rellich's invariant $R=AC-ED-B_1^2-B_2^2$. As in the $2d$ case, this parameter separates elliptic and hyperbolic cases.

Indeed, the non-zero eigenvalues of g_ω are:

$$
\lambda_1=2R
$$
$$
\lambda_2=S+\sqrt{S^2-4(1+A^2)R}
$$
$$
\lambda_3=S-\sqrt{S^2-4(1+A^2)R}
$$

with $S=R+1+A^2+B_1^2+B_2^2$. Therefore,

1. if $R>0$, then g_ω has signature $(3,0)$ and $3d$-Dritsche–Viudez equation is elliptic, equivalent in each point to $\Delta u=0$.

2. if $R < 0$, then g_ω has signature $(1, 2)$ and $3d$-Dritsche–Viudez equation is hyperbolic, equivalent in each point to $\Box u = 0$.

This equivalence is actually explicit: consider the following isomorphism P_a of $T_a M$ in each point a:

$$P_a = \begin{pmatrix} A & 0 & 0 & 0 & 0 & 0 \\ 0 & A & 0 & 0 & 0 & 0 \\ B_1 & B_2 & -T & 0 & 0 & 0 \\ 0 & 0 & 0 & 1 & 0 & B_1/T \\ 0 & 0 & 0 & 0 & 1 & B_2/T \\ 0 & 0 & 0 & 0 & 0 & -A/T \end{pmatrix}$$

with $T = \sqrt{|R|}$.

We can state from one hand side that P is a conformal symplectomorphism:

$$P^*\Omega = \alpha\Omega.$$

Moreover, if $R > 0$ then

$$P^*\omega = -A^2 T \left(dy \wedge dz \wedge dp - dx \wedge dz \wedge dq + dx \wedge dy \wedge dr + D dx \wedge dy \wedge dz \right)$$

and if $R < 0$ then

$$P^*\omega = -A^2 T \left(dy \wedge dz \wedge dp - dx \wedge dz \wedge dq - dx \wedge dy \wedge dr + D dx \wedge dy \wedge dz \right).$$

Conclusion

If A et R are non zero and if there exists a local diffeomorphism F of M such that $P = T_m F$ (the condition which is automatically satisfies when the coefficients appeared in P are constants), then our equation is

$$\begin{cases} u_{xx} + u_{yy} + u_{zz} + D = 0, & \text{if } R > 0 \\ u_{xx} - u_{yy} - u_{zz} + D = 0, & \text{if } R < 0 \end{cases}$$

! Comment

All these geostrophic models in three dimensions have a real underlying geometry and are equivalent to the model

$$\text{hess}(\phi) = cst.$$

An explanation could be that these models are constructed from two- dimensional models. It is possible that a more covariant approach would lead to a complex geometry. But we have to note that two- dimensional models correspond to the the Laplace equation which is also equivalent to the MAE hess(ϕ) = 1. It is then possible that complex geometry only arises in even dimensions. It would be then necessary to better understand this real Kähler structure and the corresponding real potential on $4n + 2$-dimensional manifolds.

Acknowledgements V.R. and I. R. express their deep thanks to organizers of the Summer School "Wisla 2018" and personally to Jerzy Szmit for a very stimulating summer school and for excellent working conditions. We are grateful to all participants for useful and fruitful discussions and inspiring atmosphere during the School.

During preparation of the material, V. R was partly supported by support of the Russian Foundation for Basic Research under the Grants RFBR 18-01-00461 and 19-51-53014 GFEN.

References

1. B. Banos: *Nondegenerate Monge-Ampère structures in six dimensions*, Letters in Mat. Phys., 62 (2002).
2. B. Banos: *Monge-AmpÃlre equations and generalized complex geometry - The two-dimensional case*, Journal of Geometry and Physics, 57 (2007).
3. H. R. Birkett, A. J. Thorpe: *Superposing semi-geostrophic potential-vorticity anomalies*, Q.J.R. Meteorol. Soc, 123 (1997).
4. R. Courant, D. Hilbert: *Methods of mathematical physics. Vol. II: Partial differential equations.* (Vol. II by R. Courant.) Interscience Publishers (a division of John Wiley & Sons), New York-London 1962 xxii+830 pp.
5. D.G. Dritschel, A. Viudez: *An explicit potential vorticity conserving approach to modelling nonlinear internal gravity waves*, submitted to Journ; Fluid Mech. (2000).
6. D.G. Dritschel, A. Viudez: *A balanced approach to modelling rotating stably-stratified geophysical flows*, submitted to Journ; Fluid Mech. (2003).
7. N. Hitchin: *The geometry of three-forms six and seven dimensions*, Journal of Differential Geometry, 56 (2001).
8. B. Hoskins: *The geostrophic Momentum Approximation and the Semi-geostrophic Equations*, Journal of the atm. sciences, vol 32, n° 2 (1975).
9. V.V Lychagin, V. Roubtsov: *On Sophus Lie Theorems for Monge-Ampère equations*, Doklady Bielorrussian Academy of Science, vol. 27, 5 (1983), p. 396–398 (in Russian).
10. V.V. Lychagin, V. Roubtsov: *Local classifications of Monge-Ampère equations*, Soviet. Math. Dokl, vol 28,2 (1983), pp. 396–398.
11. V.V Lychagin, V. Roubtsov and I.V. Chekalov: *A classification of Monge-Ampère equations*, Ann. scient. Ec. Norm. Sup, 4 ème série, t.26 (1993), pp.281–308.
12. McI M.E. McIntyre, I. Roulstone: *Are there Higher -Accuracy Analogues of Semigeostrophic Theory?*, Large-scale Atmosphere-Ocean Dynamics, II (2001).
13. J. Norbury and I. Roulstone (eds.) Large-Scale Atmosphere-Ocean Dynamics. Volume 1. Cambridge University Press (2002).
14. J. Norbury and I. Roulstone (eds.) Large-Scale Atmosphere-Ocean Dynamics. Volume 2. Cambridge University Press (2002).

15. V. Roubtsov, I. Roulstone: *Holomorphic structures in hydrodynamical models of nearly geostrophic flow*, Proc. R. Soc. Lond. A, 457 (2001).
16. C. Snyder, W. C. Skamarock, R. Rotunno: *A comparison of Primmitive-Equation and Semi-geostrophic Simulations of Baroclinic Waves*, Journal of the atm. sciences, vol 48, n° 19 (1991).

Chapter 6
Gas Flow with Phase Transitions: Thermodynamics and the Navier–Stokes Equations

Anton A. Gorinov, Valentin V. Lychagin, Mikhail D. Roop and Sergey N. Tychkov

6.1 Introduction

One-dimensional flows of gas or liquid are described by the following system of Navier–Stokes equations (see for example, [1]):

$$
\begin{cases}
\rho(u_t + u u_x) = -p_x + \eta u_{xx}, \\
\rho_t + (\rho u)_x = 0, \\
T\rho(s_t + u s_x) - kT_{xx} - \eta(u_x)^2 = 0.
\end{cases}
\tag{6.1}
$$

Here $\rho(t, x)$ is the density of the gas, $u(t, x)$ is the velocity, $p(t, x)$ is the pressure, $s(t, x)$ is the specific entropy, $T(t, x)$ is the temperature, k and η are coefficients of thermal conductivity and viscosity correspondingly, which are assumed to be constants.

The first equation of system (6.1) corresponds to the momentum conservation law of the medium, the second one is the continuity equation and the third one is the equation of heat conduction, which represents the energy conservation law. System (6.1) is incomplete. It consists of three equations for five unknown functions

A. A. Gorinov · V. V. Lychagin · M. D. Roop (✉) · S. N. Tychkov
Institute of Control Sciences of Russian Academy of Sciences,
Profsoyuznaya 65, Moscow, Russia
e-mail: mihail_roop@mail.ru

M. D. Roop
Lomonosov Moscow State University, Leninskie Gory 1, Moscow, Russia

A. A. Gorinov
e-mail: gorinov.anton@physics.msu.ru

V. V. Lychagin
e-mail: valentin.lychagin@uit.no

S. N. Tychkov
e-mail: sergey.lab06@gmail.com

© Springer Nature Switzerland AG 2019
R. A. Kycia et al. (eds.), *Nonlinear PDEs, Their Geometry, and Applications*,
Tutorials, Schools, and Workshops in the Mathematical Sciences,
https://doi.org/10.1007/978-3-030-17031-8_6

$\rho(t, x), u(t, x), p(t, x), s(t, x), T(t, x)$. To make it complete we need two additional equations describing thermodynamic properties of the gas—the equations of state.

The paper has the following structure. In Sect. 6.2, we give a geometrical description of the thermodynamic state. We consider the thermodynamic state as two-dimensional Lagrangian manifold, which can be defined by two equations with compatibility condition.

In Sect. 6.3, we study state equations and corresponding Lagrangian manifolds for van der Waals gases and get its applicable domains with a description of phase transitions.

In Sect. 6.4, we look for solutions as asymptotic expansions and analyse the zeroth and the first-order approximations.

In Sect. 6.5, we show space-time domains corresponding to different phases of the medium.

Essential computations in this paper were done in Maple with the Differential Geometry package created by I. Anderson, the corresponding files could be found in http://d-omega.org/appendices/.

6.2 Geometric Representation of Thermodynamic States

Let $\mathbb{R}^5$ be a 5-dimensional contact space equipped with the coordinates (p, ρ, e, T, s), where e represents the specific energy and the other coordinates represent thermodynamic quantities mentioned above, and the contact 1-form [2, 3]:

$$\theta = \frac{1}{T}de - ds - \frac{p}{T\rho^2}d\rho.$$

In our consideration, the thermodynamic state is a 2-dimensional Legendrian manifold $L \subset \mathbb{R}^5(p, \rho, e, T, s)$, such that

$$\theta|_L = 0.$$

The last condition means that the first law of thermodynamics holds on the manifold L.

If the specific entropy is a given function $s = s(e, \rho)$, the condition $\theta|_L = 0$ leads to the following relations, that define 2-dimensional Legendrian manifold $L \subset \mathbb{R}^5(p, \rho, e, T, s)$:

$$s = s(e, \rho), \quad p = -\rho^2 \frac{s_\rho}{s_e}, \quad T = \frac{1}{s_e}. \tag{6.2}$$

Since the equations of state usually include the specific energy and do not include the specific entropy, we shall eliminate the specific entropy s from our consideration. To this end, we consider the projection $\phi: \mathbb{R}^5 \to \mathbb{R}^4$, $\phi: (\rho, p, e, T, s) \mapsto (\rho, p, e, T)$ and symplectic space $\mathbb{R}^4$ equipped with structure 2-form

$$\Omega = d\theta = \frac{1}{T^2} de \wedge dT - \frac{1}{T\rho^2} dp \wedge d\rho + \frac{p}{T^2\rho^2} dT \wedge d\rho.$$

The restriction of the map ϕ on the Legendrian surface L is a diffeomorphism on the image $\overline{L} = \phi(L)$. The surface $\overline{L} \subset \mathbb{R}^4$ is a Lagrangian manifold. Equally, the thermodynamic state can be considered as a 2-dimensional Lagrangian manifold $\overline{L} \subset \mathbb{R}^4(\rho, p, e, T)$, i.e. $\Omega|_{\overline{L}} = 0$.

Any 2-dimensional Lagrangian manifold $\overline{L} \subset \mathbb{R}^4(\rho, p, e, T)$ is defined by the two equations

$$\begin{cases} f(\rho, p, e, T) = 0, \\ g(\rho, p, e, T) = 0, \end{cases} \tag{6.3}$$

and the condition that the surface $\overline{L}$ is Lagrangian can be written as:

$$[f, g] = 0 \text{ on } \overline{L}, \tag{6.4}$$

where $[f, g]$ is the Poisson bracket with respect to the symplectic form Ω, i.e.

$$[f, g] \, \Omega \wedge \Omega = df \wedge dg \wedge \Omega.$$

In coordinates (ρ, p, e, T) this bracket has the following form:

$$[f, g] = T\rho^2 \left(\frac{\partial f}{\partial p} \frac{\partial g}{\partial \rho} - \frac{\partial f}{\partial \rho} \frac{\partial g}{\partial p} \right) + T^2 \left(\frac{\partial f}{\partial T} \frac{\partial g}{\partial e} - \frac{\partial f}{\partial e} \frac{\partial g}{\partial T} \right) + Tp \left(\frac{\partial f}{\partial p} \frac{\partial g}{\partial e} - \frac{\partial f}{\partial e} \frac{\partial g}{\partial p} \right).$$

Condition (6.4) means the integrability of the following system of PDEs:

$$\begin{cases} f\left(\rho, -\rho^2 \frac{s_\rho}{s_e}, e, \frac{1}{s_e}\right) = 0, \\ g\left(\rho, -\rho^2 \frac{s_\rho}{s_e}, e, \frac{1}{s_e}\right) = 0. \end{cases}$$

Thus, in what follows, by the system of Navier–Stokes equations we shall understand system (6.1) together with two additional equations of state (6.3) satisfying relation (6.4).

6.3   Van der Waals Gases

6.3.1   The Equations of State

The most important class of real gases is described by the van der Waals equation:

$$f(\rho, p, e, T) = (p + a\rho^2)\left(\frac{1}{\rho} - b\right) - RT, \tag{6.5}$$

here a is a characteristic of the gas responsible for the interaction between the particles and b is particles' volume, R is the universal gas constant. To find the second equation we use the condition of compatibility, which is expressed in (6.4) and gives the following result:

Proposition 6.1 *Assuming that the specific energy is a function of density ρ and temperature T*

$$g(\rho, p, e, T) = e - \beta(\rho, T),$$

the second state equation for the van der Waals gas has to be in the form

$$\beta(\rho, T) = -a\rho + E(T),$$

where $E(T)$ is a smooth function.

Proposition 6.2 *Since the specific energy is the sum of the energy of the particles' motion and the energy of their interaction, the function $E(T)$ has to be as follows*

$$E(T) = \frac{fR}{2}T,$$

here f is a degree of freedom.

Thus, the equations of state for the van der Waals gas are

$$\begin{cases} (p + a\rho^2)\left(\frac{1}{\rho} - b\right) - RT = 0, \\ e - \frac{fRT}{2} + a\rho = 0. \end{cases} \tag{6.6}$$

To get the specific entropy s as function of the specific energy e and density ρ, we integrate system (6.6) using (6.2). We have

$$s(e, \rho) = \frac{fR}{2}\ln(e + a\rho) + R\ln\left(\frac{1}{\rho} - b\right) + s_0. \tag{6.7}$$

Thus, formulae (6.6) and (6.7) define the thermodynamic state of van der Waals gases or Legendrian manifold L.

6.3.2 Applicable Domains for the Van der Waals Gas

In this section we discuss domains where the van der Waals model is valid. We call them applicable.

Let $V = 1/\rho$ be a specific volume. First of all, we note that due to (6.5) we have restriction for volumes to consider:

$$V > b.$$

This condition is absolutely clear from the physical point of view: the volume occupied by the gas cannot be less than the particles' volume.

There is another condition for thermodynamic quantities to be applicable. The Lagrangian manifold $\overline{L}$ is equipped with quadratic differential form $\kappa|_{\overline{L}}$, which has to be negative [6]:

$$\kappa|_{\overline{L}} = d(T^{-1}) \cdot de + d(pT^{-1}) \cdot dV.$$

This allows to select domains on $\overline{L}$ where the model of van der Waals gas is applicable. In case of van der Waals gases the form $\kappa|_{\overline{L}}$ is following:

$$\kappa|_{\overline{L}} = -\frac{fR}{2(e+a/V)^2} de \cdot de + \frac{fRa}{(e+a/V)^2 V^2} de \cdot dV$$
$$+ \left(\frac{fRa}{V^3(e+a/V)} - \frac{fRa^2}{2V^4(e+a/V)^2} - \frac{R}{(V-b)^2} \right) dV \cdot dV.$$

The form $\kappa|_{\overline{L}}$ is negative if and only if its determinant is positive, which leads to the following inequality:

$$eV^3 - a(f-1)V^2 + 2abfV - ab^2 f > 0.$$

Using $e = fRT/2 - a/V$ we get:

$$\frac{1}{2}RTV^3 - aV^2 + 2abV - ab^2 > 0. \tag{6.8}$$

Let us introduce contact transformation

$$\tilde{T} = \frac{T}{T_{\text{crit}}}, \quad \tilde{V} = \frac{V}{V_{\text{crit}}}, \quad \tilde{p} = \frac{p}{p_{\text{crit}}}, \quad \tilde{e} = \frac{e}{e_{\text{crit}}}, \quad \tilde{s} = \frac{s}{s_{\text{crit}}},$$

where $T_{\text{crit}}, V_{\text{crit}}, p_{\text{crit}}, e_{\text{crit}}, s_{\text{crit}}$ are critical parameters for van der Waals gases:

$$T_{\text{crit}} = \frac{8a}{27Rb}, \quad V_{\text{crit}} = 3b, \quad p_{\text{crit}} = \frac{a}{27b^2}, \quad e_{\text{crit}} = \frac{a}{9b}, \quad s_{\text{crit}} = \frac{3R}{8}.$$

Then inequality (6.8) can be written in dimensionless variables $\tilde{T}$ and $\tilde{V}$:

$$4\tilde{V}^3\tilde{T} - 9\tilde{V}^2 + 6\tilde{V} - 1 > 0,$$

which defines the applicable domains of specific volume and temperature for the van der Waals gas. They are shown in Fig. 6.1. The picture shows that the van der Waals model is correct at any point (V, T) over the critical one. The forbidden area corresponds to phase transitions.

Fig. 6.1 Applicable domains for the van der Waals gas. White area corresponds to forbidden volumes and temperatures

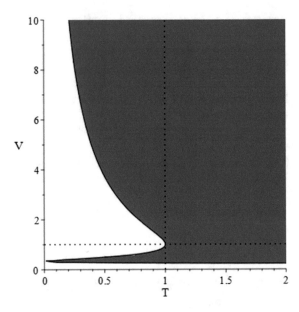

Fig. 6.2 Isotherm for the van der Waals gas

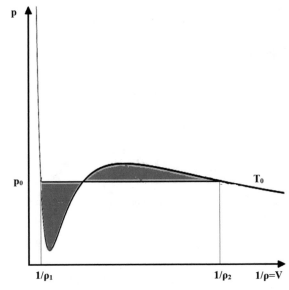

6.3.3 Phase Transitions

Phase transitions for the van der Waals gas can be described by means of Fig. 6.2. Grey domains in this picture correspond to intermediate state. At points $1/\rho_2$ and $1/\rho_1$, phase transition starts and finishes correspondingly. To find these points, we use the following condition of thermodynamic equilibrium, which claims that the chemical potential of different phases of our system is the same:

$$\mu(T_0, p_0, \rho_1) = \mu(T_0, p_0, \rho_2) = \mu_0,$$

where T_0 and p_0 are the temperature and the pressure of phase transition. The expression for the chemical potential of gases is

$$\mu = e - Ts + \frac{p}{\rho},$$

and for van der Waals gases it can be expressed in terms of pressure p, temperature T and density ρ:

$$\mu = \frac{fRT}{2} - \rho a - T\left(\frac{fR}{2}\ln\left(\frac{fR}{2}T\right) + R\ln\left(\frac{1}{\rho} - b\right)\right) + \frac{p}{\rho}.$$

Moreover, the equation of state of the gas must be satisfied at the points (T_0, p_0, ρ_1) and (T_0, p_0, ρ_2). As a result we obtain the following system of equations for ρ_1 and ρ_2:

$$\mu_0 = \frac{fRT_0}{2} - \rho_1 a - T_0\left(\frac{fR}{2}\ln\left(\frac{fR}{2}T_0\right) + R\ln\left(\frac{1}{\rho_1} - b\right)\right) + \frac{p_0}{\rho_1}, \quad (6.9)$$

$$\mu_0 = \frac{fRT_0}{2} - \rho_2 a - T_0\left(\frac{fR}{2}\ln\left(\frac{fR}{2}T_0\right) + R\ln\left(\frac{1}{\rho_2} - b\right)\right) + \frac{p_0}{\rho_2}, \quad (6.10)$$

$$p_0 - p_0\rho_1 b + a\rho_1^2 - ab\rho_1^3 - \rho_1 RT_0 = 0, \quad (6.11)$$

$$p_0 - p_0\rho_2 b + a\rho_2^2 - ab\rho_2^3 - \rho_2 RT_0 = 0. \quad (6.12)$$

Eliminating μ_0 and p_0 from (6.9)–(6.12) we get the following equations:

$$(\rho_1 - \rho_2)(RT_0 - a(\rho_1 + \rho_2)(b\rho_1 - 1)(b\rho_2 - 1)) = 0,$$

$$\rho_1 RT_0(b\rho_2 - 1)\ln\left(\frac{\rho_1(1 - b\rho_2)}{\rho_2(1 - b\rho_1)}\right) + (\rho_1 - \rho_2)(a\rho_1(1 - b\rho_2) + ab\rho_2^2 + RT_0 - a\rho_2) = 0.$$

There is the trivial solution $\rho_1 = \rho_2$, which is out of interest, because the temperature is assumed to be under the critical value. In general case the solution is given by Fig. 6.3. We can see that the straight line and the points C and D correspond to

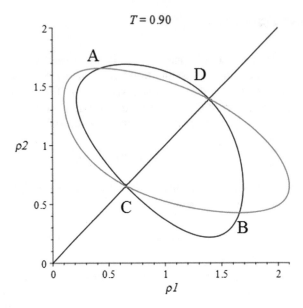

Fig. 6.3 Solution

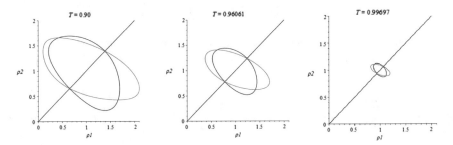

Fig. 6.4 Dynamics of the solution

trivial solution $\rho_1 = \rho_2$. The two other points A and B of intersection of the curves correspond to solution for ρ_1 and ρ_2. Since we have not specified which density is greater, the points A and B define the same solution. If we change the temperature of phase transition T_0, we can see that values ρ_1 and ρ_2 become closer and there is only one solution $\rho_1 = \rho_2$ when $T_0 = T_{crit}$ (Fig. 6.4).

6.4   Asymptotic Expansions for Solution

6.4.1   Zeroth-Order Approximation

Recall that we consider the system of equations:

$$
\begin{cases}
\rho(u_t + uu_x) = -p_x + \eta u_{xx}, \\
\rho_t + (\rho u)_x = 0, \\
T\rho(s_t + us_x) - kT_{xx} - \eta(u_x)^2 = 0,
\end{cases}
\tag{6.13}
$$

extended by the equations of state (Legendrian manifold L):

$$
\begin{cases}
s(e, \rho) = \frac{fR}{2} \ln(e + a\rho) + R \ln \left(\frac{1}{\rho} - b \right) + s_0, \\
(p + a\rho^2) \left(\frac{1}{\rho} - b \right) - RT = 0, \\
e - \frac{fRT}{2} + a\rho = 0.
\end{cases}
\tag{6.14}
$$

We are looking for asymptotical solution of system (6.13)-(6.14) with respect to van der Waals parameters a and b:

$$
u(t, x) = u_0(t, x) + au_1(t, x) + bu_2(t, x) + \cdots,
$$

$$
\rho(t, x) = \rho_0(t, x) + a\rho_1(t, x) + b\rho_2(t, x) + \cdots,
$$

$$
e(t, x) = e_0(t, x) + ae_1(t, x) + be_2(t, x) + \cdots.
$$

For simplicity, we shall continue to use $u(t, x)$, $\rho(t, x)$, $e(t, x)$ instead of $u_0(t, x)$, $\rho_0(t, x)$, $e_0(t, x)$ and get the following equations which describe the zeroth-order approximation:

$$
\rho(u_t + uu_x) + \frac{2}{f}(\rho e)_x - \eta u_{xx} = 0,
\tag{6.15}
$$

$$
\rho_t + \rho_x u + \rho u_x = 0,
\tag{6.16}
$$

$$
\rho(e_t + ue_x) - \frac{2}{f}e(\rho_t + u\rho_x) - \eta(u_x)^2 - \frac{2k}{Rf}e_{xx} = 0.
\tag{6.17}
$$

This system corresponds to equations (6.1) for the ideal gas. It defines a smooth submanifold $\mathcal{E} \subset J^2(\pi)$, here π is a 3-dimensional bundle [4, 5]:

$$
\pi : \mathbb{R}^5 \to \mathbb{R}^2, \quad \pi : (t, x, u, \rho, e) \mapsto (t, x).
$$

Proposition 6.3 *The symmetry algebra $\mathfrak{g}$ of the system $\mathcal{E}$ is solvable and generated by the following vector fields on the space $J^0(\pi)$:*

Table 6.1 The Lie algebra structure

Field	X_1	X_2	X_3	X_4	X_5
X_1	0	0	X_2	X_1	0
X_2	0	0	0	0	X_2
X_3	$-X_2$	0	0	$-X_3$	X_3
X_4	$-X_1$	0	X_3	0	0
X_5	0	$-X_2$	$-X_3$	0	0

$$X_1 = \partial_t, \quad X_2 = \partial_x, \quad X_3 = t\partial_x + \partial_u,$$

$$X_4 = t\partial_t + \rho\partial_\rho - u\partial_u - 2e\partial_e,$$

$$X_5 = x\partial_x - 2\rho\partial_\rho + u\partial_u + 2e\partial_e,$$

The Lie algebra structure is represented in Table 6.1. The table shows that the Lie algebra $\mathfrak{g}$ is solvable:

$$\mathfrak{g}^{(1)} = [\mathfrak{g}, \mathfrak{g}] = \langle X_1, X_2, X_3 \rangle$$

$$\mathfrak{g}^{(2)} = [\mathfrak{g}^{(1)}, \mathfrak{g}^{(1)}] = \langle X_2 \rangle, \quad \mathfrak{g}^{(3)} = [\mathfrak{g}^{(2)}, \mathfrak{g}^{(2)}] = 0.$$

We are going to find solutions of system (6.15)–(6.17) invariant with respect to a one-dimensional subalgebra of $\mathfrak{g}$ and also we want to get the reduced ordinary system having as many symmetries as possible. Since the symmetries in the normalizer of one-dimensional subalgebra $\mathfrak{h}$ of the Lie algebra $\mathfrak{g}$ are the symmetries of the reduced equations, we compute the normalizers of all admissible one-dimensional subalgebras in $\mathfrak{g}$:

$$N_X = \{ Y \in \mathfrak{g} \mid [X, Y] = \lambda X \}, \quad \text{where } \lambda \text{ is a parameter.}$$

One may show that in our case one-dimensional subalgebra

$$\mathfrak{h} = \langle \alpha_2 X_2 + \alpha_3 X_3 + \alpha_5 X_5 \rangle,$$

where α_j are constants, has the biggest normalizer

$$N_\mathfrak{h} = \left\langle \frac{\alpha_2}{\alpha_3} X_1 + \frac{\alpha_3}{\alpha_5} X_3 + X_5, \ -\frac{\alpha_3}{\alpha_5} X_3 + X_4, \ X_1 - \frac{\alpha_3}{\alpha_5} X_2 \right\rangle.$$

The $\mathfrak{h}$-invariant solution of system (6.15)–(6.17) has the following form:

$$e(t, x) = (\alpha_3 t + \alpha_5 x + \alpha_2)^2 F_1(t), \quad \rho(t, x) = \frac{F_2(t)}{(\alpha_3 t + \alpha_5 x + \alpha_2)^2},$$

$$u(t, x) = (\alpha_3 t + \alpha_5 x + \alpha_2) F_3(t) - \frac{\alpha_3}{\alpha_5},$$

and the reduced ordinary equations are

$$F_3' + \alpha_5 F_3^2 = 0, \quad -F_2' + \alpha_5 F_2 F_3 = 0, \tag{6.18}$$

$$F_1' F_2 R f - 2R F_1 F_2' - \alpha_5 \left(F_1 (4k\alpha_5 - 2R(f + 2) F_2 F_3) + R f \eta \alpha_5 F_3^2 \right) = 0. \tag{6.19}$$

After the integration of these equations we get the following:

$$\rho(t, x) = \frac{C_2(\alpha_5 t + C_1)}{(\alpha_3 t + \alpha_5 x + \alpha_2)^2}, \quad u(t, x) = \frac{\alpha_3 t + \alpha_5 x + \alpha_2}{\alpha_5 t + C_1} - \frac{\alpha_3}{\alpha_5}, \tag{6.20}$$

$$e(t, x) = (\alpha_3 t + \alpha_5 x + \alpha_2)^2 \left(\frac{\alpha_5 R f \eta}{2(RC_2 - 2k\alpha_5)(\alpha_5 t + C_1)^2} + C_3(\alpha_5 t + C_1)^{-2 - \frac{2}{f} + \frac{4k\alpha_5}{C_2 R f}} \right), \tag{6.21}$$

where C_1, C_2 and C_3 are constants.

This solution represents the zeroth-order approximation of the solution for the van der Waals gas. Since the flows of vector fields X_2 and X_3 are the shift and Galilean transformation correspondingly, their influence on the solution is not crucial: Galilean transformation makes the frame of reference move with constant velocity and the shift along the x-axis just changes the location of the origin. Assuming that our frame of reference does not move and the point $x = 0$ corresponds to the origin we shall take $\alpha_3 = \alpha_2 = 0$, $\alpha_5 = 1$ in (6.20)–(6.21).

6.4.2   First-Order Approximation

The equations for the first-order corrections $u_1(t, x)$, $\rho_1(t, x)$ and $e_1(t, x)$ can be written in the following form:

$$\begin{pmatrix} u_1 \\ \rho_1 \\ e_1 \end{pmatrix}_t = A(t, x) \begin{pmatrix} u_1 \\ \rho_1 \\ e_1 \end{pmatrix}_{xx} + B(t, x) \begin{pmatrix} u_1 \\ \rho_1 \\ e_1 \end{pmatrix}_x + C(t, x) \begin{pmatrix} u_1 \\ \rho_1 \\ e_1 \end{pmatrix} + D(t, x), \tag{6.22}$$

here matrixes A, B, C and D depend on the functions $u(t, x)$, $\rho(t, x)$ and $e(t, x)$ of the zeroth-order approximation found in Sect. 6.4.1:

$$A(t, x) = \frac{1}{\rho}\begin{pmatrix} \eta & 0 & 0 \\ 0 & 0 & 0 \\ 0 & 0 & \frac{2k}{fR} \end{pmatrix}, \quad B(t, x) = -\begin{pmatrix} u & \frac{2e}{f\rho} & \frac{2}{f} \\ \rho & u & 0 \\ \frac{2e}{f} & -\frac{2\eta u_x}{\rho} & 0 & u \end{pmatrix}$$

$$C(t, x) = -\begin{pmatrix} u_x & \frac{2Re_x + fR(u_t(t,x) + uu_x)}{\rho fR} & \frac{\rho_x}{\rho f} \\ \rho_x & u_x & 0 \\ e_x & \frac{2Reu_x + fR(e_t + ue_x)}{\rho fR} & -\frac{2(\rho_t + u\rho_x)}{\rho f} \end{pmatrix}, \quad D(t, x) = \begin{pmatrix} 2\rho_x\left(1 - \frac{2}{f}\right) \\ 0 \\ \frac{2k\rho_{xx} + \rho R(\rho_t + u\rho_x)(2-f)}{\rho fR} \end{pmatrix}$$

System (6.22) is linear non-homogeneous system of partial differential equations and its general solution can be represented as the sum of general solution of the corresponding homogeneous system ($D = 0$) and particular solution of non-homogeneous system.

Let us consider homogeneous system:

$$\begin{pmatrix} u_1 \\ \rho_1 \\ e_1 \end{pmatrix}_t = A(t, x)\begin{pmatrix} u_1 \\ \rho_1 \\ e_1 \end{pmatrix}_{xx} + B(t, x)\begin{pmatrix} u_1 \\ \rho_1 \\ e_1 \end{pmatrix}_x + C(t, x)\begin{pmatrix} u_1 \\ \rho_1 \\ e_1 \end{pmatrix},$$

Since its coefficients depend on the zeroth-order solution, which are invariant with respect to X_5, we are looking for the solution in the form representing the eigenfunctions of differential operator $x\partial_x = \pi_*(X_5)$:

$$\rho_1(t, x) = R(t)x^l, \quad e_1(t, x) = E(t)x^m, \quad u_1(t, x) = U(t)x^n.$$

The numbers l, m and n satisfy the linear non-homogeneous system:

$$\begin{cases} m - l = 4, \\ n - l = 3, \\ m - n = 1. \end{cases}$$

Its general solution is

$$\begin{pmatrix} l \\ m \\ n \end{pmatrix} = \begin{pmatrix} l \\ 4 + l \\ 3 + l \end{pmatrix} = \begin{pmatrix} 0 \\ 4 \\ 3 \end{pmatrix} + l\begin{pmatrix} 1 \\ 1 \\ 1 \end{pmatrix}.$$

Time-dependent part of the first-order corrections $U(t)$, $R(t)$ and $E(t)$ satisfies the following ODE system:

$$\begin{pmatrix} \dot{U} \\ \dot{R} \\ \dot{E} \end{pmatrix} = \Phi(t)\begin{pmatrix} U \\ R \\ E \end{pmatrix}.$$

Matrix $\Phi(t)$ has the following form:

Fig. 6.5 Phase picture. The thick curve at the plane (t, x) separates domains with different phases. This allows to define which phase corresponds to the medium at given point x and at given time moment t

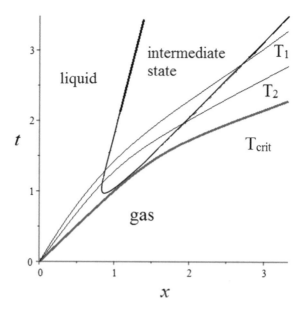

$$\Phi(t) = \begin{pmatrix} \dfrac{\eta(3+l)(2+l) - \dot{F}_2(4+l)}{F_2} & \dfrac{-2(2+l)F_1}{fF_2} & -\dfrac{2(2+l)}{f} \\ -(l+1)F_2 & -(l+1)F_3 & 0 \\ \dfrac{2(F_3\eta(3+l)f - F_1F_2(l+3+f))}{fF_2} & -\dfrac{f\dot{F}_1 + 2F_1F_3(f+1)}{fF_2} & \dfrac{2k(4+l)(3+l) - R\dot{F}_2(2+f(4+l))}{fRF_2} \end{pmatrix},$$

where functions $F_1(t)$, $F_2(t)$ and $F_3(t)$ are the solutions of reduced system (6.18)–(6.19).

If either $C_3 = 0$ or we consider gases with zero viscosity $\eta = 0$, this system can be integrated in the same way as it was done for the spatial part, because in this case components of the matrix $\Phi(t)$ are the homogeneous functions in t.

6.5   Phase Transitions Along the Gas Flow

In this section, we describe space-time domains corresponding to different phases of the medium. Since we have a solution of system (6.13)–(6.14), we can compute the corresponding set of points (t, x) of the same temperature T_0. In Sect. 6.3 we have developed a method that allows (for a given value of temperature T_0) to define the densities $\rho_1^{(0)}$ and $\rho_2^{(0)}$ of liquid and gas phases respectively, between which the phase transition occurs. The corresponding points can be found on the plane (t, x) as well. Changing the temperature, we get a set of points $(\rho_1^{(i)}, \rho_2^{(i)})$, which form a curve on the plane (t, x). This curve separates different phases of the medium. The result of this procedure for the solution obtained in Sect. 6.4 is shown in Fig. 6.5.

The curves labelled by T_1 and T_2 are the isotherms. The red curve corresponds to the critical isotherm, under which we have no phase transitions. For a given spatial coordinate x, our medium passes through three states while the time is running: gas, intermediate state and liquid.

This picture is an approximation for the real one. It can be refined by computation of further series of asymptotics.

Acknowledgements This work was supported by the Russian Foundation for Basic Research (project No 18-29-10013).

References

1. Batchelor, G.K.: An introduction to fluid dynamics. Cambridge Univ. Press, Cambridge (2000).
2. Duyunova, A.A., Lychagin, V.V.,Tychkov, S.N.: Classification of equations of state for viscous fluids. Doklady Mathematics. **95**, 172–175 (2017).
3. Duyunova, A.A., Lychagin, V.V.,Tychkov, S.N.: Differential invariants for flows of viscid fluids. J. Geom. Phys. **121**, 309–316 (2017).
4. Vinogradov, A.M., Krasilshchik, I.S.: Symmetries and Conservation Laws for Differential Equations of Mathematical Physics. Factorial, Moscow (1997).
5. Krasilshchik, I.S., Lychagin, V.V., Vinogradov, A.M.: Geometry of Jet Spaces and Nonlinear Partial Differential Equations. Gordon and Breach Science Publishers (1986).
6. Lychagin, V.V.: Contact Geometry, Measurement and Thermodynamics. Lectures at Summer School, Wisla, 2018.

Chapter 7
Differential Invariants in Thermodynamics

Eivind Schneider

7.1 Introduction

The fundamental thermodynamic relation can be formulated as $dE - (TdS - \sum_{i=2}^{n} p_i dq^i) = 0$, where E is the internal energy, S the entropy, and T the temperature while q^i and p_i are additional extensive and intensive variables, respectively. In terms of information gain I which is up to an additive constant equal to $-S$, it can be written as $dE + TdI + \sum p_i dq^i = 0$. Geometrically, we should interpret this to mean that a thermodynamic state is a Legendrian manifold, integral to a contact distribution. Let V be a vector space. We consider the contact 1-form given by $du - \sum_{i=1}^{n} \lambda_i dx^i$ on $V \times \mathbb{R} \times V^*$ (with coordinates x^i, u, λ_i). By taking $\lambda_1 = -T^{-1}, x^1 = E, \lambda_i = -T^{-1} p_i, x^i = q^i$, we can relate this to the 1-form above, which means that the Legendrian manifold is given locally over a neighborhood $D \subset V$ by

$$\{u = I(x), \lambda_i = \partial_{x^i} I(x) \mid x \in D\} \subset V \times \mathbb{R} \times V^*.$$

In Sect. 7.3, we describe two Lie group actions that appear naturally on the space $V \times \mathbb{R} \times V^*$. They arise from the fact that we may change the basis in V and change units of information. After that, in Sect. 7.4, we approach the equivalence problem of thermodynamic states under these Lie group actions by computing differential invariants of the information gain function. In Sect. 7.5, we discuss our results in the context of gases. But first, we outline how these Legendrian manifolds naturally appear in the context of measuring random vectors (following [5]), as this will help us find the natural Lie group actions acting on them.

E. Schneider (✉)
UiT The Arctic University of Norway, Tromsø, Norway
e-mail: eivind.schneider@uit.no

© Springer Nature Switzerland AG 2019
R. A. Kycia et al. (eds.), *Nonlinear PDEs, Their Geometry, and Applications*,
Tutorials, Schools, and Workshops in the Mathematical Sciences,
https://doi.org/10.1007/978-3-030-17031-8_7

7.2 Geometry of Thermodynamics

The process of measuring random vectors in a vector space V can be thought about as a map $X \colon (\Omega, \mathcal{A}, q) \to V$ from a probability space to V. This will depend on the probability measure q, which we will change in order to measure different vectors. Assume that the expected value in V is 0, i.e., $E(X) = \int_\Omega X dq = 0$. We restrict to finite-dimensional vector spaces here, even though the Bochner integral lets us treat more general Banach spaces (see [5]). Note however that the linear structure on V is important, and this will play a role later when we consider Lie group actions on V.

If we want to measure a vector $x \in V$, we choose a measure p different from q, but equivalent to it. Applying the Radon–Nikodym theorem tells us that there is a function ρ such that $dp = \rho dq$ and

$$\int_\Omega \rho dq = 1, \qquad \int_\Omega \rho X dq = x.$$

These conditions do not determine ρ uniquely. We define the information gain $I(p, q) = \int_\Omega \rho \ln \rho dq$, and add the requirement that ρ minimizes $I(p, q)$. This is the principle of minimal information gain.

As a result (see [5]), we get $\rho = \frac{1}{Z(\lambda)} e^{\langle \lambda, X \rangle}$ with $\lambda \in V^*$ where $Z(\lambda) = \int_\Omega e^{\langle \lambda, X \rangle} dq$ is called the partition function. Due to $\int_\Omega \rho X dq = x$, we get $d_\lambda Z = Z(\lambda)x$. Thus, if we define $H(\lambda) = -\ln Z(\lambda)$, we end up with $x = -d_\lambda H$. And, we also get $I(p, q) = H(\lambda) - \langle \lambda, d_\lambda H \rangle = H(\lambda) + \langle \lambda, x \rangle$. Assuming that $x = -d_\lambda H$ has a unique solution $\lambda(x)$, we may write $I = I(x) = H(\lambda(x)) + \langle \lambda(x), x \rangle$. Then, we get $d_x I = \lambda$, where the functions H and I are related by $H = I + \langle \lambda, x \rangle$.

Now, let x^i be coordinates on the vector space V and λ_i be the dual coordinates on V^*. On the space $V \times V^*$, we have the natural symplectic form $\omega = \sum_{i=1}^n d\lambda_i \wedge dx^i$. The function $H(\lambda)$ determines a submanifold $L_H = \{x^i = -\frac{\partial H}{\partial \lambda_i}\} \subset V \times V^*$ which is Lagrangian with respect to ω, meaning that $\omega|_{L_H} = 0$.

The Lagrangian manifold $L_H \subset V \times V^*$ can be extended to a manifold $\tilde{L}_H \subset V \times \mathbb{R} \times V^*$. Let u be the coordinate on $\mathbb{R}$. Then, we get a submanifold in $V \times \mathbb{R} \times V^*$ which is locally described by the function I over a neighborhood $D \subset V$:

$$\tilde{L}_H = \{u = I(x), \lambda_i = \partial_{x^i} I(x) \mid x \in D\} \in V \times \mathbb{R} \times V^*.$$

Thus, the principle of minimal information gain leads to a submanifold in $V \times \mathbb{R} \times V^*$ which is Legendrian with respect to the 1-form $\theta = du - \sum \lambda_i dx^i$.

Conversely, one can ask whether it is possible to reconstruct I, H, and Z from any Legendrian manifold $\tilde{L} \subset V \times \mathbb{R} \times V^*$. This is the case if the symmetric 2-form $(\sum d\lambda_i dx^i)|_{\tilde{L}}$ is positive definite, where the product is the symmetric one. By starting with such a Legendrian manifold $\tilde{L}$, one can recover the functions

$$I = u|_{\tilde{L}}, \quad H = I - \langle \lambda, x \rangle|_{\tilde{L}}, \quad Z = e^{-H}.$$

If $\tilde{L}$ is given by I as above, the symmetric form can be written as $\sum_{i,j} \frac{\partial^2 I}{\partial x^i \partial x^j} \, dx^i \otimes dx^j$.

7.3 Equivalence of Thermodynamical Systems

We start by describing two different Lie group actions arising naturally from the set up above, and thereby defining what it could mean for two thermodynamic states to be equivalent.

Affine Action

Since the choice of basis on V is arbitrary, we consider the Legendrian manifold $\tilde{L}$ up to linear transformations on $V \times V^*$. In addition, since the point $x = 0$ have no special significance, we add n translations to $GL(V)$ and obtain the affine group $\mathrm{Aff}(V) = V \rtimes GL(V)$ on V. After choosing a basis, the affine action on $V \times \mathbb{R} \times V^*$ is given by $(x^i, u, \lambda_k) \mapsto (\sum a^i_j x^j + c_i, u, \sum b^l_k \lambda_l)$ where the matrix (b^i_j) is the transpose of $(a^i_j)^{-1}$. The corresponding Lie algebra of vector fields is spanned by $x^i \partial_{x^j} - \lambda_j \partial_{\lambda_i}$ and ∂_{x^i}. Notice that this action does not alter the value of I (or H) at a point.

Scaling

In addition to changing basis, we may change the unit of information. The unit of information appears as the base of the logarithm we use, which in Sect. 7.2 was chosen to be e. Let $Z_a(\lambda_a)$, $H_a(\lambda_a)$, and $I_a(x_a)$ be defined in a way similar to $Z(\lambda)$, $H(\lambda)$, $I(x)$ above but with a as the base of the logarithm instead of e, and let $a = e^b$. The functions in the new units are related to the old ones in the following way.

$$Z_a(\lambda_a) = \int_\Omega a^{\langle \lambda_a, v \rangle} dq = \int_\Omega e^{\langle b\lambda_a, v \rangle} dq = Z(b\lambda_a)$$
$$H_a(\lambda_a) = -\log_a Z_a(\lambda_a) = -\log_a Z(b\lambda_a) = -\ln Z(b\lambda_a)/b = \frac{H(b\lambda_a)}{b}$$
$$I_a(x_a) = H_a(\lambda_a) + \langle \lambda_a, x_a \rangle = \frac{H(b\lambda_a) + \langle b\lambda_a, x_a \rangle}{b} = \frac{I(x_a)}{b}$$

In other words, we have the scaling transformation $(x, u, \lambda) \mapsto (x, bu, b\lambda)$ on $V \times \mathbb{R} \times V^*$. Denote by G_0 the Lie group of such transformations. The corresponding infinitesimal action is given by $\sum \lambda_i \partial_{\lambda_i} + u \partial_u$.

Remark 7.1 A natural question is where we allow b to take its values. We could restrict to $b > 0$, or to $b \neq 0$. We will discuss this in more detail when we compute the differential invariants of these Lie group actions.

7.4   Differential Invariants

The Legendrian manifold $\tilde{L}$ (the thermodynamic state) is locally determined by the information gain function I on V. We will compute differential invariants for I under the two Lie group actions $\mathrm{Aff}(V)$ and $G_0 \times \mathrm{Aff}(V)$. We consider I as local a section of the trivial bundle $V \times \mathbb{R}$ on which we continue to use coordinates $x^1, \ldots, x^n, u$. In order to study the orbit space of such sections under the action of these Lie groups, we look at their prolonged action on the jet bundles $J^k(V) \to V \times \mathbb{R} = J^0(V)$ (we will use the simplified notation J^k).

Let x^i, u_σ be canonical coordinates on J^k where $0 \leq |\sigma| \leq k$ for the multi-index $\sigma = (i_1, \ldots, i_n), i_j \geq 0$. For example, when $n = 2$ we have coordinates x^i, u_{ij} on J^2, with $0 \leq i + j \leq 2$ and $i, j \geq 0$. For $|\sigma| = 0$, we will also use the notation $u_\sigma = u$. The section $u = I(x)$ on $V \times \mathbb{R}$ prolongs to a section on J^k given by $u = I(x), u_\sigma = \frac{\partial^{|\sigma|}}{\partial x^\sigma} I(x)$. We denote this prolongation by $j^k(I)$. Since diffeomorphisms on $V \times \mathbb{R}$ transform sections of $V \times \mathbb{R}$, they lift naturally to J^k. Thus, we can consider the action of the aforementioned Lie groups on J^k. In fact we already described their action on J^1 since J^1 can be naturally identified with $V \times \mathbb{R} \times V^*$.

Differential invariants are functions on J^k that are constant on the orbits of the Lie group actions. For transitive and algebraic Lie group actions, the global Lie–Tresse theorem [4] guarantees that the algebra of rational differential invariants separates orbits in general position in J^∞, and that it is finitely generated. Since $\mathrm{Aff}(V)$ acts transitively on the base V, and not at all along the fiber, it is clear that we also in this intransitive case can separate orbits by rational invariants (the algebra of differential invariants for the $\mathrm{Aff}(V)$-action can be gotten from that of $G_0 \times \mathrm{Aff}(V)$ by adding the $G_0 \times \mathrm{Aff}(V)$-invariant u). For more thorough treatments of jet bundles and differential invariants, we refer to [2, 3, 6].

For the Lie groups $\mathrm{Aff}(V)$ and $G_0 \times \mathrm{Aff}(V)$, we give a complete description of their algebras of rational differential invariants.

7.4.1   Differential Invariants Under Aff(V)

In order to describe the field of differential invariants, we follow [1], where differential invariants under the $GL(V)$-action are found.

Theorem 7.1 *The horizontal symmetric forms $\alpha_k = k! \sum_{|\sigma|=k} \frac{u_\sigma}{\sigma!} dx^\sigma$ are Aff(V)-invariant, for $k \geq 0$.*

From these symmetric forms, we may construct (rational) scalar differential invariants in the following way. First, $\alpha_0 = u$ is a scalar differential invariant. The symmetric 2-form α_2 is nondegenerate for points in general position in J^2, so we may use it to construct the vector $v_1 = \alpha_2^{-1}(\alpha_1)$. By using this vector, we may construct a new symmetric 2-form $\alpha_{1,3} = i_{v_1}\alpha_3$. We can use α_2 to turn $\alpha_{1,3}$ into an operator $A \colon T \to T$. For a point in J^3 in general position, the vectors $v_k = A^{k-1}(v_1)$ for

$k = 1, \ldots, n$ are independent and thus define a frame. Expressing α_k in terms of v_k gives us scalar differential invariants from their coefficients. In other words, the functions $\alpha_k(v_{i_1}, \ldots, v_{i_k})$ are differential invariants. Remark that all these differential invariants will be rational functions on J^k, and affine on the fibers of $J^k \to J^3$. In particular, we get only two independent differential invariants on J^2, given by α_0 and $\alpha_1(v_1)$.

This way of generating differential invariants may not be the most convenient one. Another way to generate the field is to use invariant derivations and a finite number of differential invariants, in accordance with the Lie–Tresse theorem. As invariant derivations, we can take $v_1, \ldots, v_n$. They are of the form $\sum \alpha_i D_{x^i}$, where D_{x^i} are total derivatives and α_i are functions on J^3 (and on J^2 for v_1).

Theorem 7.2 *The field of differential invariants are generated by the invariant derivations $v_1, \ldots, v_n$ together with the first-order invariant $\alpha_0 = u$, the second-order invariant $\alpha_1(v_1)$, the third-order invariants $\alpha_3(v_{i_1}, v_{i_2}, v_{i_3})$, and the fourth-order invariants $\alpha_4(v_{i_1}, v_{i_2}, v_{i_3}, v_{i_4})$.*

It is not difficult to see that this set of invariants is sufficient for generating all differential invariants of the higher order. However, we do not necessarily need all of them.

The two-dimensional case

We take a closer look at the case when V is two-dimensional, as they are particular important when we will consider gases. We have the first-order invariant $\alpha_0 = u$ and the second-order invariant

$$\alpha_1(v_1) = \frac{u_{10}^2 u_{02} - 2u_{10}u_{01}u_{11} + u_{01}^2 u_{20}}{(u_{20}u_{02} - u_{11}^2)}.$$

The invariant derivations are given by

$$v_1 = \frac{1}{u_{20}u_{02} - u_{11}^2} \left((u_{10}u_{02} - u_{01}u_{11})D_{x^1} + (u_{01}u_{20} - u_{10}u_{11})D_{x^2} \right),$$

$$v_2 = \frac{1}{(u_{20}u_{02} - u_{11}^2)^3} \Big(\big($$
$$-(u_{10}u_{02} - u_{01}u_{11})(3u_{10}u_{11}u_{02} - 2u_{01}u_{20}u_{02} - u_{01}u_{11}^2)u_{21}$$
$$+(u_{10}u_{11} - u_{01}u_{20})(3u_{10}u_{11}u_{02} - u_{01}u_{20}u_{02} - 2u_{01}u_{11}^2)u_{12}$$
$$+u_{02}(u_{10}u_{02} - u_{01}u_{11})^2 u_{30} - u_{11}(u_{10}u_{11} - u_{01}u_{20})^2 u_{03}\big)D_{x^1}$$
$$+\big((u_{10}u_{02} - u_{01}u_{11})(u_{10}u_{20}u_{02} + 2u_{10}u_{11}^2 - 3u_{01}u_{20}u_{11})u_{21}$$
$$-(u_{10}u_{11} - u_{01}u_{20})(2u_{10}u_{20}u_{02} + u_{10}u_{11}^2 - 3u_{01}u_{20}u_{11})u_{12}$$
$$-u_{11}(u_{10}u_{02} - u_{01}u_{11})^2 u_{30} + u_{20}(u_{10}u_{11} - u_{01}u_{20})^2 u_{03}\big)D_{x^2}\Big).$$

In this case, the four third-order differential invariants $\alpha_3(v_{i_1}, v_{i_2}, v_{i_3})$ are independent, and together with v_1, v_2, and $\alpha_1(v_1)$, they generate the algebra of differential invariants.

Note that when $\tilde{L}$ is the Legendrian manifold corresponding to the information gain function I we have, in coordinates, $(\sum d\lambda_i dx^i)|_{\tilde{L}} = \sum I_{x^i x^j} dx^i \otimes dx^j = \alpha_2|_I$. In particular, since we require α_2 to be a definite symmetric 2-form, we may find differential invariants from the curvature tensor of this 2-form. For example, the Ricci scalar is a third-order differential invariant. Such invariants are, however, invariant under much more general transformations, so they don't generate all Aff(V)-invariants. What's more, they will not be invariant under $G_0 \times$ Aff(V).

7.4.2 Differential Invariants Under $G_0 \times$ Aff(V)

Now, we consider the action by the Lie group $G_0 \times$ Aff(V). It acts on $V \times \mathbb{R}$ by $(b, A) \cdot (x, u) \mapsto (Ax, bu)$. We mentioned previously that we have a choice for the G_0-parameter b, since we may take it from either $\mathbb{R} \setminus \{0\}$ or $(0, \infty)$. This corresponds to $a = e^b$ in $(0, \infty) \setminus \{1\}$ or $(1, \infty)$, respectively. Here, we will stick to the first choice $\mathbb{R} \setminus \{0\}$, with the main reason that this gives the Zariski closure of the other option. The theorems we have for the existence of rational invariants separating orbits hold for algebraic (Zariski-closed) groups. However, the structure of both orbit spaces on J^k should be clear as soon as we understand one of them. Also, the obtained orbit space will be the same as that of positive I's under the action of the topologically connected component of the Lie group (not containing $u \mapsto -u$.

In order to describe the field of differential invariants we reuse ideas from the previous section. The symmetric forms α_k are scaled by the $\mathbb{R}^*$-action. If we modify them to $\beta_k = \alpha_k/\alpha_0$, for $k \geq 1$, we obtain $G_0 \times$ Aff(V)-invariant symmetric forms. By using these instead of α_k, we can generate the algebra of differential invariants in exactly the same way as we did in the previous section. Invariant vectors can be constructed from β_k in exactly the same way as above (from α_k), and they will in fact be the exact same vectors v_i as before.

Theorem 7.3 *The field of differential invariants are generated by the invariant derivations $v_1, \ldots, v_n$ together with the second-order invariant $\beta_1(v_1)$, the third-order invariants $\beta_3(v_{i_1}, v_{i_2}, v_{i_3})$, and the fourth-order invariants $\beta_4(v_{i_1}, v_{i_2}, v_{i_3}, v_{i_4})$.*

The two-dimensional case

When V is two-dimensional, we have the second-order invariant

$$\beta_1(v_1) = \frac{u_{10}^2 u_{02} - 2u_{10}u_{01}u_{11} + u_{01}^2 u_{20}}{(u_{20}u_{02} - u_{11}^2)u}.$$

The invariant derivations from the previous section are still invariant under the current Lie group action. We have the relations $\beta_3(v_{i_1}, v_{i_2}, v_{i_3}) = \alpha_3(v_{i_1}, v_{i_2}, v_{i_3})/\alpha_0$. These four invariants are thus also independent, and together with v_1, v_2, and $\beta_1(v_1)$, they generate the algebra of differential invariants.

7.5 Application to Gases

We explain how we can use differential invariants in order to distinguish gases under the Lie group actions considered above. We use the ideal gas, and the van der Waals gases as examples. In order to keep this chapter concise, we do not go into the detailed physics but instead refer to [5] for more details about gases.

The simplest gases can be described as Legendrian manifolds of the contact form $\theta = du - (-T^{-1})d\varepsilon - (-pT^{-1})dv$ where T is temperature, p is pressure, v is specific volume, and ε is specific energy. We consider this as a contact form on $V \times \mathbb{R} \times V^*$ where V is a two-dimensional vector space. In order to relate it to our formulas above, we let $x^1 = \varepsilon, x^2 = v$ (and $\lambda_1 = -T^{-1}, \lambda_2 = -pT^{-1}$).

7.5.1 Distinguishing Gases

Integral manifolds of θ are locally determined by the information gain function I. We can use the differential invariants from above to determine when two different Legendrian manifolds of this type are equivalent under the groups $\mathrm{Aff}(V)$ and $G_0 \times \mathrm{Aff}(V)$, respectively.

We outline first how to do it for the $\mathrm{Aff}(V)$-action. For a function f on J^k, we denote by $f|_I$ the restriction of f to the section $u = I(x)$, i.e., $f|_I = f \circ j^k(I)$. For a point in general position in J^3, the differential invariants $\xi = \alpha_1(v_1)$ and $\eta = \alpha_3(v_1, v_1, v_1)$ will be horizontally independent, meaning that $\hat{d}\xi \wedge \hat{d}\eta \neq 0$. Here, $\hat{d}$ denotes the horizontal differential, which can be defined in coordinates by $\hat{d}f = D_{x^1}(f)dx^1 + D_{x^2}(f)dx^2$, so that $(\hat{d}f)|_I = d(f|_I)$. Thus for generic I, we have $d(\xi|_I) \wedge d(\eta|_I) \neq 0$, so $\xi|_I$ and $\eta|_I$ can be taken as local coordinates on V. The four invariants $h_{ij} = \alpha_3(v_i, v_j, v_2)$ and $h_0 = \alpha_0$ may also be restricted to I, and the functions $h_0|_I, h_{ij}|_I$ on V may be written in terms of $\xi|_I$ and $\eta|_I$. The four functions $h_0|_I(\xi|_I, \eta|_I), h_{ij}|_I(\xi|_I, \eta|_I)$ determine the equivalence class of the Legendrian manifold given by I.

To check equivalence under the $G_0 \times \mathrm{Aff}(V)$-action, we use the invariants $\tilde{\xi} = \xi/h_0, \tilde{\eta} = \eta/h_0, \tilde{h}_{ij} = h_{ij}/h_0$ instead, and the equivalence class of I is determined by the three functions $\tilde{h}_{ij}|_I(\tilde{\xi}|_I, \tilde{\eta}|_I)$.

Remark 7.2 The functions $\tilde{h}_{ij}|_I(\tilde{\xi}|_I, \tilde{\eta}|_I)$ are not arbitrary functions. They must satisfy a system of differential equations defined by the differential syzygies in the algebra of differential invariants.

7.5.2 Ideal Gas

As our first example we take the ideal gas, which is defined by the state equations $pv = RT$ and $\varepsilon = \frac{n}{2}RT$ where n counts the degrees of freedom. As shown in [5], they give the following information gain function:

$$I = -R \ln \left(\ln(v) + \frac{n}{2} \ln(\varepsilon) \right) + C = -R \ln \left(\ln(x^2) + \frac{n}{2} \ln(x^1) \right) + C$$

We look at the differential invariants found above restricted to the section $u = I$.

We first consider the $\mathrm{Aff}(V)$-invariants. We have $\xi|_I = R(n+2)/2$, $\eta|_I = R(n+2)$. These are constant, so the ideal gas lies in fibers over the singular set in J^3 determined by the equation $\hat{d}\xi \wedge \hat{d}\eta = 0$. The functions $h_{ij}|_I$ are also constant multiples of $R(n+2)$. The derivations v_1 and v_2 are both constant multiples of $x^1\partial_{x^1} + x^2\partial_{x^2}$ on the ideal gas. The only nonconstant function we get from the invariants is $h_0|_I$.

If we consider $G_0 \times \mathrm{Aff}(V)$-invariants instead, we still have $(\hat{d}\tilde{\xi} \wedge \hat{d}\tilde{\eta})|_I = 0$, even though $\tilde{\xi}|_I$, $\tilde{\eta}|_I$ are not constant. Thus the ideal gas is a singular Legendrian manifold, also with respect to this Lie group action.

7.5.3 Van der Waals Gas

The state equations for van der Waals gases are $\left(p + \frac{a}{v^2}\right)(v - b) = RT$ and $\varepsilon = \frac{n}{2}RT - \frac{a}{v}$, and their information gain function is given by

$$I = -R \left((v - b) \left(\frac{a}{v} + \varepsilon \right)^{n/2} \right) + C = -R \left((x^2 - b) \left(\frac{a}{x^2} + x^1 \right)^{n/2} \right) + C.$$

For van der Waals gases, we have $(\hat{d}\xi \wedge \hat{d}\eta)|_I \neq 0$, so we can use the differential invariants above to distinguish them, and it is not difficult to find the functions $h_{ij}|_I(\xi|_I, \eta|_I)$ and $h_0|_I(\xi|_I, \eta|_I)$:

$$
\begin{aligned}
h_{11} = {} & \tfrac{1}{4R^3n(Rn-2\xi)} \Big(n^2(2-n)(n\xi - 2\xi + 4\eta)R^4 \\
& + 8n\left(\left((n^2 - n - 1)\xi + 2n\eta\right)\xi - (\xi - \eta)^2\right)R^3 \\
& - 8\left((3n^2 + 2n + 2)\xi + 2n\eta\right)\xi^2 R^2 + 32\xi^4(n+1)R - 16\xi^5 \Big) \\
h_{12} = {} & \tfrac{1}{4nR^4(Rn-2\xi)^2} \Big(\left(30n^3\xi^3 - 20n^2(2\xi - \eta)\xi^2 \right. \\
& \left. - 8n\left(4\xi^2 - \xi\eta + 4\eta^2\right)\xi - 16\xi^3 + 24\xi^2\eta - 32\xi\eta^2 + 8\eta^3\right)R^4 \\
& + \left(-160n^3\xi^4 + 32n\left(2\xi^2 + \xi\eta + 2\eta^2\right)\xi^2 + 64\xi^3\eta\right)R^3 \\
& + 16\left(15\xi n^2 + 5(2\xi - \eta)n + 4\xi - 8\eta\right)\xi^4 R^2 \\
& - 64\left(3n\xi^6 + 2\xi^6 - \xi^5\eta\right)R + 64\xi^7 \Big)
\end{aligned}
$$

$$h_{22} = \frac{1}{16n^2 R^5 (Rn-2\xi)^3} \Big((2-n)^3 n^5 (3n\xi - 6\xi + 8\eta) R^8$$
$$+24 (n-2) n^4 \big(2n^3\xi^2 - n^2 (7\xi - 3\eta)\xi + n (2\xi + \eta)(2\xi - 3\eta)$$
$$+4\xi^2 - 4\xi\eta + 2\eta^2 \big) R^7 - 16 n^3 \big(21 n^4\xi^3 - n^3 (73\xi - 12\eta)\xi^2$$
$$+n^2 \big(42\xi^3 - 18\xi^2\eta - 33\xi\eta^2\big) + \big(36\xi^3 + 12\xi\eta^2 + 12\eta^3\big) n$$
$$+8\xi^3 - 24\xi^2\eta + 36\xi\eta^2 - 8\eta^3\big) R^6$$
$$+32 n^2 \big(42 n^4\xi^4 - n^3 (81\xi + 5\eta)\xi^3 + n^2 \big(-18\xi^4 - 12\xi^3\eta - 42\xi^2\eta^2\big)$$
$$+12 n (2\eta + \xi)\xi \big(2\xi^2 - 2\xi\eta + \eta^2\big) + 40\xi^3\eta - 36\xi^2\eta^2 + 24\xi\eta^3 - 4\eta^4\big) R^5$$
$$-96 n\big(35 n^4\xi^3 - 10 n^3 (3\xi + 2\eta)\xi^2 - 4n^2 \big(7\xi^2 + 8\xi\eta + 3\eta^2\big)\xi$$
$$-8 (\xi + \eta)(2\xi - \eta)\eta n + 8\xi (\xi - \eta)^2 \big)\xi^2 R^4$$
$$+128\xi^4 \big(42 n^4\xi^2 - 11\xi (\xi + 3\eta) n^3 - 3 n^2 \big(10\xi^2 + 18\xi\eta - \eta^2\big)$$
$$-12 n \big(\xi^2 + 3\xi\eta - \eta^2\big) - 8\xi^2\big) R^3$$
$$-256\xi^5 \big(21 n^3\xi^2 - n^2 (3\xi + 16\eta)\xi - 3 n \big(6\xi^2 + 6\xi\eta - \eta^2\big) - 12\xi^2\big) R^2$$
$$+1536 \big(2\xi n^2 - n (\xi + \eta) - 2\xi\big)\xi^7 R - 256 (-4 + 3n)\xi^9\Big)$$
$$h_0 = -R\Big(\tfrac{n}{2} \ln\big(\tfrac{nF^2 G}{(Rn-2\xi)^5((n+2)R-2\xi)^4}\big) + \ln\big(\tfrac{F}{G}\big) + \ln\big(\tfrac{a^{n/2}}{b^{n/2-1}}\big) + (n+1)\ln(2) - \tfrac{3\ln(3)}{2} n\Big) + C$$

where

$$F = n (n + 3n + 2) R^3 - \big(6\xi n^2 + 12 n\xi + 12\xi - 4\eta\big) R^2 + 12 (n+1)\xi^2 R - 8\xi^3,$$
$$G = n \big(n^2 - 4\big) R^3 - \big(6 n^2\xi - 24\xi + 8\eta\big) R^2 + 12 n\xi^2 R - 8\xi^3.$$

We suppressed the notation signifying restriction to I in order to simplify the equations. The functions $h_{ij}|_I(\xi|_I, \eta|_I)$ are rational functions, and we notice that they do not depend on the constants a, b, C. The expression for $h_0|_I(\xi|_I, \eta|_I)$ shows that changing a and b will only affect the constant C under the Aff(V)-action.

The $G_0 \times$ Aff(V)-invariants are more difficult to handle. In order to find the functions $\tilde{h}_{ij}|_I(\tilde{\xi}|_I, \tilde{\eta}|_I)$, we can in theory make the substitutions

$$\xi|_I = \tilde{\xi}|_I \cdot h_0|_I, \qquad \eta|_I = \tilde{\eta}|_I \cdot h_0|_I, \qquad h_{ij}|_I = \tilde{h}_{ij}|_I \cdot h_0|_I$$

and eliminate $h_0|_I$ in order to get three equations determining $\tilde{h}_{ij}|_I(\xi|_I, \eta|_I)$. However, this seems unmanageable in practice. The first three equations are polynomial in $h_0|_I$, but with degrees up to 18, while the fourth equation is not even algebraic.

It is well known that we can use $G_0 \times$ Aff(V) to normalize the constants a, b, R. In the "critical variables", in which the critical point is given by $(p, v, T) = (1, 1, 1)$, the constants are normalized to $a = 3$, $b = 1/3$, $R = 8/3$. Thus, every van der Waals gas is, under the $G_0 \times$ Aff(V)-action, equivalent to the one given by the equations

$$\Big(p + \frac{3}{v^2}\Big)\Big(v - \frac{1}{3}\Big) = \frac{8}{3} T, \qquad \varepsilon = \frac{4n}{3} T - \frac{3}{v}$$

and the information gain function

$$I = -\frac{8}{3} \ln \left(\left(v - \frac{1}{3} \right) \left(\frac{3}{v} + \varepsilon \right)^{n/2} \right) + C.$$

Notice that normalizing a, b, R in this way will affect the value of C.

Acknowledgements A part of this work was done during the summer school "Nonlinear PDEs, their geometry, and applications" in Wisła in August 2018. I would like to thank Valentin V. Lychagin for his guidance throughout this project.

References

1. P.V. Bibikov, V.V. Lychagin, *Classification of Linear Actions of Algebraic Groups on S-paces of Homogeneous Forms*, V.V. Dokl. Math. **85**, 109–112 (2012), https://doi.org/10.1134/S1064562412010383.
2. I. Krasilshchik, V. Lychagin, A. Vinogradov, *Geometry of jet spaces and nonlinear partial differential equations*, Gordon and Breach (1986).
3. B. Kruglikov, V. Lychagin, *Geometry of Differential equations*, Handbook of Global Analysis, Ed. D.Krupka, D.Saunders, Elsevier, 725–772 (2008).
4. B. Kruglikov, V. Lychagin, *Global Lie-Tresse theorem*, Selecta Mathematica **22**, 1357–1411 (2016).
5. V.V. Lychagin, *Contact Geometry, Measurement and Thermodynamics*, to appear in the same book as this paper.
6. P. Olver, *Equivalence, Invariants, and Symmetry*, Cambridge University Press, Cambridge (1995).
7. M. Rosenlicht, *Some basic theorems on algebraic groups*, Amer. J. Math. **78**, 401–443 (1956).

Chapter 8
Monge–Ampère Grassmannians, Characteristic Classes and All That

Valentin V. Lychagin and Volodya Roubtsov

Introduction

1.1. It is known (since Maslov observations) that the analogues of Bohr–Sommerfeld conditions in the asymptotic quantisation (know also as the Maslov's canonical operator method) have a topological nature.

This condition has a form of annihilation of some cohomology classes (the Maslov–Arnold classes). The Maslov–Arnold classes are, in fact, the examples of *characteristic classes*, which are completely defined by a universal construction of a 'classifying map' into a 'classifying space' phase space T^*M. The topology of this Lagrangian Grassmannian and its $\mathbb{Z}_2$−cohomology ring $H^*(LG, \mathbb{Z}_2)$ are well known (A. Borel, D. Fuchs).

The classes of Maslov–Arnold contain an important information about singularities for the Lagrangian projections.

In this paper, we review and describe one of the generalisations of Maslov–Arnold classes associated with a topological study of Monge–Ampère equations and their solutions. This important tool for studies of Monge–Ampère solution singularities, topological properties of discontinuous solutions were almost out of the scope of the

V. V. Lychagin
V. A. Trapeznikov Institute of Control Sciences of Russian Academy of Sciences, 65 Profsoyuznaya street, Moscow 117997, Russia
e-mail: valentin.lychagin@uit.no

UiT Norges Arktiske Universitet, Postboks 6050, 9037 Langnes, Tromso, Norway

V. Roubtsov (✉)
LAREMA UMR 6093 du CNRS, Département de Mathématiques,
Université d'Angers, Angers, France
e-mail: roubtsov.vladimir@gmail.com; volodya@univ-angers.fr

Theory Division of ITEP, Moscow, Russia

© Springer Nature Switzerland AG 2019
R. A. Kycia et al. (eds.), *Nonlinear PDEs, Their Geometry, and Applications*,
Tutorials, Schools, and Workshops in the Mathematical Sciences,
https://doi.org/10.1007/978-3-030-17031-8_8

lectures during the Summer School 'Wisla-2018' and we hope that this note will fill this gap.

The main object of our interest is the Monge–Ampère Grassmannian, constructed by the tangent hyperplanes to generalised solutions. We will study their topological and geometric properties with the aim of (via the analogue of the 'universal construction') to define some cohomology characteristic classes. One-cohomology classes enter in the construction of the discontinuous solutions for the given 'multi-valued' generalised solutions. The co-dimension one-cohomology classes are obstructed to a solvability of the boundary value problem. The higher order co-dimension classes are responsible for the solvability of more complicated boundary problems (when the boundary may have the components of different co-dimensions).

1.2. A Grassman variety $Gr_k(n)$ of k-dimensional subspaces in n-dimensional vector space V has long history and is used in different branches of mathematics. Among them are the universal G-bundle construction and characteristic classes in algebraic topology, Schubert and Plücker calculus in algebraic geometry, Gauss tangential mappings in differential geometry, Maslov index and its generalisation in symplectic geometry, etc.

We will use all this incarnation of the Grassmannian to apply them to our study of the Monge–Ampère equations and their solutions. Roughly speaking, the Grassmannian associated to a Monge–Ampère equation $\mathcal{E}_\omega \subset J^2(1)$ at a point $x \in J^1(1)$ is an approximation of the order 1 (a 'linearization') to a generalized solution $\mathcal{L}$ passing through the point x. In other words, it is a set of the tangent spaces at x of solutions passing through this point.

We will denote, following the tradition, this Grassmannian by $I\mathcal{E}_\omega(x)$ there, as above $\omega \in \Omega^n_\varepsilon(J^1(1))$ is an effective n-form defining the Monge–Ampère operator

$$\Delta_\omega : \mathcal{E}_\omega = \{\Delta_\omega = 0\}$$

and the generalized solution $\mathcal{L}$ satisfies the 'integrability' conditions

$$\omega \mid_{\mathcal{L}} = U_1 \mid_{\mathcal{L}} = 0.$$

We will restrict in this paper our attention to the class of symplectic Monge–Ampère operators and equations such that for any $x \in T^*\mathbb{R}^n$

$$I\mathcal{E}_\omega(x) = \left\{ L \subset T_x(T^*\mathbb{R}^n) \mid \omega_x \mid_L = \Omega_x \mid_L = 0 \right\}$$

is a subvariety in the Lagrangian Grassmannian

$$LG(x) = \left\{ L \subset T_x(T^*\mathbb{R}^n) \mid \Omega_x \mid_L = 0 \right\},$$

where L denotes an oriented Lagrangian subspace.

The well-known result [1] is that the space $LG(x)$ is a homogeneous space of the group $U(n)$

$$LG(x) = U(n)/SO(n), \qquad \forall x \in T^*\mathbb{R}^n.$$

We will give in this paper a description of the integral Monge–Ampère Grassmannians $I\mathcal{E}_\omega(x)$ for $n = 2, 3$ for ω being a non-degenerate effective form. We refer to [2] for an algebro-geometric description of this grassmannian for $n = 4$ if Δ_ω is a Monge–Ampère pluriharmonic operator and where it had been proven that this Grassmannian is a real algebraic submanifold in $\mathbb{CP}^4$.

Finally, we will give an account to the theory of the characteristic classes based on the integral Grassmannians $I\mathcal{E}_\omega(x)$. We show that even for $n = 3$ the topological structure of such Grassmannians is interesting and complicated. Their topological structure (decomposition to various 'regularity' strata) is different for even (easy case) and for odd (highly non-trivial) values of n. We have corrected some results of [2] and find a relation with some exciting object (Cayley affine cubic surface) which deserves future explorations.

The main sources of bibliographic references for this chapter are [2–4].

8.1   Grassmannians, Associated with the Lagrangian and Legendrian Planes

2.1. Let $f \in C^\infty(M)$ be a smooth function on a smooth $n-$ dimensional variety M. Then the section

$$\sigma_f : M \longrightarrow J^1 M$$

is given by

$$\sigma_f(a) = \left[f\right]_a^1 = j_1(f)(a).$$

We will consider the image $\sigma_f(M)$ as a submanifold in $J^1 M$:

$$\sigma_f(M) \subset J^1 M$$

and consider the tangent plane $T_{a_1}(\sigma_f(M))$ at the point $a_1 \in J^1 M$ such that

$$a_1 = j_1(f)(a) = [f]_a^1.$$

This plane is

(a) n - dimensional;
(b) 'Legendrian', i.e. the restriction of the Cartan universal 1-form on this tangent plane is zero: $U_{1,a_1} \mid T_{a_1}(\sigma_f(M)) \equiv 0$,
(c) 'Lagrangian'(i.e. the natural 2-form dU_{1,a_1} annihilates on $T_{a_1}(\sigma_f(M))$).

The set of all such planes

$$\{L \subset T_{a_1}(J^1(M)), U_1 \mid_L = dU_1 \mid L = 0\}$$

at the point $a_1 \in \sigma^1 M$ is isomorphic to the Lagrangian grassmannian $LG(x_1)$, where $x_1 = \pi(a_1)$ under the natural projection

$$\pi : J^1 M \longrightarrow T^* M.$$

To describe a topological structure of this grassmannian, we should remark that being considered as a complex manifold the Lagrangian grassmannian may be represented as a homogeneous space in two different ways:

(a) $LG^{\mathbb{C}}(x_1) = Sp(n, \mathbb{C})/G$, where the stabilizer subgroup G consist of the triangular block-matrices:

$$G = \begin{pmatrix} A & B \\ 0 & (A^t)^{-1} \end{pmatrix}$$

with the matrix entries $A \in GL(n, \mathbb{C})$ and $B \in Mat(n, \mathbb{C})$ such that

$$AB^t = BA^t;$$

(b) $LG^{\mathbb{C}}(x) = Sp(n)/U(n)$, where we denote by $Sp(n)$ the intersection of $Sp(n, \mathbb{C})$ with the unitary $2n \times 2n$ matrices:

$$Sp(n) = Sp(n, \mathbb{C}) \cap U(2n).$$

In the real case, the Lagrangian grassmannian $LG^{\mathbb{R}}(x_1)$ is isomorphic to the quotient

$$LG^{\mathbb{R}}(x_1) \simeq U(n)/O(n),$$

or (when we consider the oriented planes), we have

$$LG^{\mathbb{R}}_+(x_1) \simeq U(n)/SO(n).$$

The topology of such homogeneous spaces was studied in the papers of A. Borel and D. Fuchs [5, 6] from which we get the following

Theorem 8.1 $\mathbb{Z}_2$-cohomology ring of the Lagrangian grassmannian $LG^{\mathbb{R}}_+(x_1)$ is isomorphic (as a graded ring up to degree n) to the quotient

$$\mathbb{Z}_2[w_1, \dots, w_n]/(w_1^2, \dots, w_n^2),$$

where $w_1, \dots, w_n$ are the Stiefel–Whitney classes of the tautological bundle over the grassmannian $LG^{\mathbb{R}}_+(x_1)$: the fibre of this bundle in the point $L \in LG^{\mathbb{R}}_+(x_1)$ is the same vector space L.

8.2 Integral or Monge–Ampère Grassmannians

Now, we remind the notion of *integral* Grassmannian which assigns to the symplectic Monge–Ampère operators on a smooth n−dimensional manifolds M.

Our considerations are (basically) local so far we shall write regularly $\mathbb{R}^n$ instead of M and $\mathbb{R}^{2n} = T^*(\mathbb{R}^n)$ instead of T^*M.

Define for an effective n−form $\omega \in \Omega_\epsilon(T^*\mathbb{R}^n)$ and any $x \in T^*\mathbb{R}^n$ the set

$$I\mathcal{E}_\omega(x) := \left\{ L \subset T_x(T^*\mathbb{R}^n) \mid \omega_x \mid_L = \Omega_x \mid_L = 0 \right\}$$

as a subset in the Lagrangian Grassmannian

$$LG_+^{\mathbb{R}}(x) = \left\{ L \subset T_x(T^*\mathbb{R}^n) \mid \Omega_x \mid_L = 0 \right\},$$

where L denotes a Lagrangian subspace with a fixed orientation.

We will suppose that the natural projection of

$$I\mathcal{E}_\omega \overset{def}{=} \underset{x_1 \in T^*M}{U} I\mathcal{E}_\omega(x_1)$$

on the cotangent space T^*M is a smooth bundle:

$$\alpha : I\mathcal{E}_\omega \to T^*M.$$

8.3 Grassmannians for 2− and 3− Effective Forms

Now, we specify grassmannians $I\mathcal{E}_\omega(x_1) \subset LG_+(x_1)$ associated with the effective 2- and 3-forms.

8.3.1 Integral Grassmannians for Monge–Ampère Equations in Dimension 2

Grassmannians of a Monge–Ampère equation in dimension 2 could be characterised by the pfaffian $Pf(\omega_{x_1})$ ([7, 8]). We will restrict ourselves to the case of general position, or to the *non-degenerated* Monge–Ampère operators with $Pf(\omega_{x_1}) \neq 0$.

Theorem 8.2 *If the form ω_{x_1} defines an elliptic Monge–Ampère operator such that $(Pf(\omega_{x_1}) > 0)$, then the grassmannian $I\mathcal{E}_\omega(x_1)$ is homeomorphic to the projective line $\mathbb{CP}^1$. If the Monge–Ampère operator is hyperbolic at x_1, i.e. $(Pf(\omega_{x_1}) < 0)$ then the integral grassmannian $I\mathcal{E}_\omega(x_1)$ is homeomorphic to a torus $\mathbb{T}^2$.*

Proof ([8])

1. If the form $\omega \in \Omega^2(T^*M)$ belongs to the 'elliptic orbit' at the point $x_1 \in T^*M$, then we can introduce the field of endomorphisms $\{A_{x_1}\} \in End\,(T_{x_1}(T^*M))$ such that $-Pf(\omega_{x_1}) = A_x^2$, or, after a proper normalisation

$$\omega_{x_1} \to \frac{\omega_{x_1}}{\sqrt{|Pf(\omega_{x_1})|}},$$

we obtain the field $\{A_{x_1}\}x_1 \in T^*M$ such that $A_{x_1}^2 = -1$ and as it was mentioned in the cited classification theorem, the operator A_{x_1} gives an almost-complex structure on T^*M. So we have that there is a one-to-one correspondence between the choice of the complex line and the choice of a point in the integral grassmannian (a choice of Lagrangian plane L with the condition $\omega_{x_1}\,|_L= 0$). Given almost-complex structure A_{x_2} defines an isomorphism

$$T_{x_1}(T^*\mathbb{R}^2) \simeq \mathbb{C}^2,$$

and therefore $I\mathcal{E}_\omega(x_1) = \mathbb{C}P^1$

2. In the case of the hyperbolic Monge–Ampère integral grassmannian instead of almost-complex structures, we have the field of almost-product structures given by the operators A with the eigenvalue ± 1, i.e. $A^2 = 1$. The almost-product structure dictates the choice of coordinates on $T_{x_1}(T^1\mathbb{C}^2)$ such that the equations $\omega_{x_1}\,|_L= \Omega_{x_1}\,|_L= 0$ at point x_1 are read as

$$dp_1 \wedge dq_1\,|_L= dp_2 \wedge dq_2\,|_L= 0.$$

This equation shows that such Lagrangian planes are direct sums of two 1−dimensional subspaces obtained from the eigenspaces of A, i.e. $I\mathcal{E}_\omega(x_1)$ is isomorphic to the product $S^1 \times S^1 = \mathbb{T}^2$. $\square$

8.3.1.1   Digression: Plücker Embedding and Homotopy Type of Generic 2d−MA Grassmannians

Here, we propose a pure algebraic derivation of the previously proved results. This approach is very much in the spirit of F. Klein.

Denote by V a tangent space to T^*M at some point. Then fixing the Liouville volume form Vol on V, we obtain a symmetric bilinear pairing as above:

$$g : \Lambda^2(V) \otimes \Lambda^2(V) \mapsto \mathbb{R}, \quad g(\alpha, \beta) = \frac{(\alpha \wedge \beta)}{Vol}, \alpha, \beta \in \Lambda^2(V).$$

The form g is non-degenerate and (as it is easy to check) has signature (3,3). A bivector α is g-isotropic iff α is a decomposable: $\alpha \wedge \alpha = 0$.

There is a bijective correspondence between the points of the grassmannian $Gr_2(4, V)$ and the points of the projective quadric in $\mathbb{R}P^5$ given by the image of Plücker embedding

$$Gr_2(4, V) \to \mathbb{P}(\Lambda^2(V)) = \mathbb{R}P^5$$

which is in coordinates is written as

$$p_{12}p_{34} - p_{13}p_{24} + p_{14}p_{23} = 0,$$

where $\alpha \in \Lambda^2(V)$ is represented as $\alpha = 1/2 \sum_{i,j} p_{ij}e_i \wedge e_j$ for some base e_i of V and $p_{ij} + p_{ji} = 0$.

To describe in the Plücker terms the Lagrangian grassmannians and the Monge–Ampère grassmannians, we need to choose the volume form $Vol = \Omega \wedge \Omega$ and to suppose that the symplectic bivector $\alpha_\Omega = e_1 \wedge e_3 + e_2 \wedge e_4$. The effectivity condition gives the relation $p_{13} + p_{24} = 0$ and the relations, which are singled out. The planes tangent to the elliptic MA and to the hyperbolic MA are correspondingly read as $p_{14} + p_{23} = 0$ and as $p_{14} - p_{23} = 0$. Then the Plücker embedding is restricted to the embeddings of special Lagrangian grassmannian

$$SLG_2(4, V)_+ = SU(2)/SO(2) = S^2 (= I\mathcal{E}_{\omega_+})$$

in $\mathbb{R}P^3$(which is exactly the elliptic $2d$−MA grassmannian) such that the images of it are projective quadrics, which are in appropriate homogeneous coordinates $(x_1 : x_2 : x_3 : x_4) \in \mathbb{R}P^3$ given by the equations

$$q_+ : x_1^2 - x_2^2 - x_3^2 - x_4^2 = 0$$

and in the case $SLG_2(4, V)_- (= I\mathcal{E}_{\omega_-})$

$$q_- : x_1^2 - x_2^2 - x_3^2 + x_4^2 = 0.$$

Now to precise the topological type of the quadrics $q_\pm$, let us identify $\mathbb{R}P^3$ as the three-sphere $x_1^2 + x_2^2 + x_3^2 + x_4^2 = 2$ with identified poles : $\mathbb{R}P^3 = S^3/\mathbb{Z}_2$.

Then, the q_+ quadric reduces to the relations $x_1 = 1, x_2^2 + x_3^2 + x_4^2 = 1$, which are singled out to a two-sphere S^2. The q_- quadric reduces to the two-dimensional torus $\mathbb{T}^2 = S^1 \times S^1$, which is given by the relations $x_1^2 + x_4^2 = 1, x_2 = x_3 = 0$ and $x_2^2 + x_3^2 = 1, x_1 = x_4 = 0$.

Corollary 8.1 $\pi_1(I\mathcal{E}_{\omega_+}) = 0, \pi_1(I\mathcal{E}_{\omega_-}) = \mathbb{Z} \oplus \mathbb{Z}$.

8.3.2 Geometric Structure Associated with 3d – MA Equations

Let us consider the case when $\dim M = 3$. Now, we describe the grassmannians associated with the non-degenerate Monge–Ampère operators in dimension 3. The adequate language to do it was proposed in the Ph.D. thesis of B. Banos [2, 9], who had generalised and extended the correspondence between the Monge–Ampère operators with constant coefficients and flat integrable geometric structure on dimension 3. We will give a review to remind his approach.

We are interested here in a local description of MA structures, so we can assume in the subsection $T^*M = \mathbb{R}^6$.

Let ω be an effective 3-form on $\mathbb{R}^6$, $\omega \in \Omega_\varepsilon^3(\mathbb{R}^6)$, and $q_\omega \in S^2(\mathbb{R}^6)$ is the corresponding quadratic form. We will suppose that the form ω is non-degenerate in Hitchin sense ([]Hit) (remind that, in this case, the signature of q_ω may be equal to $(3, 3)$, or to $(0, 6)$, or to $(4, 2)$). If Ω denotes the standard canonical 2-form on $\mathbb{R}^6$ we will call a *generalised Calabi–Yau structure* on $\mathbb{R}^6$ the quintuple $(g, \Omega, K, \alpha, \beta)$ where g is a (pseudo) metric on $\mathbb{R}^6$ (a non-degenerate quadratic form, possibly indefinite), $K \in \mathrm{End}\,\mathbb{R}^6$ such that $K^2 = \pm 1$ and

$$g(x, y) = \Omega(Kx, y)$$

for any vector $x, y \in \mathbb{R}^6$, and the α, β are 3-forms on $\mathbb{R}^6$ which are decomposable, i.e.

$$\alpha = \lambda_1 \wedge \lambda_2 \wedge \lambda_3, \beta = \mu_1 \wedge \mu_2 \wedge \mu_3$$

for $\lambda_i, \mu_i \in \Omega^1(\mathbb{R}^6)$, $1 \leqslant i \leqslant 3$ and such that

$$\alpha \wedge \Omega = \beta \wedge \Omega = 0$$

and

$$\alpha \wedge \beta = \mathrm{const} \cdot \Omega^3.$$

The subspaces associated to α and β are the eigenspaces of the endomorphism K:

$$C_\alpha \stackrel{def}{=} \{x \in \mathbb{R}^6 \mid Kx = -x\} \iff \{\iota_x \alpha = 0\}$$

$$C_\beta \stackrel{def}{=} \{x \in \mathbb{R}^6 \mid Kx = x\} \iff \{\iota_x \beta = 0\}$$

or

$$C_\alpha \stackrel{def}{=} \{x \in \mathbb{R}^6 \mid Kx = -ix\} \iff \{\iota_x \alpha = 0\}$$

$$C_\beta \stackrel{def}{=} \{x \in \mathbb{R}^6 \mid Kx = ix\} \iff \{\iota_x \beta = 0\}$$

We remark that for the 'usual' Calabi–Yau structure on $\mathbb{R}^6$, we should take $K = I$, $I^2 = -1$, $g = \Omega(I\cdot, \cdot)$ and by $(\alpha, \beta) = (\alpha, \overline{\alpha})$, i.e. when α is a $(3,0)$-form for this complex structure. Usually, the standard Calabi–Yau structure imposes the positivity condition on g. These (generalized) Calabi–Yau structures are in one-to-one correspondence with the Monge–Ampère constant coefficient operators.

Example 3 The standard Calabi–Yau structure on $\mathbb{R}^6$ is equivalent to *Special Kähler structure* on $\mathbb{C}^3$ (for details see in the lectures of one of the authors in the same volume [10]), is given by the quadruple (g, I, Ω, α) (in this case $\beta = \overline{\alpha}$), and defines the special Lagrangian Monge–Ampère operator

$$\Delta u - \mathrm{Hess}(u) = 0$$

with coordinates $q_1, q_2, q_3, p_1, p_2, p_3$ on $T\mathbb{R}^3$ as

$$g = -(dq_1)^2 - (dq_2)^2 - (dq_3)^2 + (dp_1)^2 + (dp_2)^2 + (dp_3)^2,$$

$$I = \frac{\partial}{\partial p_1} \otimes dq_1 + \frac{\partial}{\partial p_2} \otimes dq_2 +$$

$$\frac{\partial}{\partial p_3} \otimes dq_3 - \frac{\partial}{\partial q_1} \otimes dp_1 - \frac{\partial}{\partial q_2} \otimes dp_2 - \frac{\partial}{\partial q_3} \otimes dp_3 \otimes dp_3,$$

$$\Omega = dq_1 \wedge dp_1 + dq_2 \wedge dp_2 + dq_3 \wedge dp_3$$

and

$$\alpha = dz_1 \wedge dz_2 \wedge dz_3 = d(q_1 + ip_1) \wedge d(q_2 + ip_2) \wedge d(q_3 + ip_3).$$

Example 4 The following (pseudo)-special Lagrangian operator

$$\Box u + \mathrm{Hess}(u) = 0$$

(where $\Box u = \dfrac{\partial^2 u}{\partial q_1} - \Delta_{q_2, q_3} v$ is the $3d$-wave operator) is associated with the generalised (pseudo) Calabi–Yau structure (g, I, Ω, α), where

$$g = (dq_1)^2 - (dq_2)^2 + (dq_3)^2 + (dp_1)^2 - (dp_2)^2 + (dp_3)^2,$$

$$I = \frac{\partial}{\partial q_1} \otimes dp_1 - \frac{\partial}{\partial p_1} \otimes dq_1 - \frac{\partial}{\partial q_2} \otimes dp_2 + \frac{\partial}{\partial p_2} \otimes dq_2 + + \frac{\partial}{\partial q_3} \otimes dp_3 - \frac{\partial}{\partial p_3} \otimes dq_3,$$

$$\Omega = dq_1 \wedge dp_1 + dq_2 \wedge dp_2 + dq_3 \wedge dp_3,$$

$$\alpha = (dq_1 + idp_1) \wedge (dq_2 + idp_2) \wedge (dq_3 + idp_3)$$

Example 5 The 'real' Calabi–Yau structure $(g, S, \Omega, \alpha, \beta)$ is associated with the Hessian constant coefficient equation $\mathrm{Hess}(u) = 1$, such that

$$g = dq_1 \otimes dp_1 + dq_2 \otimes dp_2 + dq_3 \otimes dp_3,$$

$$S = \frac{\partial}{\partial q_1} \otimes dq_1 - \frac{\partial}{\partial p_1} \otimes dp_1 + \frac{\partial}{\partial q_2} \otimes dq_2 - \frac{\partial}{\partial p_2} \otimes dp_2 + + \frac{\partial}{\partial q_3} \otimes dq_3 - \frac{\partial}{\partial p_3} \otimes dp_3,$$

and $S^2 = Id$;

$$\Omega = dq_1 \wedge dp_1 + dq_2 \wedge dp_2 + dq_3 \wedge dp_3,$$
$$\alpha = dq_1 \wedge dq_2 \wedge dq_3,$$
$$\beta = dp_1 \wedge dp_2 \wedge dp_3.$$

8.3.3 Integrability and MA Grassmannians in 3d

Now, we remind the $3d-$analogue of the classification theorem in geometric form. More exactly, we will show that these are similar integrability conditions for the non-degenerated Monge–Ampère operators, which are nothing but the integrability conditions of the corresponding Calabi–Yau generalized structures.

Recall that in $2d$-case, the constant coefficient reducibility condition of a non-degenerate (in the given open neighbourhood) effective 2-form ω is written as the closeness of the normalised form

$$d\left(\frac{\omega}{\sqrt{|Pf(\omega)|}} \right) = 0,$$

where $Pf(\omega) \neq 0$. Its geometric counterpart could be written in the form Nijenhuis tensor torsion-free condition $N_A = 0$ for the endomorphism A.

Let $\omega_0 \in \Omega^3_\epsilon(T^*M)$, for $3d-$dimensional M be an effective non-degenerate differential form and we will denote by q_ω the quadratic form, associated with the form ω by the following formula for the corresponding bilinear symmetric form (see the details in [8]):

$$q_\omega(X, Y) = \frac{\iota_X(\omega) \wedge \iota_Y(\omega)}{\Omega \wedge \Omega \wedge \Omega}.$$

The Hitchin linear operator at any $x \in T^*M$ acts as

$$K_\omega : T_x(T^*M) \rightarrow T_x(T^*M)$$
$$K_\omega(X) \stackrel{def}{=} \frac{A(\iota_X\omega \wedge \omega)}{\Omega \wedge \Omega \wedge \Omega},$$

where

$$A : \Omega^5(T^*M) \rightarrow T(T^*M) \otimes \Omega^6(T^*M).$$

is the isomorphism which is induced by an exterior product.

Then the following proposition is valid:

Proposition 8.1 ($3d-$ 'splitting construction') *The MA effective form can be split in the sum of two 3—forms: $\omega = \alpha + \beta$ for*

$$\alpha = \frac{1}{2}\left(\omega + \left|\frac{1}{6}TrK_\omega^2\right|^{-\frac{3}{2}} K_\omega^*(\omega)\right), \quad \beta = \frac{1}{2}\left(\omega - \left|\frac{1}{6}TrK_\omega^2\right|^{-\frac{3}{2}} K_\omega^*(\omega)\right) \tag{8.1}$$

and the quintuple $(\Omega, q_\omega, K_\omega, \alpha, \beta)$ is a generalised almost-Calabi–Yau structure on T^*M in the sense of previous section.

Proof It was observed by Banos [2] that $q_\omega = \Omega(K_\omega\cdot, \cdot)$. Then we should only check that α and β are effective and the ratio $\frac{\alpha \wedge \beta}{\Omega^3}$ is a constant. But it is clear from

$$K^*(\Omega \wedge \omega) = K^*(\Omega) \wedge K^*(\omega) = \pm\Omega \wedge K^*(\omega) = 0$$

that the form $K^*(\omega)$ is effective if ω is. The effectiveness of α and β is evident from 8.1. From the normalisation condition, we obtain

$$\frac{1}{6}TrK_\omega^2 = \pm 1$$

and hence,

$$\frac{\alpha \wedge \beta}{\Omega^3}$$

is a constant. □

Theorem 8.3 *There are three different models of the integral grassmannian in the dimension 3, corresponding to the three different non-degenerated Monge–Ampère operators with constant coefficients:*
1. If $\omega \in \Omega_\mathcal{E}^3(\mathbb{R}^6)$ is such that $\frac{1}{6}TrK_\omega^2 = 1$, then

$$I\mathcal{E}_\omega = SL(3)/SO(3) \coprod SL(3)/SO(1, 2),$$

where $\coprod$ means the disjoint sum.
2. If $\omega \in \Omega_\mathcal{E}^3(\mathbb{R}^6)$ is such that $\frac{1}{6}TrK_\omega^2 = -1$, and the signature of q_ω is (0, 6), then

$$I\mathcal{E}_\omega = SU(3)/SO(3);$$

3. If $\omega \in \Omega_\mathcal{E}^3(\mathbb{R}^6)$ is such that $\frac{1}{6}TrK_\omega^2 = -1$, then and the signature of q_ω is (4, 2), then

$$I\mathcal{E}_\omega = SU(2, 1)/SO(2, 1).$$

To prove this Theorem, we need some technical results about the grassmannians of integral planes.

Lemma 8.1 *The set $I\mathcal{E}_\omega^* \subset I\mathcal{E}_\omega$ of the integral grassmannians having the non-degenerated quadratic form q_ω is an open subset in all integral grassmannians.*

Proof Easy.

□

We denote the signature of q_φ by (p, q) and by $I\mathcal{E}_\omega^{(p,q)}$. We denote the subset in $I\mathcal{E}_\omega$ with q_φ of signature (p, q). Then, taking the unifications

$$I\mathcal{E}_\omega^k = \coprod_{p+q\leqslant k} I\mathcal{E}_\omega^{(p,q)}$$

for $p + q \leq 3$ we obtain

$$I\mathcal{E}_\omega^{-1} = \varnothing \subset I\mathcal{E}_\omega^0 \subset I\mathcal{E}_\omega^1 \subset I\mathcal{E}_\omega^2 \subset I\mathcal{E}_\omega^3 = I\mathcal{E}_\omega$$

a filtration of the topological space $I\mathcal{E}_\omega$.

Proof (*Proof of the Theorem*) Let $V \in I\mathcal{E}_\omega^*$ be an integral plane of the non-degenerate Monge–Ampère operator Δ_ω. The Hitchin operator K_ω transforms V to the orthogonal space $V_0 = K_\omega(V)$: we have $q_\omega(V, V_0) \equiv 0$, because

$$q_\omega(V, V_0) = q_\omega(V, K_\omega V) = \Omega(KV, K_\omega^2 V) = \pm c\,\Omega(KV, V) = 0$$

(V is a Lagrangian!). So $K_\omega(V) \subseteq V^0$, but $q_\omega|_V$ is non-degenerate hence the exact coincidence ($dim V_0 = 3$),

$$K_\omega(V) \oplus V = T_x(T^*M) \simeq T^*(\mathbb{R}^3) = \mathbb{R}^6.$$

1. Consider the case when the signature $\epsilon(q_\omega) = (3, 3)$. Let $\frac{1}{6} Tr K_\omega^2 = 1$. Taking the base of $\mathbb{R}^6(e_1, e_2, e_3, f_1, f_2, f_3)$ as a canonical one with

$$\begin{aligned}
\Omega &= e_1^* \wedge f_1^* + e_2^* \wedge f_2^* + e_3^* \wedge f_3^*, \\
q &= e_1^* f_1^* + e_2^* f_2^* + e_3^* f_3^*; \\
K_\omega &= e_1 \otimes e_1^* + e_2 \otimes e_2^* + e_3 \otimes e_3^* - f_1 \otimes f_1^* - f_2 \otimes f_2^* - f_3 \otimes f_3^*, \\
\omega &= e_1^* \wedge e_2^* \wedge e_3^* - f_1^* \wedge f_2^* \wedge f_3^*, \\
\bar{\omega} &= e_1^* \wedge e_2^* \wedge e_3^* + f_1^* \wedge f_2^* \wedge f_3^*.
\end{aligned}$$

We introduce the sets

$$V_{(3,0)} \stackrel{def}{=} \{(v, v)|v \in \mathbb{R}^3\} \in I\mathcal{E}_\omega^{(3,0)}$$

and

$$V_{(1,2)} \stackrel{def}{=} \{(x_1, x_2, x_3, -x_1, -x_2, -x_3)\} \in I\mathcal{E}_\omega^{(1,2)}.$$

Choose an orthonormal base a_1, a_2, a_3 of q_ω on V:

$$q_\omega(a_i, a_j) = \varepsilon_i\, \delta_{ij}, \quad \varepsilon_i = \pm 1$$

$$\langle e_1^* \wedge e_2^* \wedge e_3^* a_1 \wedge a_2 \wedge a_3 \rangle \geqslant 0$$

an let

$$u_i \overset{def}{=} \frac{a_i + K_\omega(a_i)}{\sqrt{2}}, \quad w_i \overset{def}{=} \frac{a_i - K_\omega(a_i)}{\sqrt{2}}$$

for $i = 1, 2, 3$. Let A be the matrix of u_i in the base $\{e_1, e_2, e_3\}$ and B be the matrix of v_i in the base $\{f_1, f_2, f_3\}$. Then taking the matrix

$$C = \begin{pmatrix} A & 0 \\ 0 & B \end{pmatrix}$$

we obtain $V = CV_0$. Moreover, we have

$$\begin{pmatrix} A^t & 0 \\ 0 & B^t \end{pmatrix} \begin{pmatrix} 0 & I \\ I & 0 \end{pmatrix} \begin{pmatrix} A & 0 \\ 0 & B \end{pmatrix} = \begin{pmatrix} 0 & I_{(p,q)} \\ I_{(p,q)} & 0 \end{pmatrix},$$

with

$$I_{(p,q)} \overset{def}{=} \begin{pmatrix} \varepsilon_1 & 0 & 0 \\ 0 & \varepsilon_2 & 0 \\ 0 & 0 & \varepsilon_3 \end{pmatrix}$$

and $A^t B = I_{(p,q)}$. $\det B = \frac{1}{\det A}$, if $\varepsilon(q_\omega) = (3, 0)$ or $(1, 2)$

$\det B = -\frac{1}{\det A}$, if $\varepsilon(q_\omega) = (0, 3)$ or $(1, 2)$

Now, using that

$$\langle \omega, a_1 \wedge a_2 \wedge a_3 \rangle = 0,$$

we obtain

$$\det A = \det B.$$

Thus, we have

$$I\mathcal{E}_\omega^{(3,0)} = \left\{ \begin{pmatrix} A & 0 \\ 0 & (A^t)^{-1} \end{pmatrix} V_{(3,0)} \middle| A \in SL_3 \right\};$$

$$I\mathcal{E}_\omega^{(0,3)} = \left\{ \begin{pmatrix} A & 0 \\ 0 & (A^t)^{-1} \end{pmatrix} V_{(1,2)} \middle| A \in SL_3 \right\};$$

$$I\mathcal{E}_\omega^{(0,3)} = I\mathcal{E}_\omega^{(2,1)} = \varnothing.$$

But

$$\begin{pmatrix} A & 0 \\ 0 & (A^t)^{-1} \end{pmatrix} V_{(p,q)} = V_{(p,q)} \begin{pmatrix} A & 0 \\ 0 & (A^t)^{-1} \end{pmatrix}$$

if and only if

$$(A^t)^{-1}I_{(p,q)}X = I_{(p,q)}AX$$

for any $X \in \mathbb{R}^3$. It means that we have to choose the matrix A from $SO(p, q)$:

$$I\mathcal{E}_\omega^{(3,0)} = SL(3)/SO(3), \quad I\mathcal{E}_\omega^{(1,2)} = SL(3)/SO(1, 2).$$

2. Similarly, in the case $TrK^2 = -6$ and $\varepsilon(q_\omega) = (4, 2)$, we can take canonical base $e_1, e_2, e_3, f_1, f_2, f_3$ in $\mathbb{R}^6$ in a such way that

$$
\begin{aligned}
\Omega &= e_1^* \wedge f_1^* + e_2^* \wedge f_2^* + e_3^* \wedge f_3^*, \\
q_\omega &= e_1^* e_1^* - e_2^* e_2^* + e_3^* e_3^* + f_1^* f_1^* - f_2^* f_2^* + f_3^* f_3^*, \\
K_\omega &= e_1 \otimes f_1^* - e_2 \otimes f_2^* + e_3 \otimes f_3^* - f_1 \otimes e_1^* + f_2 \otimes e_2^* - f_3 \otimes e_3^*, \\
\omega &= Re((e_1^* + if_1^*) \wedge (e_2^* - if_2^*) \wedge (e_3^* + if_3^*)).
\end{aligned}
$$

Denote by $V_0 \in I\mathcal{E}_\omega^{(2,1)}$ the subspace generated by f_1, f_2, f_3. Let $V \in I\mathcal{E}_\omega$ and a_1, a_2, a_3 is an orthonormal base of V :

$$q_\omega(a_i, a_j) = \varepsilon_i \delta_{ij}, \quad \varepsilon_i = \pm 1$$

We introduce the $\mathbb{C}$-linear endomorphism by

$$Af_i = a_i, \quad i = 1, 2, 3.$$

Then

$$\overline{A}^t I_{(2,1)} A = \begin{pmatrix} \varepsilon_1 & 0 & 0 \\ 0 & \varepsilon_2 & 0 \\ 0 & 0 & \varepsilon_3 \end{pmatrix}$$

and we can similarly deduce that

$$I_{(2,1)} = \begin{pmatrix} \varepsilon_1 & 0 & 0 \\ 0 & \varepsilon_2 & 0 \\ 0 & 0 & \varepsilon_3 \end{pmatrix},$$

which implies that $A \in SU(2, 1)$ (because of the condition that $\omega/V = 0$ and $\det A \in \mathbb{C}$). So we obtain

$$I\mathcal{E}_\omega^* = \{AV_0 | A \in SU(2, 1)\},$$

But $AV_0 = V_0$ if and only if A is real; hence the result:

$$I\mathcal{E}_\omega^* = SU(1, 2)/SO(2, 1).$$

3. The case $TrK^2 = -6$ and $\varepsilon(q_\omega) = (0, 6)$ corresponds to the standard Calabi–Yau structure on the space $\mathbb{C}^3 = \mathbb{R}^6$ known also as *special Lagrangian*.

It is well known after Harvey and Lawson pioneering paper [11] that the special Lagrangian grassmannian

$$I\mathcal{E}_\omega(x) := \left\{ L \subset T_x(T^*\mathbb{R}^3) \mid \omega_x \mid_L = \Omega_x \mid_L = 0 \right\} = SU(3)/SO(3).$$

8.4 Multidimensional Generalisation of Splitting Construction

8.4.1 Non-degenerate $2k + 1-$ Forms in Sense of Hitchin

Let V be a real vector space of dimension $4k + 2$ and $\Lambda^p(V^*)$ the space of exterior $p-$forms on V. Let us fix a volume form $\Theta \in \Lambda^{4k+2}(V^*)$ on V. Denote by $A : \Lambda^{4k+1}(V^*) \to V \otimes \Lambda^{4k+2}(V^*)$, the isomorphism which is induced by the exterior product: $A = \tilde{A} \otimes \Theta$, where

$$\langle \tilde{A}(\theta), \alpha \rangle = \frac{\theta \wedge \alpha}{\Theta}, \quad \alpha \in V^*.$$

Now and in what follows in this section $V = T_x(T^*M)$ at some point $x \in T^*M$ and the volume form Θ, we shall identify with the Liouville symplectic volume form $\Omega^{\wedge k}$.

We shall use an invariant operator $K_\omega^\Theta : V \to V$ defined by

$$K_\omega^\Theta(X) = \frac{A(\iota_X \omega \wedge \omega)}{\Theta}, \quad X \in V.$$

Definition 8.1 The Hitchin Pfaffian of a $2k + 1-$form $\omega \in \Lambda^{2k+1}(V^*)$ is

$$\lambda_\Theta(\omega) := \frac{1}{4k + 2} \mathrm{Tr}(\mathrm{K}_\omega^{\Theta^2}).$$

A $2k + 1-$form ω is called non-degenerate in Hitchin sense if and only if $\lambda_\Theta(\omega) \neq 0$.

Lemma 8.2 Let $\omega \in \Lambda^{2k+1}(V^*)$ be a $2k + 1-$form which is non-degenerate in Hitchin sense. Then
$$K_\omega^{\Theta^2} = \lambda_\Theta(\omega)\mathrm{Id}.$$

Proposition 8.2 Let $\omega \in \Lambda^{2k+1}(V^*)$ be a $2k + 1-$form, which is non-degenerate in Hitchin sense.

1. $\lambda_\Theta(\omega) > 0$ iff $\omega = \alpha + \beta$ where α, β are decomposable $2k + 1-$forms on V. Moreover, if $\frac{\alpha \wedge \beta}{\Theta} > 0$ then α and β are uniquely defined:

$$\begin{cases} 2\alpha = \omega + |\lambda_\Theta(\omega)|^{-k-\frac{1}{2}} (K_\omega^\Theta)^* \omega, \\ 2\beta = \omega - |\lambda_\Theta(\omega)|^{-k-\frac{1}{2}} (K_\omega^\Theta)^* \omega. \end{cases}$$

2. $\lambda_\Theta(\omega) < 0$ iff $\omega = \alpha + \bar\alpha$, where $\alpha \in \Lambda^{2k+1}(V^* \otimes \mathbb{C})$ is a decomposable over $\mathbb{C}$ $2k+1-$form on V. Moreover, if $\frac{\alpha \wedge \bar\alpha}{i\Theta} > 0$ then then α is uniquely defined:

$$\alpha = \omega + i|\lambda_\Theta(\omega)|^{-k-\frac{1}{2}} (K_\omega^\Theta)^* \omega.$$

Remark 8.1 Let us fix a basis $(e_1, \dots, e_{4k+2})$ in V and the dual one $(e_1^*, \dots, e_{4k+2}^*)$ in V^*. We pose $\Theta = e_1^* \wedge, \dots, \wedge e_{4k+2}^*$.

1. $\lambda_\Theta(\omega) > 0$ iff ω is in the $GL(V)-$orbit of

$$e_1^* \wedge, \dots, \wedge e_{2k+1}^* + e_{2k+2}^* \wedge, \dots, \wedge e_{4k+2}^*$$

2. $\lambda_\Theta(\omega) < 0$ iff ω is in the $GL(V)-$orbit of

$$(e_1^* + ie_{2k+2}^*) \wedge (e_2^* + ie_{2k+3}^*) \wedge \dots \wedge (e_{2k+1}^* + ie_{4k+2}^*)$$
$$+ (e_1^* - ie_{2k+2}^*) \wedge (e_2^* - ie_{2k+3}^*) \wedge \dots \wedge (e_{2k+1}^* - ie_{4k+2}^*).$$

The action of $GL(V)$ on $\Lambda^{2k+1}(V^*)$ has two open orbits splited by the hypersurface

$$\{\lambda_\Theta = 0\}.$$

The unicity of this decomposition of a non-degenerate exterior form ω in a sum of two decomposable forms (up to a choice of orientation) provides a construction of a dual form $\check\omega$:

Definition 8.2 (*Hitchin*)

1. If $\lambda_\Theta(\omega) > 0$ and $\omega = \alpha + \beta$, then $\check\omega = \alpha - \beta$.
2. If $\lambda_\Theta(\omega) < 0$ and $\omega = \alpha + \bar\alpha$, then $\check\omega = i(\bar\alpha - \alpha)$.

8.4.1.1 Hamiltonian Approach

One can define a non-degenerate exterior two-form (a symplectic structure) on $\Lambda^{2k+1}(V^*)$. More precisely,

$$\mathfrak{D}_\Theta(\omega, \omega') = \frac{\omega \wedge \omega'}{\Theta}$$

is a symplectic form on $\Lambda^{2k+1}(V^*)$. Hitchin has proven the following

Proposition 8.3 *The action of $SL(4k+2)$ on the symplectic space $(\Lambda^{2k+1}(V^*), \mathfrak{D}_\Theta)$ is a hamiltonian with the moment map*

$$K^\Theta : \Lambda^{2k+1}(V^*) \to sl(4k+2, \mathbb{R}).$$

We have identified here the Lie algebra $sl(4k+2, \mathbb{R})$ with its dual $sl(4k+2, \mathbb{R})^*$ with a help of the Killing form $(X, Y) \to \frac{1}{4k+2}\mathrm{Tr}(XY)$.

Let suppose now that in our $4k+2$–dimensional vector space, V is a symplectic with a symplectic form Ω. The Liouville volume form $\Theta := -\Omega^{\wedge, 2k+1}$ is fixed and it is denoted $\lambda := \lambda_\Theta$, $K := K^\Theta$, $\mathfrak{D} := \mathfrak{D}_\Theta$. The subspace of effective $2k+1$–forms $\Lambda_\epsilon^{2k+1}(V^*)$ is a symplectic subspace in $(\Lambda^{2k+1}(V^*), \mathfrak{D})$. In fact, due to Hodge–Lepage–Lychagin theorem, each form $\omega \in \Lambda^{2k+1}(V^*)$ can be written as

$$\omega = \omega_0 + \omega_1 \wedge \mathfrak{D}$$

with effective ω_0. The exterior product is non-degenerate on $\Lambda_\epsilon^{2k+1}(V^*)$. Moreover the action of $Sp(2k+1)$ preserve $\mathfrak{D}$. So it is natural to ask this action to be hamiltonian.

Lemma 8.3 *A* $2k+1$*– form on* V *is effective iff* $K_\omega \in sp(2k+1)$

Proof Take a symplectic base $(e_1, e_2, \ldots, e_{2k+1}, f_1, f_2, \ldots, f_{2k+1})$, then K_ω is written

$$K_\omega(X)\Theta = \sum_{j=1}^{2k+1}(\iota_X\omega \wedge \omega e_j^*) \otimes e_j + \sum_{j=1}^{2k+1}(\iota_X\omega \wedge \omega f_j^*) \otimes f_j.$$

Let

$$K_\omega = \left(\begin{array}{c|c} A & B \\ \hline C & D \end{array}\right)$$

then

$$\begin{cases} A_{jl}\Theta = \iota_{e_j}\omega \wedge \omega \wedge e_l^* \\ B_{jl}\Theta = \iota_{f_j}\omega \wedge \omega \wedge e_l^* \\ C_{jl}\Theta = \iota_{e_j}\omega \wedge \omega \wedge f_l^* \\ D_{jl}\Theta = \iota_{f_j}\omega \wedge \omega \wedge f_l^*. \end{cases}$$

So ω is effective iff the following relations are satisfied for $l = 1, \ldots, 2k+1$:

$$\begin{cases} \iota_{e_l}\omega \wedge \mathfrak{D} = \mathfrak{D} \wedge \iota_{e_l}\omega = \omega \wedge f_l^* \\ \iota_{f_l}\omega \wedge \mathfrak{D} = \mathfrak{D} \wedge \iota_{f_l}\omega = -\omega \wedge e_l^*. \end{cases}$$

and in the matrix form $D = -A^t$, $B^t = B$ and $C = C^t$, in other words, iff $K_\omega \in sp(2k+1)$.

Corollary 8.2 *The action* $Sp(2k+1)$ *on the symplectic space* $(\Lambda_\epsilon^{2k+1}(V^*), \mathfrak{D})$ *is hamiltonian with the moment map* $K : \Lambda_\epsilon^{2k+1}(V^*) \to sp(2k+1)$.

8.5 Characteristic Classes of Monge–Ampère Equations on a 3-Dimensional Manifolds

Here, we discuss the most interesting application of the theory of integral Grassmannians—the theory of the characteristic classes for solutions of the Monge–Ampère equations.

The Grassmannians, which we had studied before, play the role of the universal bundles and their cohomology are defined some characteristic classes via the Gauss tangential mappings. To be more precise, we consider the cotangent bundle $\pi : T^*M \to M$ and a solutions Z_ω of a Monge–Ampère symplectic equation $\triangle_\omega = 0$ given by an effective n-form $\omega \in \Omega^n(T^*M)$.

We consider the bundle of the integral Grassmannians $I\mathcal{E}_\omega \to T^*M$. The fibre of this bundle at the point $x \in T^*M$ is the Grassmannian $I\mathcal{E}_{\omega,x}$ of all Lagrangian n-planes such that they are integral to $\omega (\omega|_V = 0$ for any $V \in I\mathcal{E}_{\omega,x})$. If we consider now the cohomology space $H^*(I\mathcal{E}_\omega, \mathbb{Z}_2)$ and any cohomology class $C_\omega \in H^*(I\mathcal{E}_\omega, \mathbb{Z}_2)$ then the natural analogue of the Gauss map associated with a multivalued solution $Z_\omega \subset T^*M$

$$J_{Z_\omega} : Z_\omega \to I\mathcal{E}_\omega, \ J_{Z_\omega}(x) = T_x(Z),$$

induced the map of the cohomology

$$J^*_{Z_\omega} : H^*(I\mathcal{E}_\omega, \mathbb{Z}_2) \to H^*(Z_\omega, \mathbb{Z}_2).$$

and a cohomology class $J^*_{Z_\omega}(C_\omega) \in H^*(Z_\omega, \mathbb{Z}_2)$.

Taking C_ω from the set of generators in $H^*(H^*(I\mathcal{E}_\omega, \mathbb{Z}_2)$, we obtain some cohomology characteristic invariants of the multivalued solution Z_ω.

8.5.1 Special Lagrangian Monge–Ampère Characteristic Classes

We shall realise the program outlined above for the case of the most interesting low-dimensional Monge–Ampère operator

$$\triangle_\omega = \triangle - \text{Hess}.$$

Here $M = \mathbb{R}^3, T^*M = \mathbb{R}^6$,

$$\Omega = dq_1 \wedge dp_1 + dq_2 \wedge dp_2 + dq_3 \wedge dp_3$$

and $\pi : T^*\mathbb{R}^3 \to \mathbb{R}^3$ is given by $\pi(q, p) = q$. We denote(for a chosen subspace $V \subset T^*\mathbb{R}^6$) by $V|_p := V \cap \ker(\pi)$ which is parameterised by $(p, 0)$.

Let $LG_+(\mathbb{R}^6) = U(3)/SO(3)$ be the Grassmannian of oriented Lagrangian subspaces in $T^*\mathbb{R}^3$, and the filtration of it is defined by V_p:

$$F_n = \{V \in LG_+(\mathbb{R}^6)|\dim V_p \geqslant 3 - n\},$$

$$\varnothing = F_{-1} \subset F_0 \subset F_1 \subset F_2 \subset F_3 = LG_+(\mathbb{R}^6),\ 0 \leqslant n \leqslant 3. \tag{8.2}$$

The cohomology of this Grassmannian are well known and its calculation is an application of the following classical A. Borel's theorem:

Theorem 8.4 *There is a convergent spectral sequence* $\{E_r^{(p,q)}, d_r^{(p,q)}\}$, $r \geqslant 0$, *constructed by the filtration 8.2 such that* $E_1^{(p,q)} = H^{(p+q)}(F_p/F_{p-1}, \mathbb{Z}_2)$ *is given by the following picture.*

0	0	0	0
0	$\mathbb{Z}_2$	$\mathbb{Z}_2$	0
0	$\mathbb{Z}_2$	$\mathbb{Z}_2$	0
$\mathbb{Z}_2$	$\mathbb{Z}_2$	0	0

The sequence is stabilized on the first step $(E_2^{(p,q)} = \ldots = (E_\infty^{(p,q)})$ and converge to the cohomology $H^*(LG_+(\mathbb{R}_6), \mathbb{Z}_2)$, which are isomorphic, (as graded algebras), up to the order 3 to the graded ring $\mathbb{Z}_2[w_1, w_2, w_3]/(w_{1,1}^2, w_{1,2}^2, w_{1,3}^2)$ generated by the Stiefel–Whitney's classes of the tautological bundle over the Grassmannian $LG_+(\mathbb{R}^6)$.

Now, we will consider the integral sub-Grassmannian $I\mathcal{E}_\omega$, associated with the special Lagrangian Monge–Ampère operator $\triangle_\omega = \triangle - \text{Hess}$ and the effective form ω is

$$dp_1 \wedge dp_2 \wedge dq_3 - dp_1 \wedge dp_3 \wedge dq_2 + dp_2 \wedge dp_3 \wedge dp_1 - dq_1 \wedge dq_2 \wedge dq_3.$$

The subspaces $V \in LG_+(\mathbb{R}^6)$ such that $\omega|_V = 0$ are formed the sub-Grassmannian $I\mathcal{E}_\varepsilon \subset LG_+(\mathbb{R}^6)$

Taking the filtration as above, one can define

$$\varnothing = F_\omega^{-1} \subset F_\omega^0 \subset F_\omega^1 \subset F_\omega^2 \subset F_\omega^3 = I\mathcal{E}_\omega, \tag{8.3}$$

where $F_\omega^1 = F_i \cap I\mathcal{E}_\omega$, and, then, will compute the spectral sequence corresponding to the filtration (8.3). Take the decomposition $\mathbb{R}^6 = T\mathbb{R}^3 = L_p \oplus L_q$, where

$$L_p = \left\langle \frac{\partial}{\partial p_1}, \frac{\partial}{\partial p_2}, \frac{\partial}{\partial p_3} \right\rangle, \tag{8.4}$$

$$L_q = \left\langle \frac{\partial}{\partial q_1}, \frac{\partial}{\partial q_2}, \frac{\partial}{\partial q_3} \right\rangle. \tag{8.5}$$

and identify the tangent $T(T^*M)$ with the direct sum $TM \oplus T^*M$.
The following theorem based on some unpublished computations of B. Banos [2]:

Theorem 8.5 *1. The quotient $F_\omega^0/F_\omega^{-1} \simeq \emptyset$;*
2. *The quotient $F_\omega^1/F_\omega^0 \simeq \tau_1$,, where τ_1 is the tautological $1d$−dimensional bundle over $\mathbb{P}(T)$ understanding as $Gr_1(3)$;*
3. *The quotient $F_\omega^2/F_\omega^1 \simeq \tau_2$, where τ_2 is the tautological $2d$−dimensional bundle over $\mathbb{P}(T^*)$ understanding as $Gr_2(3)$;*
4. *The quotient $F_\omega^3/F_\omega^2 \simeq \mathbb{S} \subset \mathbb{R}^6$, where the set $\mathbb{S}$ is the $5d$−dimensional singular cubic hypersurface with two connected components, whose regular part is a covering (with discrete fibres $\mathbb{Z}_2^3$) over*

$$x_1 x_2 x_3 - x_1 - x_2 - x_3 = 0$$

modulo symmetric group $\mathfrak{S}_3$.

Proof 1. L_p is the only Lagrangian space, which are belonged to ker π. From the condition

$$\left\langle \omega, \frac{\partial}{\partial p_1} \wedge \frac{\partial}{\partial p_2} \wedge \frac{\partial}{\partial p_3} \right\rangle \neq 0 \tag{8.6}$$

it follows that

$$F_\omega^0/F_\omega^{-1} \simeq \emptyset; \tag{8.7}$$

2. Let $L \in I\mathcal{E}_\omega$ be such a space that $L \in F_\omega^1$. Then one can represent L as the direct sum

$$L = V_p \oplus \langle \xi_1 + \xi_1^\nu, \xi_2 + \xi_2^\nu \rangle, \tag{8.8}$$

where V_p is generated by a vector $\hat{\xi} \in T*$, say,

$$\hat{\xi} = \alpha \partial_{p_1} + \beta \partial_{p_2} + \gamma \partial_{p_3}.$$

Vectors $\xi_1, \xi_2 \in T$ are a basis in the ker $\hat{\xi} \subset T$, i.e.

$$\xi_1 = \beta \partial_{q_1} - \alpha \partial_{q_2}, \xi_2 = \gamma \partial_{q_2} - \beta \partial_{q_3}.$$

In its turn, $\xi_1^\nu, \xi_2^\nu \in T^*$, can be written as

$$\xi_1^\nu = a_1\partial_{p_1} + b_1\partial_{p_2} + c_1\partial_{p_3},$$
$$\xi_2^\nu = a_2\partial_{p_1} + b_2\partial_{p_2} + c_2\partial_{p_3}$$

and

$$\iota_\xi\omega = \alpha\,dp_2 \wedge dp_3 - \beta\,dp_1 \wedge dp_3 + \gamma\,dp_1 \wedge dp_2 -$$
$$- \alpha\,dq_2 \wedge dq_3 + \beta\,dq_1 \wedge dq_3 - \gamma\,d_1 \wedge dq_2.$$

Making the contraction with $\xi_1 + \xi_1^\nu$, obtain

$$\iota_{\xi_1+\xi_1^\nu}(\iota_\xi\omega) = -(\alpha\gamma\,dq_1 + \beta\gamma\,dq_2 + (\alpha^2 - \beta^2)dq_3 +$$
$$+(-\gamma b_1 + \beta c_1)dp_1 + (\gamma a_1 - \alpha c_1)dp_2 + (\alpha b_1 - \beta a_1)dp_3.$$

and further, contracting with $\xi_2 + \xi_2^\nu$, obtain

$$\iota_{\xi_2+\xi_2^\nu}\iota_{\xi_1+\xi_1^\nu}(\iota_\xi\omega) = -\beta(\alpha^2 + \gamma^2 + \beta^2) +$$
$$+\gamma(a_1 b_2 - a_2 b_1) + \alpha(b_1 c_2 - b_2 c_1) + \beta(a_2 c_1 - a_1 c_2).$$

We observe that $\beta \neq 0$ (this is a consequence of the linear independence ξ_1, ξ_2) and hence we get a linear equation on $\xi_1^\nu \wedge \xi_2^\nu$ and conclude that the associate graded F_ω^1/F_ω^0 is isomorphic to the tautological $1d$−dimensional bundle over Grassmannian $Gr_3(1) = \mathbb{P}(T)$.

3. Let us describe the associate graded quotient F_ω^2/F_ω^1. Similarly to previous description (8.8), $L \in F_\omega^2$ implies

$$L = V_p \oplus \langle \xi + \xi^\nu \rangle, \tag{8.9}$$

where $V_p \subset T^*$ is the kernel $\ker\xi$ of $\xi \in T$, and $\xi^\nu \in T^*$. We remind that this a description of a Lagrangian space with $2d$−dimensional kernel of the projection, when we identifies the tangent space $T(T^*M)$ with the direct sum $T \oplus T^*$. If

$$\xi = \alpha\partial_{p_1} + \beta\partial_{p_2} + \gamma\partial_{p_3},$$

then V_p is generated by vectors $\xi_1 = \alpha\partial_{p_2} - \beta\partial_{p_1}$ and $\xi_2 = \beta\partial_{p_3} - \gamma\partial_{p_2}$ and if

$$\xi^\nu = a\partial_{p_1} + b\partial_{p_2} + c\partial_{p_3},$$

then

$$\xi_1 \wedge \xi_2 = \beta\gamma\,\partial_{p_1} \wedge \partial_{p_2} + \beta\alpha\,\partial_{p_3} \wedge \partial_{p_2} - \beta^2\partial_{p_1} \wedge \partial_{p_3},$$

and

$$\theta = \iota_{\xi_1\wedge\xi_2}\omega = \beta\gamma\,dp_3 - \beta\alpha\,dp_1 - \beta^2 dp_2.$$

Therefore $\omega|_L = 0$ if and only if $\theta(\xi^v) = 0$. This condition implies the relation (here again, $\beta \neq 0$)

$$\alpha a + \beta b + \gamma c = 0,$$

which gives an isomorphism of F_ω^2/F_ω^1 with the total space of the tautological $2d$–dimensional bundle over the Grassmannian $Gr_3(3) = \mathbb{P}(T^*)$.

4. The most interesting and difficult topologically case of F_ω^3/F_ω^2 admits a nice and straightforward algebraic description.

In this case $V_p = 0$ and

$$L = \langle \partial_{q_i} + \sum_{j=1}^{3} a_{ij} \partial_{p_j} \rangle, \quad i = 1, 2, 3.$$

One can conclude that $a_{ij} = a_{ji}$ (because L is a Lagrangian) and the condition $\omega|_L = 0$ is reduced to the algebraic equation

$$\det(\|a_{ij}\|) = \operatorname{tr}(\|a_{ij}\|).$$

The associated graded F_ω^3/F_ω^2 is identified with

$$\{A = \|a_{ij}\|, A = A^* \mid \det(A) = \operatorname{tr}(A), \quad 1 \le i, j \le 3.\}$$

This is a $5d-$ dimensional singular cubic hypersurface $\mathbb{S}$ in $\mathbb{R}^6$ with highly non-trivial topology. The regular part (which one can obtain by removing multiplicities) can be identified with a covering with discrete fibres $\mathbb{Z}_2^3$ over the cubic orbifold

$$\{x_1 x_2 x_3 - x_1 - x_2 - x_3 = 0\}/\mathfrak{S}_3$$

under the automorphism actions of the symmetric group $\mathfrak{S}_3$. $\square$

8.5.2 *Remarks and Speculations About* $\mathbb{S}$

The 'regular' part of $\mathbb{S}$ can be interpreted in the following way:

Complexifying the tangent space V (going to $V \otimes \mathbb{C}$), we identify T^* with $\mathbb{C}^3$ and the cubic $\mathbb{S}$ can be considered like a singular affine complex $2d-$ surface in $\mathbb{C}^3$:

$$x_1 x_2 x_3 - x_1 - x_2 - x_3 = 0. \tag{8.10}$$

This is the so-called Cayley surface. The 'infinite part' of the cubic consists of three lines. The three pair-wise intersection points of this lines are singularities in infinity. There is one extra 'finite' singular point $x_1 = x_2 = x_3 = 1$. Singularities has A_1-type singularities (verified by direct change of variables, see, i.e. the Appendix B in [12]). The homotopy type of this variety is the bouquet of two-dimensional spheres ('Milnor spheres') [13]. The Poincaré–Hilbert polynomial of $\mathbb{S}$ over $\mathbb{C}$ is $P_{\mathbb{S}}(t) = 1 + 2t^2$. [14]

The 'regular' part of $\mathbb{S}$ can be interpreted in following way: we consider the interpretation of the effective form ω entering in the definition of special Lagrangian MA operator

$$\Delta_\omega = \Delta - \text{Hess},$$

as an imaginary part $\Im(\alpha)$ of the holomorphic $(3, 0)$−form $\alpha = dz_1 \wedge dz_2 \wedge dz_3$ where $dz_i = dq_i + \sqrt{-1}dp_i$, $i = 1, 2, 3$.

We shall identify $T^*\mathbb{R}^3$ with $\mathbb{R}^3 \times \mathbb{R}^3$ and denote by $\pi_{1,2}$ the projections $T^*\mathbb{R}^3 \to \mathbb{R}^3$ such that $\pi_1(q, p) = q$ and $\pi_2(q, p) = p$.

Let $V \subset T^*\mathbb{R}^3$ be a Lagrangian linear subspace ($\Omega|_V = 0$) such that $\omega|_V = \Im(\alpha)|_V = 0.$.

There are two different possibilities:

1. V is transversal to both projections π_1 and π_2 simultaneously.
2. V is neither transversal to π_1 nor to π_2 and codim$\pi_1(V) = $ codim$\pi_2(V)$

Let $A : \mathbb{R}^3 \to \mathbb{R}^3$ be a linear map and the graph$(A) \subset T^*\mathbb{R}^3$ is a linear Lagrangian subspace.

Then, the first type of the special Lagrangian hyperplanes belongs to the regular part of the Grassmannian $I\mathcal{E}_\omega$ (be represented by graphs of some self-adjoint linear map A and expressed in terms on non-multiple eigenvalues x_1, x_2, x_3 as above.)

The second type forms the singular part of the special Lagrangian Grassmannian and only simple (conic type) singularity in C^3 is relatively studied (see [15]).

We postpone the study of full special Lagrangian Grassmannian and hope to come back to it in our future work. The interest in this study is not restricted to the geometry of non-linear PDEs, but there are many other avatars of this cubic and the links between them are still unclear. Among them the following:

Moduli space or isomorphism classes of Sklyanin algebras with 4 generators which are in one-to-one correspondence with the orbifold $\mathbb{S}/\mathfrak{S}_3$ [16]. Here, $\mathbb{S}$ is given a cubic relation

$$F(x, y, z) = xyz + x + y + z = 0 \qquad (8.11)$$

which geometrically describes a 2-dimensional affine variety in $\mathbb{S} \subset \mathbb{C}^3$. We suppose that (x, y, z) is a 'generic' point in $\mathbb{S}$. The symmetric group $\mathfrak{S}_3$ of order 3 isomorphically acts by cyclic permutations of (x, y, z). The coordinates has explicit expression via four Jacobi theta functions θ_{ij} and the equation can be interpreted like a classical 'quartic identity' for them [17].

Another reincarnation of (8.10) is the variety of monodromy date for the first-order linear system of complex ODEs associated with Painlevé II equation [12].

Recently [18], the same cubic had appeared in a description of cyclically ordered 6−tuples of Lagrangian subspaces $(L_1, L_2, \ldots L_6)$ in $\mathbb{C}^4$ such that every two consecutive subspaces L_i and L_i' are 'maximally non-transversal'. These configurations are in some sense dual to Lagrangian configurations, and the Maslov index may be applied to study them.

The moduli space $\mathcal{L}(2, 6)$ (under $Sp(4, \mathbb{C})$ of generic $(2, 6)$-Lagrangian configurations, i.e. Legendrian hexagons in $\mathbb{C}P^3$ can be described in terms of the so-called 'symplectic cross-ratio' and each hexagon defines a diametric symplectic cross-ratio. There are three non-independent diametric cross-ratio in $\mathcal{L}(2, 6)$. They satisfy the cubic relation (8.10).

We address to the study and to a clarification of all this intriguing coincidences in future publications.

Acknowledgements V. L. and V. R. express their deep thanks to the organisers of the Summer School 'Wisla 2018' and personally to Jerzy Szmit for a very stimulating summer school and for excellent working conditions. We are grateful to all participants for useful and fruitful discussions and inspiring atmosphere during the School.1
During preparation of the material, V. R. was partly supported by the project IPaDEGAN (H2020-MSCA-RISE-2017), Grant Number 778010, and by the Russian Foundation for Basic Research under the Grants RFBR 18-01-00461 and 19-51-53014 GFEN. The research of V. L. was partly supported by the Russian Foundation for Basic Research under RFBR Grant 18-29-10013.

References

1. V. Guillemin, S. Sternberg: Geometric Asymptotics, AMS, Series: Mathematical Surveys and Monographs Number 14 , revised ed. 1977, ISBN-13: 978-0821816332.
2. B. Banos, Opérateurs Monge-Ampère Symplectiques en dimensions 3 et 4, Thèse de Doctorat, Université d'Angers, (Novembre, 2002) 1–128.
3. V. V. Lychagin, Geometric theory of singularities of solutions of Nonlinear Differential Equations, J. Soviet Math., vol 51, 1990, n. 6, 2735–2357 (English transl. of russian ed. "Itogi Nauki VINITI", 1988).
4. L. V. Zilbergleit, Characteristic classes of Monge-Ampère equations, "The interplay between differential geometry and differential equation", AMS Transl. Ser.2, 167, 279–294, Providence, RI 1995.
5. A. Borel: La cohomologie mod 2 de certains espaces homogènes, Comment. Math. Helv., vol. 27(1953), p 165–197.
6. D. B. Fuchs: About Maslov-Arnold characterstic classes, Soviet. Math. Dokl., vol. 178 (1968), n. 2, 301–306 (russian), English transl. Soviet. Math. Dokl., 9 (1968).
7. V. Lychagin, V. Rubtsov and I. Chekalov : A classification of Monge-Ampère equations, Ann. scient. Ec. Norm. Sup., 4 ème série, t.26, 1993, 281–308.
8. A. Kushner, V. Lychagin and V. Rubtsov, Contact geometry and Non-linear Differential Equations, Cambridge University Press, 2007.
9. B. Banos, Non-degenerate Monge-Ampère Structures in dimension 3, Letters in Mathematical Physics, 62 (2002) 1–15.
10. V. Rubtsov: Geometry of Monge-Ampère structures, Lectures delivered on the Summer School "Nonlinear PDEs, their Geometry and Applications -Wisla 2018", this vol. to appear, ed. Birkhäuser, 2019.
11. R. Harvey and H. B. Lawson, Calibrated geometries, Acta. Math. 148, p. 47–157, 1982.
12. L.Chekhov, M. Mazzocco, V. Rubtsov : Painlevé monodromy manifolds, decorated character varieties, and cluster algebras. Int. Math. Res. Not. IMRN., 2017, no. 24, 7639–7691.
13. M. Tibar, Polynomials and vanishing cycles., Cambridge Tracts in Mathematics, 170. Cambridge University Press, Cambridge, 2007. xii+253 pp.

14. S. Szilárd: Perversity equals weight for Painlevé spaces, arXiv:1802.03798, 2018.
15. D. D. Joyce, Singularities of special Lagrangian submanifolds, math.DG/0310460, 2003.
16. R. Bocklandt, T. Schedler, M. Wemyss, Superpotentials and higher order derivations., J. Pure
    Appl. Algebra , 214 (2010), no. 9, 1501–1522.
17. Whittaker, E. T.; Watson, G. N.: A course of modern analysis. An introduction to the general
    theory of infinite processes and of analytic functions; with an account of the principal tran-
    scendental functions , Reprint of the fourth (1927) edition, Cambridge Mathematical Library.
    Cambridge University Press, Cambridge, 1996. vi+608 pp. ISBN: 0-521-58807-3.
18. Ch. Conley, V. Ovsienko Lagrangian configurations and symplectic cross-ratios,
    arXiv:1812.04271, 2018.

Chapter 9
Weak Inverse Problem of Calculus of Variations for Geodesic Mappings and Relation to Harmonic Maps

Stanislav Hronek

9.1 Geodesic Mappings and Basic Setting

Let us start with geodesic mappings of manifolds with affine connections. For the theory of geodesic mappings, we refer to [1]. Because geodesics on manifolds are characterized by the symmetric part of the connection only, we can restrict ourselves to torsion-free manifolds with affine connections, i.e. from now on, we assume that all connections under consideration are symmetric. We will also only consider naturally parametrized geodesics. Consider manifolds with affine connections $(M, {}^{M}\nabla)$ and $(N, {}^{N}\nabla)$ and a map between them $\phi : (M, {}^{M}\nabla) \longrightarrow (N, {}^{N}\nabla)$. This map is said to be a geodesic map if

1. ϕ is a diffeomorphism of M onto N; and
2. the image under ϕ of any geodesic arc in M is a geodesic arc in N; and
3. the image under the inverse function ϕ^{-1} of any geodesic arc in N is a geodesic arc in M.

The usual example of a geodesic mapping would be an isometry of Euclidean surfaces. In our paper, we will generalize this definition a little, we will give up the assumptions 1. and 3. meaning instead of diffeomorphisms we will be working with immersion. Mathematically speaking, a mapping $\phi : (M, {}^{M}\nabla) \longrightarrow (N, {}^{N}\nabla)$, $(dim(M) \leq dim(N))$ is geodesic if for every geodesic curve $x(t)$ on $(M, {}^{M}\nabla)$, $\phi \circ x(t)$ is a geodesic curve on $(N, {}^{N}\nabla)$. For solving the inverse problem of calculus of variation, we would like to use the formalism of calculus of variations on fibred manifolds. The fibred space for the problem will be $(M \times N, \pi, M)$, which has dimension $m + n$ and the mapping ϕ now serves as a fibre coordinate. We also suppose geodesics on M and N are parametrized by the same parameter t. From the simple definition of geodesic mappings, one can derive a set of geodesic equations which will serve as conditions for the mapping ϕ to be a geodesic mapping. Let us write out the equations in coordinate systems (x^i, ϕ^σ) on the total space and

S. Hronek (✉)
Faculty of Natural Sciences, Masaryk University,
Kotlarska 2, 602 00 Brno, Czech Republic
e-mail: standahronek@gmail.com

© Springer Nature Switzerland AG 2019
R. A. Kycia et al. (eds.), *Nonlinear PDEs, Their Geometry, and Applications*,
Tutorials, Schools, and Workshops in the Mathematical Sciences,
https://doi.org/10.1007/978-3-030-17031-8_9

adapted system on the basis (x^i). In these coordinate systems, the affine connections $^M\nabla$, respectively, $^N\nabla$ have components denoted by $^M\Gamma^h_{ij}$, respectively, $^N\Gamma^\mu_{\nu\lambda}$. The geodesic equations for a curve $x(t)$ in M and a curve $y(t)$ in N are

$$\ddot{x}^h + {}^M\Gamma^h_{ij}\dot{x}^i\dot{x}^j = 0,$$

$$\ddot{y}^\sigma + {}^N\Gamma^\sigma_{\mu\nu}\dot{y}^\mu\dot{y}^\nu = 0,$$

$$i, j, l = 1, \ldots, m = dim(M) \quad \mu, \nu, \lambda = 1, \ldots, n = dim(N),$$

For a geodesic mapping ϕ, the geodesic curve $y(t)$ is the image of $x(t)$ by ϕ, substituting $y(t) = \phi(x(t))$ in the second equation we get

$$\frac{d}{dt}\left(\phi^\mu_l\dot{x}^l\right) + {}^N\Gamma^\mu_{\nu\lambda}\phi^\nu_i\dot{x}^i\phi^\lambda_j\dot{x}^j = 0,$$

where we used the chain rule in the second equation $\frac{d}{dt}(\phi^\mu(x^l(t))) = \phi^\mu_l\dot{x}^l$. Computing the derivative in the second equation and then substituting for the second derivative from the first we get

$$\frac{d}{dt}\left(\phi^\mu_l\dot{x}^l\right) + {}^N\Gamma^\mu_{\nu\lambda}\phi^\nu_i\dot{x}^i\phi^\lambda_j\dot{x}^j = \phi^\mu_{kl}\dot{x}^k\dot{x}^l + \phi^\mu_l\ddot{x}^l + {}^N\Gamma^\mu_{\nu\lambda}\phi^\nu_i\dot{x}^i\phi^\lambda_j\dot{x}^j = 0,$$

$$\phi^\mu_{kl}\dot{x}^k\dot{x}^l - {}^M\Gamma^h_{ij}\dot{x}^i\dot{x}^j\phi^\mu_h + {}^N\Gamma^\mu_{\nu\lambda}\phi^\nu_i\dot{x}^i\phi^\lambda_j\dot{x}^j = \dot{x}^i\dot{x}^j\left(\phi^\mu_{ij} - {}^M\Gamma^h_{ij}\phi^\mu_h + {}^N\Gamma^\mu_{\nu\lambda}\phi^\nu_i\phi^\lambda_j\right),$$

$$\phi^\sigma_{ij} - {}^M\Gamma^k_{ij}\phi^\sigma_k + {}^N\Gamma^\sigma_{\alpha\lambda}\phi^\alpha_i\phi^\lambda_j = 0. \tag{9.1}$$

What we get is a sufficient condition for ϕ to be a geodesic mapping. The second part of interest is harmonic mappings.

9.2 Harmonic Mappings

The basics of harmonic mappings can be found in [2, 3]. Their applications to physics, which include string theory, sigma models, and general relativity, are presented in papers [4, 5]. The main change from geodesic mappings is in the setting. Harmonic mappings are defined on Riemannian manifolds, i.e. manifolds endowed with a metric tensor. We say that a mapping ϕ between two Riemannian manifolds (M, g) and (N, h) is harmonic if it is a stationary (extremal) point of the energy functional.

$$E(\phi) = \int_M \frac{1}{2}Tr_g(\phi^*h)\omega_0,$$

where ω_0 is the volume element on M corresponding to the metric tensor g. Euler–Lagrange equations of this functional yield similar equations as for geodesic mappings (9.1), with the difference that for harmonic mappings the connections are metric connections. The Lagrange function in the chosen coordinate system (x^i, ϕ^σ) has the following form:

$$L = \frac{1}{2} g^{ij} h_{\alpha\lambda} \phi_i^\alpha \phi_j^\lambda$$

Its Euler–Lagrange equations are as follows:

$$d_k \frac{\partial L}{\partial \phi_k^\sigma} = \frac{\partial L}{\partial \phi^\sigma}$$

After computing the derivatives and arranging the terms, we get

$$g^{ij} h_{\sigma\nu} \phi_{ij}^\sigma + g^{ij} \phi_i^\alpha \phi_j^\lambda \left(\frac{1}{2} h_{\sigma\lambda,\alpha} + \frac{1}{2} h_{\sigma\alpha,\lambda} - \frac{1}{2} h_{\alpha\lambda,\sigma} \right) + g^{kj}_{,k} h_{\sigma\nu} \phi_j^\sigma = 0$$

$$g^{ij} h_{\sigma\nu} \left(\phi_{ij}^\sigma - {}^M \Gamma_{ij}^k \phi_k^\sigma + {}^N \Gamma_{\alpha\lambda}^\sigma \phi_i^\alpha \phi_j^\lambda \right) = 0, \qquad (9.2)$$

where in the last step we used an expression for the metric trace of Christoffel symbols $g^{ij} \Gamma_{ij}^k = -g^{kl}_{,l}$. These equations seem rather complicated, let us give some examples of harmonic mappings. Constant maps are harmonic. If the source manifold (M, g) was $\mathbb{R}$ with the natural metric, the mapping ϕ would be harmonic if and only if it was a geodesic. On the other hand, if $\mathbb{R}$ with the natural metric was the target space, Eq. (9.2) would be the Laplace equation and its solutions, called harmonic functions, are a special case of harmonic mappings. As mentioned above we get similar equations, more concretely we get a trace of the geodesic equations (9.1). There is also an additional difference being affine connections on one hand and metric on the other. The similarities suggest the following question: what is the connection between variationality of Eq. (9.1) and the corresponding connections being metric?

9.3   Weak Inverse Problem of Calculus of Variations

To answer this question, we will start with Eq. (9.1) assuming that the connections are affine and not necessary metric and solve the inverse problem of calculus of variations. From the equations, it is obvious that they are not variational by themselves, meaning we need to impose and solve the weak inverse problem. To do that we multiply the equation by a multiplier $B_{\sigma\nu}^{ij}(x^k, \phi^\mu)$ and assume it does not depend on the derivatives of the basis and fibre coordinates, which is usually the case, but this also means our conclusions will be only sufficient and possibly not necessary. By multiplying Eq. (9.1) with a multiplier B, we get new equations which do not need to be equivalent to (9.1) but are a differential consequence of (9.1).

We associate a dynamical $(m + 1)$-form with the equation, this form is the following $E = E_\nu \omega^\nu \wedge \omega_0$, where ω_0 is the volume element on M and $\omega^\nu = d\phi^\nu - \phi_i^\nu dx^i$ is the contact 1-form on $M \times N$.

$$E_\nu = B_{\sigma\nu}^{ij} \left(\phi_{ij}^\sigma -^M \Gamma_{ij}^k \phi_k^\sigma +^N \Gamma_{\alpha\lambda}^\sigma \phi_i^\alpha \phi_j^\lambda \right). \tag{9.3}$$

Again using the logic that we are studying geodesic equations and the connections and their components are supposed to be symmetric we can assume the multiplier is also symmetric in the upper indices i, j.

Instead of solving variationality Eq. (9.1), we study the variationality of the associated form E using the tools of calculus of variations on fibred manifolds, which can be found in [6, 7]. The conditions for this form to be variational are called Helmholtz conditions of variationality, which are of the following form (derivation can be found in [7]):

$$\frac{\partial E_\nu}{\partial \phi_{lp}^\mu} - \frac{\partial E_\mu}{\partial \phi_{lp}^\nu} = 0, \tag{9.4}$$

$$\frac{\partial E_\nu}{\partial \phi_l^\mu} + \frac{\partial E_\mu}{\partial \phi_l^\nu} - 2d_p \frac{\partial E_\mu}{\partial \phi_{lp}^\nu} = 0, \tag{9.5}$$

$$\frac{\partial E_\nu}{\partial \phi^\mu} - \frac{\partial E_\mu}{\partial \phi^\nu} + d_l \frac{\partial E_\mu}{\partial \phi_l^\nu} - d_l d_p \frac{\partial E_\mu}{\partial \phi_{lp}^\nu} = 0. \tag{9.6}$$

From the first condition, we get symmetry of the multiplier B in the lower indices

$$B_{\sigma\nu}^{ij} = B_{\nu\sigma}^{ij}.$$

Because in the beginning, we assumed B does not depend on derivatives the equations for the second condition split into a polynomial form in the derivatives of ϕ. Setting each coefficient to zero results in two conditions.

$$-B_{\mu\nu}^{ij} {}^M \Gamma_{ij}^l = \frac{\partial B_{\mu\nu}^{lp}}{\partial x^p}. \tag{9.7}$$

$$^N \Gamma_{\mu\lambda}^\sigma B_{\sigma\nu}^{ij} + {}^N \Gamma_{\nu\lambda}^\sigma B_{\sigma\mu}^{ij} = \frac{\partial B_{\mu\nu}^{ij}}{\partial \phi^\lambda}, \tag{9.8}$$

These conditions already tell us something about the form of the multiplier B. The second equation is the condition for a connection $^N \Gamma_{\mu\lambda}^\sigma$ to be compatible with metric tensors B^{ij}, with components $B_{\sigma\nu}^{ij}$ for any choice of indices i, j. Noticing that the equations separate, in the sense that in the Eq. (9.7) there is a derivative with respect to x^p and in the Eq. (9.8) with respect to ϕ^λ, we can guess that the multiplier B separates also into the following form

$$B_{\sigma\nu}^{ij} = g^{ij}(x^k) h_{\sigma\nu}(\phi^\mu),$$

This would be the simplest choice, in particular, we know this choice is correct (is a solution to the inverse problem) because it gives the harmonic mappings Eq. (9.2). We can be more general and allow the functions $h_{\sigma\nu}(\phi^\mu)$ to also depend on x^k. The reasoning is the following. The second equation tells us that $h_{\sigma\nu}$ are components of a metric tensor of the connection ${}^N\Gamma^\sigma_{\mu\lambda}$ at one particular fibre $\pi^{-1}(x^k)$. If we move to another fibre the connection ${}^N\Gamma^\sigma_{\mu\lambda}$ is also metric but the metric can be different. We can have a different metric tensor h in each fibre and we allow the functions $h_{\sigma\nu}$ to also depend on x^k for this very reason. Therefore, we choose the multiplier in the following form:

$$B^{ij}_{\sigma\nu} = g^{ij}(x^k)h_{\sigma\nu}(x^k, \phi^\mu). \tag{9.9}$$

To justify calling the functions g^{ij} and $h_{\sigma\nu}$ components of metric tensors, we need to check if they are symmetric and regular. Their symmetry follows from symmetry of the multiplier B, symmetry of B in upper indices we assumed and in lower indices we got from Helmholtz conditions, the regularity also follows from the regularity of the multiplier B. In coordinate-free form, we have $B = g \otimes h$.

We know that Eq. (9.8) assures that the connection ${}^N\Gamma^\sigma_{\mu\lambda}$ comes from a metric h, which can be different in each fibre (for different x). We can calculate how it changes between fibres from Eq. (9.7). We substitute for B into Eq. (9.7) and simplify

$$-B^{ij}_{\mu\nu}{}^M\Gamma^l_{ij} = \frac{\partial B^{lp}_{\mu\nu}}{\partial x^p}$$
$$-g^{ij}h_{\mu\nu}{}^M\Gamma^l_{ij} = g^{lp}_{,p}h_{\mu\nu} + h_{\mu\nu,p}g^{lp}$$
$$h_{\mu\nu}\left(g^{ij}{}^M\Gamma^l_{ij} + g^{lp}_{,p}\right) = -h_{\mu\nu,p}g^{lp}$$

where $g^{lp}_{,p}$ is actually a trace of a connection induced by the metric tensor g and let us denote it by ${}^M\bar{\nabla}$, meaning the equation is a difference of traces of two connections. Originally, we assumed the space M is only endowed with an affine connection, however we see there also supposedly exists a metric g, but that is no surprise because we know every smooth manifold admits a metric. We see that the dependency of metric tensor h on the basis coordinate x is given by the difference of traces of connections on the space M. We can then express the relation between the connection components

$$^M\Gamma^l_{ij} = {}^M\bar{\Gamma}^l_{ij} + S^l_{ij},$$

where S^l_{ij} is a tensor which satisfies

$$g^{ij}S^l_{ij} = -\frac{1}{n}h^{\mu\nu}h_{\mu\nu,p}g^{lp}.$$

The last remaining Helmholtz condition unfortunately brings no new information after rearranging it into a polynomial form and from requiring that all the polynomial coefficients vanish we get four conditions

$$\partial_\nu B^{ij}_{\sigma\mu} - \partial_\mu B^{ij}_{\sigma\nu} - \partial_\sigma B^{ij}_{\mu\nu} + B^{ij}_{\alpha\nu}{}^N\Gamma^\alpha_{\mu\sigma} + B^{ij}_{\alpha\nu}{}^N\Gamma^\alpha_{\sigma\mu} = 0 \tag{9.10}$$

$$\partial_\nu(B^{ij}_{\sigma\mu})^N\Gamma^\sigma_{\alpha\lambda} + B^{ij}_{\sigma\mu}\partial_\nu^N\Gamma^\sigma_{\alpha\lambda} - \partial_\mu(B^{ij}_{\sigma\nu})^N\Gamma^\sigma_{\alpha\lambda} - B^{ij}_{\sigma\nu}\partial_\mu^N\Gamma^\sigma_{\alpha\lambda} + \partial_\alpha(B^{ij}_{\sigma\nu})^N\Gamma^\sigma_{\mu\lambda} + B^{ij}_{\sigma\nu}\partial_\alpha^N\Gamma^\sigma_{\mu\lambda}$$

$$+ \partial_\lambda(B^{ij}_{\sigma\nu})^N\Gamma^\sigma_{\alpha\mu} + B^{ij}_{\sigma\nu}\partial_\lambda{}^N\Gamma^\sigma_{\alpha\mu} - \partial_\lambda\partial_\alpha B^{ij}_{\mu\nu} = 0 \tag{9.11}$$

$$^M\Gamma^k_{ij}\left(\partial_\mu B^{ij}_{\sigma\nu} - \partial_\nu B^{ij}_{\sigma\mu} - \partial_\sigma B^{ij}_{\mu\nu}\right) + 2\partial_l\left(B^{lk}_{\lambda\nu}\right)^N\Gamma^\lambda_{\mu\sigma} - 2\partial_p\partial_\sigma B^{kp}_{\mu\nu} = 0 \tag{9.12}$$

$$- \partial_l(B^{ij}_{\mu\nu})^M\Gamma^l_{ij} - B^{ij}_{\mu\nu}\partial_l^M\Gamma^l_{ij} - \partial_l\partial_p B^{lp}_{\mu\nu} = 0. \tag{9.13}$$

Equation (9.10) provides us with the same information as (9.8) that being, the connections $^N\nabla$ is metric with the metric tensor h.

We can also see that the remaining equations are dependent. Equation (9.13) is just a derivative of (9.7) with respect to x^l and Eq. (9.12) is a multiple of Eq. (9.10) by $^M\Gamma^k_{ij}$. The only equation that does not depend on the previous ones is Eq. (9.11) but after some calculations it results in an identity for the Riemann curvature tensor R.

$$h_{\beta\mu}R^\beta_{\lambda\nu\alpha} = h_{\sigma\nu}R^\sigma_{\alpha\mu\lambda} \longrightarrow R_{\mu\lambda\nu\alpha} = R_{\nu\alpha\mu\lambda}.$$

9.4 Summary and Conclusions

We now summarize everything we discovered from the Helmholtz conditions. We have Eq. (9.1) for a geodesic mapping ϕ. We ask the following question: what is the connection between variationality of this equation and the corresponding connections being metric? Therefore, we are solving an inverse problem for the associated dynamical form $E = E_\nu\,\omega^\nu \wedge \omega_0$.

$$E_\nu = B^{ij}_{\sigma\nu}\left(\phi^\sigma_{ij} - {}^M\Gamma^k_{ij}\phi^\sigma_k + {}^N\Gamma^\sigma_{\alpha\lambda}\phi^\alpha_i\phi^\lambda_j\right)$$

We choose a specific form (9.9) of the variational multiplier B. The conditions for variationality are

1. Connection $^N\nabla$ is metric and is fibre-wise induced by the metric h. (Metric h is generally different in each fibre $h_{\sigma\nu} = h_{\sigma\nu}(x^k, \phi^\mu)$. The way in which this metric changes in x^k is given by the connection $^M\nabla$ and metric g)
2. Connection $^M\nabla$ does not need to be metric but is related to a metric connection by

$$\Gamma^k_{ij} = \bar{\Gamma}^k_{ij} + S^k_{ij},$$

where S^k_{ij} is a tensor whose metric trace by the tensor g relates to the changes in the metric h

$$g^{ij}S^l_{ij} = -h^{\mu\nu}h_{\mu\nu,p}g^{lp}.$$

The form (9.9) of the variational multiplier B is a solution to the inverse problem if the corresponding metrics satisfy the above conditions. The results suggest that both spaces are Riemannian, where h is compatible with the connection $^N\nabla$ on N

but g is not necessarily compatible with $^M\nabla$ on M. The interesting result is that the metric h can be different in different fibres and the changes are related to structures on the base manifold M. This conclusion mostly results from using the formalism of fibred manifolds and finding a non-trivial solution to the weak inverse problem, but it suggests more complicated fibred structure for the problem of geodesic and harmonic mappings. There remains the question of finding more general forms of the multiplier B, which could be a part of further research.

Acknowledgements I would like to thank Olga Rossi for the help and guidance.

References

1. Josef, Mikes: *Differential geometry of special mappings*. Palacky university, Olomouc 2015.
2. Eells, James, Jr.; Sampson, J. H. *Harmonic mappings of Riemannian manifolds*. Amer. J. Math. 86 1964 109–160.
3. Hajime, Urakawa, *Calculus of Variations and Harmonic Maps*. American Mathematical Society 1993.
4. Charles W. Misner, *Harmonic maps as models for physical theories*. Phys. Rev. D 18, 4510 1978.
5. Sanchez, N., *Harmonic maps in general relativity and quantum field theory, CERN-TH. 4583/86*. Invited lecture at the Meeting, "Applications harmoniques", CIRM, Luminy, June 1986.
6. Krupka, Demeter: *Introduction to global variational geometry*. Atlantis Press 2015.
7. O. Krupkova: *The Geometry of Ordinary Variational Equations*. Lecture Notes in Mathematics 1678. Springer Verlag, Berlin, Heidelberg 1997.

Chapter 10
Integrability of Geodesics of Totally Geodesic Metrics

Radosław A. Kycia and Maria Ułan

10.1 Introduction

In [11], a class of totally geodesic metrics were given. For convenience, we outline here the main steps referring interested reader to the paper for details.

The starting point is to decompose the Weyl tensor in the base of 2-forms, which are eigenvectors of the corresponding Weyl operator. Then it results that the space-time contains totally geodesic distributions [11] of hyperbolic (H) and elliptic (E) tangent planes. This induces the solutions of the Einstein's equations with the cosmological constant Λ in the form

$$g = g^H \oplus g^E,$$
$$g^H = e^{\alpha(x_0,x_1)}(dx_0^2 - dx_1^2), \quad g^E = -e^{\beta(x_2,x_3)}(dx_2^2 + dx_3^2). \tag{10.1}$$

The functions α and β are the solutions of the hyperbolic and elliptic Liouville equations, correspondingly, [4]

$$\begin{cases} \frac{\partial^2 \alpha(x_0,x_1)}{\partial^2 x_0} - \frac{\partial^2 \alpha(x_0,x_1)}{\partial^2 x_1} + 2\Lambda e^{\alpha(x_0,x_1)} = 0, \\ \frac{\partial^2 \beta(x_2,x_3)}{\partial^2 x_2} + \frac{\partial^2 \beta(x_2,x_3)}{\partial^2 x_3} - 2\Lambda e^{\beta(x_2,x_3)} = 0. \end{cases} \tag{10.2}$$

The solutions are as follows:

R. A. Kycia (✉)
The Faculty of Science, Masaryk University, Kotlářská 2, 602 00 Brno, Czech Republic
e-mail: kycia.radoslaw@gmail.com

Faculty of Physics Mathematics and Computer Science, Cracow University of Technology, 31155 Kraków, Poland

M. Ułan
Baltic Institute of Mathematics, Wałbrzyska 11/85, 02-739 Warszawa, Poland
e-mail: maria.ulan@baltinmat.eu

© Springer Nature Switzerland AG 2019
R. A. Kycia et al. (eds.), *Nonlinear PDEs, Their Geometry, and Applications*,
Tutorials, Schools, and Workshops in the Mathematical Sciences,
https://doi.org/10.1007/978-3-030-17031-8_10

$$\alpha(x_0, x_1) = ln(h_1(v)(v_{x_0}^2 - v_{x_1}^2)),$$
$$\beta(x_2, x_3) = ln(h_2(u)(u_{x_2}^2 + u_{x_3}^2)), \tag{10.3}$$

where u and v are the solutions of the two-dimensional hyperbolic and elliptic equations:

$$v_{x_0 x_0} - v_{x_1 x_1} = 0,$$
$$u_{x_2 x_2} + u_{x_3 x_3} = 0, \tag{10.4}$$

and where h_1 and h_2 are the solutions of a second-order ODEs. Full list of the solutions is presented [11].

In this paper, we analyse the geodesic governed by (10.1). Computations of the geodesic equations were performed using the Mathematica package CCGRG, see [14, 18, 19], and symmetries were computed using the *Differential Geometry* Maple package.

This paper is organized as follows: In the next section, it is shown that there is no true singularities of geodesics in the model of [11], i.e. the space-time is totally geodesic. Then the analysis of Liouville integrability [1] of the geodesics equations is provided. Finally, the analogous model with additional coupling to the electromagnetic field described in [12] is considered in the terms of integrability of geodesics.

The presentation starts with the analysis of the singularities of geodesics.

10.2 Singularities

The metric tensors described in [11] have obvious singularities. Generally, the singularities in the General Relativity have two origins [17]:

- singularities of the coordinates which results from the fact that in the coordinate patch ill-defined coordinate functions are used over regular points of manifold;
- true singularities which indicate geodesic incompleteness of the manifold;

True singularities are usually visible as the singularities of some invariants of curvature. The simplest second-order one is the square of the Riemann curvature (called Kretschmann scalar [3])

$$K = R_{abcd} R^{abcd}. \tag{10.5}$$

For (10.1) that are solutions of (10.2), this invariant is constant

$$K = 8\Lambda^2, \tag{10.6}$$

which suggests no singularities, i.e. completeness of the pseudo-Riemannian manifold. The answer is affirmative as it is provided by the following Lemma[1]

---

[1]RK would like to thank Igor Khavkine for discussion on this subject and suggestions of the outline of the proof.

Lemma 10.1 *The pseudoriemannian manifold (10.1) with (10.2) is complete.*

Proof From the metric decomposition (10.1) and the fact that

$$R^H = \sum_{i,j=0}^{1} R^{ij}_{..ij} = 2\Lambda, \quad R^E = \sum_{i,j=2}^{3} R^{ij}_{..ij} = 2\Lambda, \tag{10.7}$$

it results that the space factorizes into two-dimensional subspaces of constant curvature. These subspaces are isometric to spaces with no singularities according to the well-known Killing–Hopf theorem (see, e.g. Theorem 6.3 in [2]). □

The lemma states that any singularity of (10.1), (10.2) is an artificial singularity only and can be removed by a suitable change of coordinates.

10.3 Geodesics

In this section, the analysis of the geodesic equations will be provided. In the first part, the canonical form of the geodesic equations and their symmetries will be presented. Then the (Liouville) integrable cases will be singled out.

10.3.1 Geodesic Equations

Since the tangent space decomposes into two-dimensional subspaces, therefore, the geodesic equations consist of two pairs of two coupled ODEs for $\gamma(s) = (x_0(s), x_1(s), x_2(s), x_3(s))$, namely,

$$\begin{cases} x_0'' + x_0' x_1' \alpha_{x_1} + \frac{1}{2}\left(x_0'\right)^2 \alpha_{x_0} + \frac{1}{2}\left(x_1'\right)^2 \alpha_{x_0} = 0 \\ x_1'' + \frac{1}{2}\left(x_0'\right)^2 \alpha_{x_1} + \frac{1}{2}\left(x_1'\right)^2 \alpha_{x_1} + x_0' x_1' \alpha_{x_0} = 0, \end{cases} \tag{10.8}$$

$$\begin{cases} x_2'' + x_2' x_3' \beta_{x_3} + \frac{1}{2}\left(x_2'\right)^2 \beta_{x_2} - \frac{1}{2}\left(x_3'\right)^2 \beta_{x_2} = 0 \\ x_3'' - \frac{1}{2}\left(x_2'\right)^2 \beta_{x_3} + \frac{1}{2}\left(x_3'\right)^2 \beta_{x_3} + x_2' x_3' \beta_{x_2} = 0, \end{cases} \tag{10.9}$$

where $' = \frac{d}{ds}$, $\alpha_{x_i} = \frac{\partial \alpha}{\partial x_i}$ and $\beta_{x_i} = \frac{\partial \beta}{\partial x_i}$.

These equations can be significantly simplified. Adding and subtracting Eq. (10.8) and then introducing the light-cone variables (characteristics of the wave equation): $x_0 = \frac{z_0 + z_1}{2}$ and $x_1 = \frac{z_0 - z_1}{2}$ one gets

$$\Delta_1(z_0, z_1) : \begin{cases} z_0'' + \frac{\partial \alpha(z_0, z_1)}{\partial z_0}\left(z_0'\right)^2 = 0 \\ z_1'' + \frac{\partial \alpha(z_0, z_1)}{\partial z_1}\left(z_1'\right)^2 = 0. \end{cases} \tag{10.10}$$

Symmetries of (10.10) can be found by assuming that the generator of a symmetry is of the form: $X = f(s, z_0, z_1)\partial_s + g(s, z_0, z_1)\partial_{z_0} + h(s, z_0, z_1)\partial_{z_1}$, and solving the following system of PDEs:

$$\pounds_{X^{(2)}} \Delta_1(z_0, z_1)|_{\Delta_1(z_0, z_1)} = 0, \tag{10.11}$$

where $\pounds$ is the Lie derivative along $X^{(2)}$—the second prolongation of X to the jet space [5, 10, 15, 16]. The result is

$$X_1 = (As + B)\partial_s, \tag{10.12}$$

where A and B are constants. This gives a scaling and a translation symmetry of s variable, and it results from the fact that (10.8) does not depends explicitly on s. The symmetry reflects the fact that the geodesics should not depend on re-parametrization in s and is also connected with the fact that the geodesic equations are variational and should posses such symmetries.

The same procedure can be applied to the second system of (10.9). In this case we have positively defined ('elliptic') metric, which suggests complex characteristics. It is, therefore, more appropriate to use complex-valued characteristics of an elliptic equation, i.e. the substitution $x_2 = \frac{z_2+z_3}{2i}$ and $x_3 = \frac{z_2-z_3}{2}$, where $i = \sqrt{-1}$. Then adding and subtracting from the first equation of (10.9) multiplied by the imaginary unity the second one one gets the system which resembles (10.10), namely,

$$\Delta_1(z_2, z_3) : \begin{cases} z_2'' + \frac{\partial \beta(z_2, z_3)}{\partial z_2} (z_2')^2 = 0 \\ z_3'' + \frac{\partial \beta(z_2, z_3)}{\partial z_3} (z_3')^2 = 0. \end{cases} \tag{10.13}$$

Since the equations are the same as in the previous case, symmetry analysis indicates, as above, the following generator:

$$X_2 = (Cs + D)\partial_s, \tag{10.14}$$

where C and D are some constants.

In the next section, integrability of geodesics equations will be investigated.

10.3.2 Integrability of Geodesic Equations

First, let us consider the hyperbolic part of the metric, namely define the Hamiltonian

$$H_{0,\alpha} = e^{\alpha(x_0, x_1)}(p_0^2 - p_1^2), \tag{10.15}$$

which surfaces of constant value determine the movement of the particles (positive-massive particles, zero-massless particles). Since the submanifold dimension is 2, therefore in order to find its foliation, according the the Liouville theorem [1], one additional function that the Poisson brackets with $H_{0,\alpha}$ vanishes, is needed. It is assumed in the polynomial form in p_0 and p_1, namely,

$$H_{1,\alpha} = \sum_{k=0}^{n} f_i(x_0, x_1) p_0^k p_1^{n-k}, \tag{10.16}$$

where n is natural number that is fixed degree. Complete integrability is equivalent to the existence of a solution of

$$\{H_{0,\alpha}, H_{1,\alpha}\}_{PB} = 0, \tag{10.17}$$

where $\{.,.\}_{PB}$ is the standard Poisson bracket. Equation (10.17) gives the set of PDEs.[2] In order to check closeness of this system the Kruglikov–Lychagin multi-bracket [6–9, 13] is used. When applied on the system (10.17), it gives compatibility condition in terms of PDEs for $\alpha(x, y)$, which solutions up to $n = 5$ are

1. $n = 1, 2$:

$$\alpha(x_0, x_1) = F \tanh(B(y - x) + A)^3 + E \tanh(B(y - x) + A)^2$$
$$+ D \tanh(B(y - x) + A) + C; \tag{10.18}$$

where A, B, C, D, E, F are the constants of integration and parametrize α.

2. $n = 3, 4$:

$$\alpha(x_0, x_1) = Ax + By + C; \tag{10.19}$$

where A, B, C are the constants of integration and parametrize α.

Surprisingly, these solutions fulfil the first equation of (10.2) only when the cosmological constant $\Lambda = 0$. This is a very prominent example of the role of the cosmological constant in integrability of geodesic equations.

For the case (10.18), integration can be easily performed using (10.10) and gives

$$\begin{cases} z_0(s) = As + J, \\ \int_0^{z_1(s)} \exp(F \tanh(Ba + A)^3 + E \tanh(Ba + A)^2 + D \tanh(Ba + A) + C) da + Gs + H = 0, \end{cases} \tag{10.20}$$

where the second solution is expressed in the implicit form, $A, B, \ldots, F$ are as in (10.18) and G, H, J are the constants dependent on initial data.

The second case (10.19) can be explicitly expressed in terms of elementary functions, namely,

$$z_0(s) = -2 \frac{\ln\left(\frac{2}{(Ds+E)(A+B)}\right)}{A+B},$$
$$z_1(s) = -2 \frac{\ln\left(\frac{2}{(Fs+G)(A+B)}\right)}{A+B}, \tag{10.21}$$

---

[2] All calculations for this section are available as Maple files on: https://github.com/rkycia/GeodesicsIntegrability.

where D, E, F, G are the constants depending on initial data.

Similar analysis performed for the elliptic part of the metric by taking

$$H_{0,\beta} = e^{\beta(x_2,x_3)}(p_2^2 + p_3^2), \tag{10.22}$$

and

$$H_{1,\beta} = \sum_{k=0}^{n} f_i(x_2, x_3) p_2^k p_3^{n-k}, \tag{10.23}$$

and checking when

$$\{H_{0,\beta}, H_{1,\beta}\}_{PB} = 0. \tag{10.24}$$

The two solutions for β are obtained up to the degree $n = 5$, namely:

1. $n = 1, 2$:

$$\beta(x_0, x_1) = F \tanh(B(y - xi) + A)^3 + E \tanh(B(y - xi) + A)^2 \\ + D \tanh(B(y - xi) + A) + C; \tag{10.25}$$

where A, B, C, D, E, F are the constants of integration and parametrize β, and i is the imaginary unit.

2. $n = 3, 4$:

$$\beta(x_0, x_1) = Ax + By + C; \tag{10.26}$$

where A, B, C are the constants of integration and parametrize β.

As in the previous case, these βs solve (10.2) only when the cosmological constant $\Lambda = 0$.

For (10.25) the solution of (10.13) is

$$\begin{cases} z_2(s) = As + J, \\ \int_0^{z_3(s)} \exp(F \tanh(Ba + A)^3 + E \tanh(Ba + A)^2 + D \tanh(Ba + A) + C) da + Gs + H = 0, \end{cases} \tag{10.27}$$

where, as before, G, H, J are the integration constants depending on initial data.

For (10.26), the solution of (10.13) is

$$z_2(s) = -2 \frac{\ln\left(\frac{2}{(A-iB)(Ds+E)}\right)}{A - iB}, \\ z_3(s) = -2 \frac{\ln\left(\frac{2}{(A+iB)(Fs+G)}\right)}{A + iB}, \tag{10.28}$$

where D, E, F, G are again the constants depending on initial data. These solutions are complex-valued, however, since x_2 and x_3 fulfil real equations for geodesic, therefore, transforming to the original variables one gets real solutions.

In general, the geodesic solutions can be constructed by selecting the solution (10.20) or (10.21) for the hyperbolic part of the subspace, and (10.27) or (10.28) for the elliptic subspace. Therefore, in total $4 = 2 \times 2$ integrable solutions were obtained.

10.4   Einstein–Maxwell Solutions

The results from the previous section can be used for analysis of the geodesics of the solutions for coupled the Einstein and Maxwell equations described in [12]. In this model the totally geodesic solutions, the same as the solution of (10.1) for the metric, were obtained. However, now α and β are solutions of

$$\begin{cases} \frac{\partial^2 \alpha(x_0,x_1)}{\partial^2 x_0} - \frac{\partial^2 \alpha(x_0,x_1)}{\partial^2 x_1} + k_1 e^{\alpha(x_0,x_1)} = 0, \\ \frac{\partial^2 \beta(x_2,x_3)}{\partial^2 x_2} + \frac{\partial^2 \beta(x_2,x_3)}{\partial^2 x_3} + k_2 e^{\beta(x_2,x_3)} = 0, \end{cases} \tag{10.29}$$

where

$$k_1 = 2\left(\frac{kJ}{c^4} + \Lambda\right), \quad k_2 = \left(\frac{kJ}{c^4} - \Lambda\right), \tag{10.30}$$

where Λ is the cosmological constant, k is the gravitational constant, and the new parameter J is connected with the solution for the Faraday tensor of electromagnetic field

$$F = -2l e^{\alpha(x_0,x_1)} dx_0 \wedge dx_1 + 2m e^{\beta(x_2,x_3)} dx_2 \wedge dx_3, \tag{10.31}$$

where

$$l^2 = \frac{J - I_1}{2}, \quad m^2 = \frac{J + I_1}{2}, \tag{10.32}$$

and where I_1 is the invariant of the characteristic polynomial of the skew symmetric operator $\hat{F}$ (associated to F by $g(\hat{F}X, Y) = F(X, Y)$), namely, its determinant. The parameters $\pm l$ and $\pm im$, where $l, m \in \mathbb{R}$, are the eigenvalues of the hyperbolic and the elliptic parts of the operator $\hat{F}$.

The straightforward result from (10.32) is that

$$J = l^2 + m^2, \quad I_1 = m^2 - l^2. \tag{10.33}$$

From our previous considerations, the geodesic equations are (Liouville) integrable when $k_1 = 0 = k_2$, i.e., when $J = 0$ and $\Lambda = 0$. And therefore, since $l, m \in \mathbb{R}$, from the first equation of (10.33) it results that $l = 0$ and $m = 0$, and therefore, the Faraday tensor vanishes. This shows that the integrable solutions for geodesics exist when no electromagnetic field and no cosmological constant is present in this model. The solutions for geodesics are exactly the same as in the previous section for the Einstein equations only, since the electromagnetic field vanishes.

10.5 Discussion

The semi-Riemmanian metric of [11] describes anisotropic space-time, which distinguished the space direction x_1, and therefore cannot describe the observed space-time where assumption on spherical symmetry is imposed. The presence of this distinguished space direction resembles the phenomena from the phase transitions in solid state physics, and therefore it suggests that the model can be applied in some phenomena that occur when the universe undertake some kind of phase transition, e.g., in the early state of the universe. A similar description applies also to the coupled Einstein–Maxwell system.

Intriguing correspondence between vanishing of the cosmological constant and integrability of the geodesic equations was noted. In the case of electromagnetism for integrability also electromagnetic field must vanish.

10.6 Conclusions

In this paper, analysis of the geodesic of the solution of the Einstein vacuum equations resulting from the Weyl tensor bivector structure was provided. In particular, integrable geodesic equations of special solutions of the Einstein vacuum equation were found and described. A similar analysis was also performed for the Einstein–Maxwell system.

Acknowledgements We would like to thank Prof. Valentin V. Lychagin and Igor Khavkine for enlightening discussions. We would also like thank Sergey N. Tychkov for helping to master Maple. RK participation was supported by the GACR Grant 17-19437S, and MUNI/A/1138/2017 Grant of Masaryk University.

References

1. V.I. Arnold, *Mathematical Methods of Classical Mechanics*, Springer; 2nd edition (1997)
2. W.M. Boothby, *An Introduction to Differentiable Manifolds and Riemannian Geometry*, Academic Press; 2nd edition 2002
3. Ch. Cherubini, D. Bini, S. Capozziello, R. Ruffini, *Second Order Scalar Invariants of the Riemann Tensor: Applications to Black Hole Spacetimes*. International Journal of Modern Physics D. 11 (06): 827–841 (2002); arXiv:gr-qc/0302095v1; https://doi.org/10.1142/S0218271802002037
4. D.G. Crowdy, *General Solutions to the 2D Liouville equations*, International Journal of Engineering Science, 35 2 141–149 (1997)
5. I.S. Krasilshchik, A.M. Vinogradov, *Symmetries and Conservation Laws for Differential Equations of Mathematical Physics*, American Mathematical Society 1999
6. B. Kruglikov, *Note on two compatibility criteria: Jacobi-Mayer bracket vs. differential Groöbner basis*, Lobachevskii J. Math., 23, 2006, 57–70
7. B. Kruglikov, V. Lychagin, *Mayer brackets and solvability of PDEs–I*, Differential Geometry and its Applications, Elsevier BV, 17, 251–272 (2002)

8. B. Kruglikov, V. Lychagin, *Mayer brackets and solvability of PDEs–II*, Transactions of the American Mathematical Society, 358, 3, 1077–1103 (2006)
9. B. Kruglikov, V. Lychagin, *Compatibility, Multi-brackets and Integrability of Systems of PDEs*, Acta Applicandae Mathematicae, Springer, 109, 151 (2010)
10. A. Kushner, V. Lychagin, V. Rubtsov, *Contact Geometry and Nonlinear Differential Equations*, Cambridge University Press; 1 edition 2007
11. V. Lychagin, V. Yumaguzhin, *Differential invariants and exact solutions of the Einstein equations*, Anal.Math.Phys. 1664-235X 1–9 (2016); https://doi.org/10.1007/s13324-016-0130-z
12. V. Lychagin, V. Yumaguzhi, *Differential invariants and exact solutions of the Einstein–Maxwell equation*, Anal.Math.Phys. 1, 19–29, (2017); https://doi.org/10.1007/s13324-016-0127-7
13. Maple package for the Mayer and the Kruglikov-Lychagin brackets calculations can be downloaded from http://d-omega.org/brackets/
14. Mathematica package CCGRG for tensor computations can be downloaded from http://library.wolfram.com/infocenter/MathSource/8848/
15. P.J. Olver, *Applications of Lie Groups to Differential Equations*, Springer; 2nd edition 2000
16. P.J. Olver, *Equivalence, Invariants, and Symmetry*, Cambridge University Press; 1 edition 2009
17. R.M. Wald, *General Relativity*, Chicago University Press 1984
18. A. Woszczyna, R.A. Kycia, Z.A. Golda, *Functional Programming in Symbolic Tensor Analysis*, Computer Algebra Systems in Teaching and Research, IV 1 100–106 (2013)
19. A. Woszczyna, P. Plaszczyk, W. Czaja, Z.A. Golda, *Symbolic tensor calculus - functional and dynamic approach*, Technical Transactions, Y. 112 61–70 (2015); https://doi.org/10.4467/2353737XCT.15.110.4147

Index

© Springer Nature Switzerland AG 2019
R. A. Kycia et al. (eds.), *Nonlinear PDEs, Their Geometry, and Applications,*
Tutorials, Schools, and Workshops in the Mathematical Sciences,
https://doi.org/10.1007/978-3-030-17031-8